ICGSCE 2014

Editor
Md Amin Hashim
Faculty of Chemical Engineering
Universiti Teknologi MARA (UiTM)
Selangor
Malaysia

ISBN 978-981-10-1366-9 ISBN 978-981-287-505-1 (eBook)
DOI 10.1007/978-981-287-505-1

Springer Singapore Heidelberg New York Dordrecht London

Softcover reprint of the hardcover 1st edition 2015

Printed on acid-free paper

Springer Science+Business Media Singapore Pte Ltd. is part of Springer Science+Business Media
(www.springer.com)

Preface

On behalf of the editorial members, I would like to convey my heartiest congratulations and thank you to all authors who had successfully completed their original work piece, participated and hence contributed to the conference. The conference organisers also would like to acknowledge and appreciate all internal and external reviewers who had spent their precious time in reading and suggesting improvements on the selected manuscripts. Selected manuscripts were broadly inclined to the conference theme which was "Synergising Sustainable Technologies for Future Generation", and again these subjects are either directly or indirectly related to the general chemical engineering subjects.

Several selected and edited papers will be published in the "ICGSCE 2014" conference proceedings with approval from established publisher, such as Springer. Other selected and edited paper will be published in the Malaysian Journal of Analytical Sciences, and in the Materials Science and Engineering Journal, published by Transtech publications. Finally, and again, on behalf of editors, I would like to thanks the conference organisers for their trust and confidence in the editorial committee. The acknowledgment is also due to the Research and Management Institute (RMI), UiTM for their valuable advice on publication matters.

Md Amin Hashim

List of Reviewers

Abdul Aziz Ishak, Universiti Teknologi MARA
Amin Azdarpour, Universiti Teknologi MARA
Amizon Azizan, Universiti Teknologi MARA
Arina Sauki, Universiti Teknologi MARA
Assoc. Prof. Dr. Md Amin Hashim, Universiti Teknologi MARA
Assoc. Prof. Dr. Norazah Abd Rahman, Universiti Teknologi MARA
Assoc. Prof. Dr. Rahmah Mohamed, Universiti Teknologi MARA
Assoc. Prof. Dr. Ruzitah Mohd Salleh, Universiti Teknologi MARA
Assoc. Prof. Hasnora Jafri, Universiti Teknologi MARA
Assoc. Prof. Dr. Ayub Md Som, Universiti Teknologi MARA
Assoc. Prof. Zulkafli Hassan, Universiti Teknologi MARA
Che Mohd Som Said, Universiti Teknologi MARA
Dr. Abdul Hadi, Universiti Teknologi MARA
Dr. Ahmad Rafizan Mohamad Daud, Universiti Teknologi MARA
Dr. Alawi Sulaiman, Universiti Teknologi MARA
Dr. Atikah Kadri, Universiti Teknologi MARA
Dr. Azil Bahari Alias, Universiti Teknologi MARA
Dr. Fazlena Hamzah, Universiti Teknologi MARA
Dr. Hazlina Hussin, Universiti Teknologi MARA
Dr. Ibtisam Abdul Wahab, Universiti Teknologi MARA
Dr. Istikamah Subuki, Universiti Teknologi MARA
Dr. Jagannathan Krishnan, Universiti Teknologi MARA
Dr. Jefri Jaapar, Universiti Teknologi MARA
Dr. Junaidah Jai, Universiti Teknologi MARA
Dr. Kamariah Noor Ismail, Universiti Teknologi MARA
Dr. Neni Kartini Che Mohd Ramli, Universiti Teknologi MARA
Dr. Noor Fitrah Abu Bakar, Universiti Teknologi MARA
Dr. Norhuda Ismail, Universiti Teknologi MARA
Dr. Norliza Ibrahim, Universiti Teknologi MARA
Dr. Nornizar Anuar, Universiti Teknologi MARA
Dr. Norun Farihah, International Islamic University Malaysia

Dr. Nurul Fadhilah Kamalul Aripin, Universiti Teknologi MARA
Dr. Rafiziana Kasmani, Universiti Teknologi Malaysia
Dr. Rahida Wati Sharudin, Universiti Teknologi MARA
Dr. Safari Zainal, Universiti Teknologi MARA
Dr. Sherif Abdul Bari Ali, Universiti Teknologi MARA
Dr. Syed Shatir Asghrar Syed Hassan, Universiti Teknologi MARA
Dr. Tan Huey Ling, Universiti Teknologi MARA
Dr. Zulkifli Abdul Rashid, Universiti Teknologi MARA
Hanafiah Zainal Abidin, Universiti Teknologi MARA
Lim Ying Pei, Universiti Teknologi MARA
Miradatul Najwa Muhd Rodhi, Universiti Teknologi MARA
Mohibah Musa, Universiti Teknologi MARA
Dr. Nik Raikhan Nik Him, Universiti Teknologi MAR1A
Noorhaliza Aziz, Universiti Teknologi MARA
Nor Faeqah Idrus, Universiti Teknologi MARA
Norhasyimi Rahmat, Universiti Teknologi MARA
Norrulhuda Mohd Taib, Universiti Teknologi MARA
Nur Fadzeelah Abu Kassim, Universiti Teknologi MARA
Nurhaslina Che Radzi, Universiti Teknologi MARA
Prof. Dr. Ku Halim Ku Hamid, Universiti Teknologi MARA
Prof. Ir. Dr. Jailani Salihon, Universiti Teknologi MARA
Rasyidah Alrozi, Universiti Teknologi MARA
Rozaimee, Universiti Teknologi MARA
Sakinah Mohd Alauddin, Universiti Teknologi MARA
Sharmeela Matali, Universiti Teknologi MARA
Siti Shawalliah Idris, Universiti Teknologi MARA
Zalizawati Abdullah, Universiti Teknologi MARA

Contents

Part I
Green and Sustainable Technologies

Overview on Chemical-Based, Bio-based and Natural-Based Surfactants in EOR Applications

Hazlina Husin, Muhammad Najmi Ibrahim, Zulkafli Hassan, Norrulhuda Mohd Taib, Ku Halim Ku Hamid, Nik Khairul Irfan Nik Ab Lah and Muhammad Shafiq Mat Shayuti

Abstract Energy consumption in the world is increasing every year while expanding technology trying to catch up with the demand of oil production. One of the technologies which has been widely used to enhance oil recovery is surfactant flooding. A variety of surfactants promote oil recovery are reviewed in this paper, with particular emphasis on the ability: (i) to lower the interfacial tension (ii) to adjust the pH and (iii) to stand the high salinity conditions of reservoirs. To date, surfactants studied in enhanced oil recovery applications can be categorized as chemical-based, bio-based and natural-based surfactants.

Keywords Surfactant · Interfacial tension · EOR

H. Husin (✉) · M.N. Ibrahim · ZulkafliHassan · N.M. Taib · K.H. Ku Hamid · N.K.I. Nik Ab Lah · M.S. Shayuti
Faculty of Chemical Engineering, Universiti Teknologi MARA, 40450 Shah Alam, Selangor Darul Ehsan, Malaysia
e-mail: hazlina858@salam.uitm.edu.my

M.N. Ibrahim
e-mail: mdnajmi91@gmail.com

ZulkafliHassan
e-mail: zulkafli1660@salam.uitm.edu.my

N.M. Taib
e-mail: norrulhuda4670@salam.uitm.edu.my

K.H. Ku Hamid
e-mail: kuhalim@salam.uitm.edu.my

N.K.I. Nik Ab Lah
e-mail: khairulirfan@salam.uitm.edu.my

M.S. Shayuti
e-mail: mshafiq5779@salam.uitm.edu.my

H. Husin
CoRe of Frontier Materials & Industry Applications, Universiti Teknologi MARA, 40450 Shah Alam, Selangor Darul Ehsan, Malaysia

M.A. Hashim (ed.), *ICGSCE 2014*, DOI 10.1007/978-981-287-505-1_1

Introduction

Oil recovery using primary recovery methods which used reservoir pressure to bring the oil up to the surface is only ~20 %. When secondary recovery methods are used, the oil recovery would only be increased up to ~40 % maximum and thus more oil residual is still left behind. Enhanced oil recovery (EOR) is another method known to maximize oil recovery up to ~60 % of the original oil-in-placed remains in the reservoirs (also known as tertiary recovery methods).

In EOR, surfactant flooding is classified under chemical flooding methods. Surfactant flooding is when a surfactant is employed in order to lower the interfacial tension (IFT) between oil and water so that the oil becomes easier to flow. Surfactant also alters the rock wettability (from oil-wet to water-wet) and adjusts pH and salinity. This review focuses on chemical-based, bio-based and natural-based surfactants. These surfactants are unique from one another depending on composition and chemical properties.

Surfactants are compounds that have both hydrophilic groups (heads) and hydrophobic groups (tails). It means that the surfactant have both water soluble component and oil soluble component. Surfactant will disperse in water and adsorb between oil and water. The hydrophilic head will expand into the water phase while the hydrophobic tail will extend into the oil phase.

Chemical-Based Surfactants

Surfactants manufactured from chemical synthesis processes are called chemical-based surfactant. For example, ethoxylated sulfonates and petroleum sulfonates. Ethoxylated sulfonates have a strong hydrophilic group and have a very high electrolyte tolerance [1]. However, ethoxylated sulfonates have a tendency to form liquid crystals. These liquid crystals can be avoided by an addition of co-solvent or elevating the temperature. In contrast, petroleum sulfonates have a lower tolerance to salt with optimal salinity of 1–2 % [2].

Recently, ionic liquids are reported to shown the ability as chemical-based surfactants. Researchers have described that ionic liquid is able to tolerate harsh conditions of salinity. For example, 1-dodecyl-3-methylimidazolium chloride ([C_{12}mim][Cl]) and octylmethylnaphthalene sulfonate surfactants. Besides both surfactants can survive harsh conditions of salinity, the first surfactant can lower the IFT up to 70 mN m^{-1} difference at ~1000 ppm [3] while the second surfactant can lower the IFT up to 10^{-6} mN m^{-1} at a very low concentration [4].

Biosurfactant

Bio-surfactant is an EOR method employed under microbial enhance oil recovery (MEOR) techniques. Bio-surfactant is produced from the excretion of microbial cells that grown on hydrocarbon. It is complex bio-polymer and has the prospective in moving heavy crude oil from the reservoir. According to Al-Sulaimani et al. [5], bio-surfactant produced by *Bacillus licheniformis* was able to lower the IFT from 46.6 to 3.28 mN m^{-1} in incubation for 16 h. *Bacillus licheniformis* are genetically stable and can cause significant IFT reduction in different mediums [6]. Some other bio-surfactants are able to reduce the surface tension of water from 72 to 27 mN m^{-1} [7]. The effectiveness of bio-surfactant during MEOR techniques is contributed by its ability to lower the IFT, can withstand temperature up to 75 °C and pH values of up to 12 [8] and the microorganism that produced the bio-surfactant can also improve the sweep efficiency and selectively plug high permeability zones.

Natural-Based Surfactant

Recently, studies are inclined towards surfactants derived from natural resources, which also are categorized as natural-based surfactants [9]. These natural-based surfactants are biodegradable, low toxicity, cheap and renewable. Due to their unique properties, these natural-based surfactants are preferable to be used in EOR as compared to the chemical-based surfactants. Furthermore, the impact to the environment can be reduced without lowering the oil production. Although natural-based surfactants have never been produced commercially, large scale industrial production has been considered [10].

Studies [11–14] shown that sugar-based surfactants are effective surfactant because it can lower the IFT and have excellent adsorption at solid-liquid interface [15]. Additionally, glucose-based and sucrose-based surfactants have a range of beneficial physical and performance properties, including high levels of surfactancy (surface and interfacial activity), rapid biodegradability, low toxicity, effective emulsification properties and effective surface interactions [16]. In terms of adsorption mechanism, sugar-based surfactant cannot be classified as only one simple chain of surfactant. This is because there are a lot of combination of hydrophilic group and hydrophobic alkyl chain. The hydrophilic group might consist of various sugar units while the hydrophobic group might consist of one or more alkyl chains [17]. Figures 1 and 2 show some sugar-based surfactants that have two or more hydrophobic chain.

Saponin can be classified as amphipathic glycosides with a distinctive foaming characteristic. It can be found in various plant species. Saponins are surface active substances because they contain both hydrophilic regions (such as rhamnose, xylose, arabinose, galactose, fucose and glucuronic acid) and hydrophobic regions

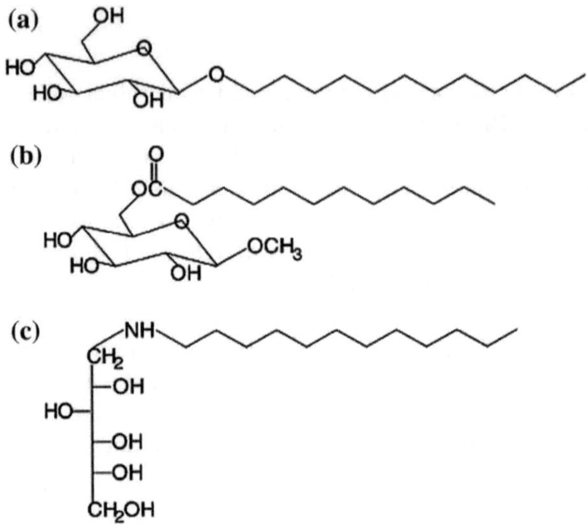

Fig. 1 Linkages between the hydrophilic and the hydrophobic in sugar surfactants: **a** ether bond, dodecyl-β-D-glucoside; **b** ester bond, β-methyl-6-*O*-dodecanoyl-D-glucoside; **c** amine bond, *N*-dodecyl-glucamine [17]

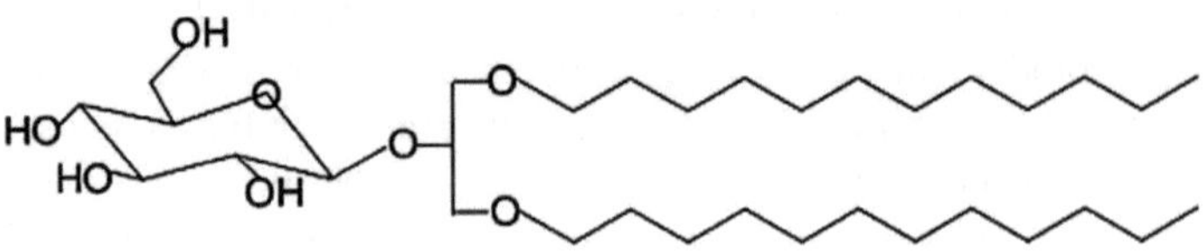

Fig. 2 A glycolipid: 1,3-di-*O*-dodecyl-2-*O*-β-D-glucosylglycerol [17]

(such as quillaic acid and gypsogenic acid) on the same molecule [18]. Example of saponin surfactant that has been extracted from the *Quillaja saponaria* is shown in Fig. 3 [18, 19]. Yang et al. [18] conducted a comparable study between a saponin surfactant (Q-Naturale®) and a synthetic surfactant (Tween 80). They reported that the saponin surfactant can lower the IFT as effective as the synthetic surfactant (as in Fig. 4). Another research on saponin which was conducted by Shahri et al. [20] have shown that the IFT decreases as the saponin concentrations increase. In summary, saponin is a suitable candidate to replace the chemical surfactant because it can lower the IFT as low as the chemical surfactant can perform, biodegradable and can be produced at low cost. It does not have any adverse effect to the environment as compared to chemical surfactant. This is a very important characteristic as the reservoir can be preserved from any pollution that might change the condition of hydrocarbon while increasing the oil recovery [20].

Malic acid-based surfactant is another suitable candidate to act as natural-based surfactant for EOR because they contain active functional groups. Malic acid is an

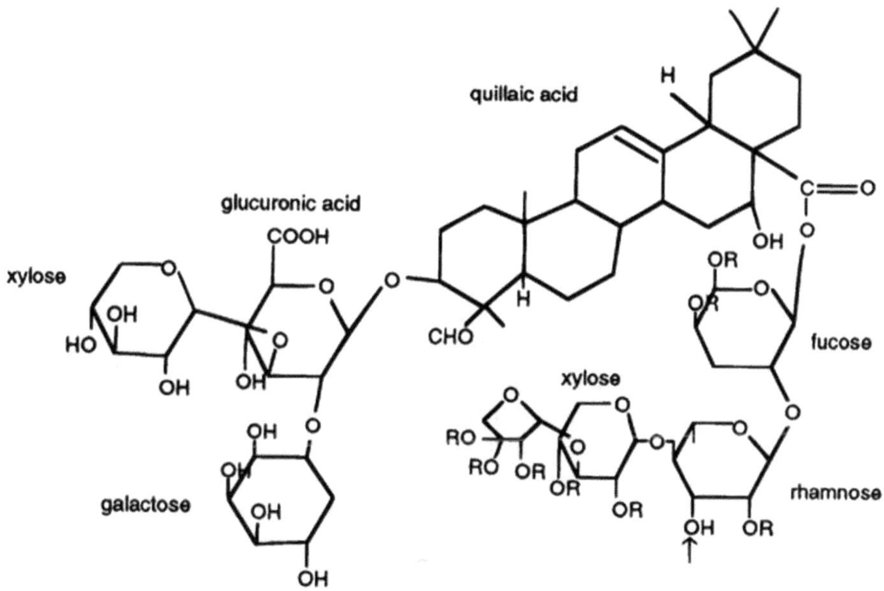

Fig. 3 Structural representations of Quillaja saponin. This structure is presented in [19]

Fig. 4 The influence of saponin (Q-Naturale®) and synthetic (Tween 80) surfactants at different concentration on the interfacial tension [18]

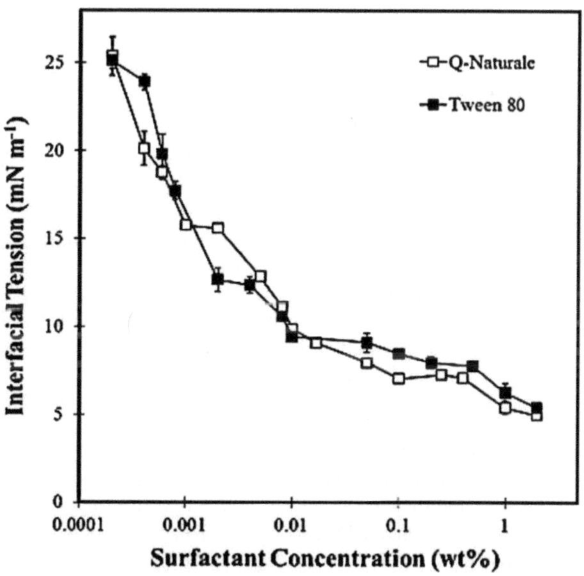

organic acid with chemical formula of $HO_2CCH_2CHOHCO_2H$. Methods to produce malic acid are described in many studies [21–25]. Lassila and Slone [26] investigated on malic acid diester surfactant in waterborne coating, inks, adhesive, fountain solution and agricultural applications. Malic acid diester surfactant was found to lower the IFT in water based systems. The adsorption mechanism study [27, 28] showed the stable conformational molecular structure of malic acid subject to particle surface. A presence of two different types of hydrogen bonds in the racemic malic acid and only one type in the pure enantiomeric form was reported.

Conclusion

Although the use of surfactant for EOR has been studied and tested for many years, the specific mechanisms involved in oil recovery are not understood fully or known specifically for a wide variety of reservoir conditions, crude oils, and surfactant-type systems. If MEOR or natural-based surfactant is to become an industrially accepted alternative EOR process in the industry, the oil recovery mechanisms such as IFT, phase and volume diagrams, and adsorptionmechanism need to be clarified. The natural-based surfactant offers cost-effective processes as compared to existing chemical-based surfactants.

Acknowledgments The author gratefully acknowledge the financial support from the Ministry of Education, Malaysia (600-RMI/FRGS 5/3 (93/2013)) and for the continuous research supports provided by the Universiti Teknologi MARA (UiTM). The author wish to thank the referee for making this a better paper.

References

1. Carmona I, Schecter RS, Wade WH, Weerasooriya U (1985) Ethoxylated oleyl sulfonates as model compounds for enhanced oil recovery. Soc Petrol Eng J 25:351–357
2. Skauge A, Palmgren O (1989) Phase behaviour and solution properties of ethoxylated anionic surfactants. In: SPE international symposium on oilfield chemistry, 1989 Copyright 1989, Society of Petroleum Engineers, Houston, Texas
3. Hezave AZ, Dorostkar S, Ayatollahi S, Nabipour M, Hemmateenejad B (2013) Dynamic interfacial tension behaviour between heavy crude oil and ionic liquid solution (1-dodecyl-3-methylimidazolium chloride ([C_{12}mim][Cl]) + distilled or saline water/heavy crude oil)) as a new surfactant. J Mol Liq 187:83–89
4. Zhao Z, Li Z, Qiao W, Cheng L (2005) Dynamic interfacial behaviour between crude oil and octylmethylnaphthalene sulfonate surfactant flooding systems. Colloids Surf A 259:71–80
5. Al-Sulaimani H, Al-Wahaibi Y, Al-Bahry S, Elshafie A, Al-Bemani A, Joshi S, Zargari S (2011) Optimization and partial characterization of biosurfactants produced by Bacillus Species and their potential for ex-situ enhanced oil recovery. Soc Petrol Eng J 16:672–682
6. Thomas CP, Duvall ML, Robertson EP, Barrett KB, Bala GA (1993) Surfactant-based EOR mediated by naturally occurring microorganisms. SPE Reservoir Eng 8:285–291
7. Illias RM, Ooi SW, Idris AK, Rahman WAWA (1999) Production of biosurfactant and biopolymer from Malaysian oil fields isolated microorganisms. In: SPE Asia Pacific improved oil recovery conference, Society of Petroleum Engineers, Kuala Lumpur, Malaysia
8. Gabitto JF (2006) Combined microbial surfactant-polymer system for improved oil mobility and conformance control. In: SPE annual technical conference and exhibition, Society of Petroleum Engineers, San Antonio, Texas, USA

9. Deymeh H, Shadizadeh SR, Motafakkerfard R (2012) Experimental investigation of *Seidlitzia rosmarinus* effect on oil–water interfacial tension: usable for chemical enhanced oil recovery. Scientia Iranica 19:1661–1664
10. Piispanen P, Persson M, Claesson P, Norin T (2004) Surface properties of surfactants derived from natural products. Part 1: syntheses and structure/property relationships—solubility and emulsification. J Surfact Deterg 7:147–159
11. Jarek E, Wydro P, Warszyński P, Paluch M (2006) Surface properties of mixtures of surface-active sugar derivatives with ionic surfactants: theoretical and experimental investigations. J Colloid Interface Sci 293:194–202
12. Rico-Lattes I, Lattes A (1997) Synthesis of new sugar-based surfactants having biological applications: key role of their self-association. Colloids Surf A 123–124:37–48
13. Shi J, Li QD, Xia XC (2006) Synthesis and properties of sugar-based gemini surfactants. Tenside Surfactants Deterg 43:204–209
14. Söderberg I, Drummond CJ, Furlong DN, Godkin S, Matthews B (1995) Non-ionic sugar-based surfactants: self assembly and air/water interfacial activity. Colloids Surf A 102:91–97
15. Zhang L, Somasundaran P (2006) Adsorption of mixtures of nonionic sugar-based surfactants with other surfactants at solid/liquid interfaces: I. Adsorption of n-dodecyl-β-d-maltoside with anionic sodium dodecyl sulfate on alumina. J Colloid Interface Sci 302:20–24
16. Rojas OJ, Stubenrauch C, Lucia LA, Habibi Y (2009) Interfacial properties of sugar-based surfactants
17. Stubenrauch C (2001) Sugar surfactants—aggregation, interfacial, and adsorption phenomena. Curr Opin Colloid Interface Sci 6:160–170
18. Yang Y, Leser ME, Sher AA, McClements DJ (2013) Formation and stability of emulsions using a natural small molecule surfactant: Quillaja saponin (Q-Naturale®). Food Hydrocolloids 30:589–596
19. Mitra S, Dungan SR (1997) Micellar properties of Quillaja Saponin. 1. Effects of temperature, salt, and pH on solution properties. J Agric Food Chem 45:1587–1595
20. Shahri MP, Shadizadeh SR, Jamialahmadi M (2012) Applicability test of new surfactant produced from Zizyphus Spina-Christi leaves for enhanced oil recovery in carbonate reservoirs. J Jpn Petrol Inst 55:27–32
21. Miyama M, Nakayama K (1993) D-malic acid production from DL-malic acid by microbial assimilation of L-malic acid: screening of microorganism. Biotechnol Tech 7:105–110
22. Chibata I, Tosa T, Yamamoto K (1978) Industrial production of L-malic acid by immobilized microbial cells. In: Pye EK, Weetall H (eds) Enzyme engineering. Springer, US, pp 463–468
23. Ado Y, Kawamoto T, Masunaga I, Takayama K, Takasawa S, Kimura K (1982) Production of 1-malic acid with immobilized thermophilic bacterium, *Thermus rubens* Nov. sp. In: Chibata I, Fukui S, Wingard L Jr (eds) Enzyme Engineering. Springer, US, pp 303–304
24. Gong CS, Cao N, Sun Y, Tsao GT (1996) Production of L-malic acid from fumaric acid by resting cells of Brevibacterium sp. In: Wyman C, Davison B (eds) 17th symposium on biotechnology for fuels and chemicals. Humana Press, New York, pp 481–487
25. West T (2011) Malic acid production from thin stillage by Aspergillus species. Biotechnol Lett 33:2463–2467
26. Lassila KR, Slone CS (1999) Malic acid diester surfactants. Air Products and Chemicals
27. Husin H, Leong YK, Liu J (2012) Molecular attributes of an effective steric agent: yield stress of dispersions in the presence of pure enantiomeric and racemate malic acids. Adv Powder Technol 23:459–464
28. Husin H, Leong YK, Liu J (2011) Conformational molecular structure-interaction force correlation in influencing effective steric between alumina and malic acid. In: CHEMECA 2011: the 41th Australasian chemical engineering conference, Engineers Australia, Sydney, Australia, p 109

OPEFB Filler from Biomass in Superabsorbent Polymer Composite for Agriculture Application: A Comparative Study

Wan Siti Nadiah Wan Yaacob, Saidatul Shima Jamari and Suriati Ghazali

Abstract An agriculture activity is one of the major sectors which contribute a significant share to Malaysia economy. Therefore it is important to maintain the soil fertility. Superabsorbent polymer composite (SAP'c) was produced using graft polymerization technique with the addition of the oil palm empty fruit bunch (OPEFB) filler to enhance the absorbance properties and decomposition process. Effect of filler addition at different size to the SAP'c was investigated. Water absorbency test was conducted to determine the absorbency and swelling properties while Fourier Transform Infrared (FTIR) and Scanning Electron Microscopic (SEM) analysis was conducted to investigate the functional groups and morphology of the SAP'c. Water absorb highest for SAP'c with coarse OPEFB while fine particle size showed a decrement in swelling behaviour. FTIR spectra shows that –C ≡ C–H with C–H bend existed in all samples. SEM showed that the addition of the OPEFB give a fibrous morphology and build a surface contact. In conclusion the addition of biomass filler in the superabsorbent polymer (SAP) will increase the water absorbency and swelling properties of the hydrogel.

Keywords Superabsorbent polymer · Superabsorbent polymer composite · Oil palm empty fruit bunch

W.S.N.W. Yaacob · S.S. Jamari (✉) · S. Ghazali
Faculty of Chemical and Natural Resources Engineering, Universiti Malaysia Pahang, Lebuhraya Tun Razak, 26300 Kuantan, Pahang, Malaysia
e-mail: sshima@ump.edu.my

W.S.N.W. Yaacob
e-mail: wan_nad90@yahoo.com

S. Ghazali
e-mail: suriati@ump.edu.my

M.A. Hashim (ed.), *ICGSCE 2014*, DOI 10.1007/978-981-287-505-1_2

Introduction

Nowadays, agricultural activities are becoming a major growing sector and contribute a lot of income to the Malaysia economy. In addition, this activity also provides employment opportunities to the citizen. In order to ensure that agricultural activities continues to contribute significantly, measures should be taken to safeguard the sustainability of this sector, for example, to maintain the soil fertility. Fertilizers made from chemicals are beneficial towards the plant however, offer adverse effects to the neutrality of the soil pH. Therefore, this research was proposed to synthesis SAP'c as a soil conditioning mechanism.

In 1960s, the United States Department of Agriculture has developed superabsorbent polymer. The early attempt for the commercialization of superabsorbent polymer was in early 1970s as it is commercialized in the form of starch/acrylonitrile/acrylamide based polymer (superslupers). At first, this material was focused to be used in agriculture/horticulture sector. In 1982, superabsorbent polymer for the diapers application were introduced to the Japanese market by Unicharm, while in 1983 by KAO and Procter and Gamble introduced in 1985, [1, 2]. In general, the definition of superabsorbent polymer is a material that can absorb huge amount of water while retain its original shape without loss the water that had been absorb. The example applications that used this type of materials were in hygiene product such as diapers. The most common technique to synthesize this material was by graft polymerization method. In graft polymerization method, usually acrylic acid or methacrylic acid was used as monomers and sometimes acrylamide and methacrylamide also can be used. The addition of acrylamide into acrylic acid can help to increase the absorbency of superabsorbent polymer besides induce the graft polymerization [3].

Currently, the world major concern is towards the environmental problems. In relative to the problem, superabsorbent polymers composite, which produce throughout the research, were synthesized by adding the OPEFB from biomass waste as filler. The addition of the OPEFB will enhance the biodegradable ability thus it will be an eco-friendly product. Employing a biomass waste as filler will reduce the land pollutant cause by the waste besides reducing the cost needed to decompose the waste and production cost of the product. In recent years, Malaysia has become the second largest (after Indonesia) palm oil exporter thus contributes up to 39 wt% of the total world palm oil production. Due to the demand of the palm oil industries around the world, few countries such as Indonesia, Colombia, Ivory Coast and Papua New Guinea had identified other good potential of the palm oil resulting in increasing of the fresh fruit bunch (FFB) production that leads to the increasing of waste streams [4].

As the OPEFB is one of the main waste biomass products in Malaysia, Malaysian Palm Oil Board (MPOB) in 2005 has record that about 20–25 % residue of OPEFB produced for every tonne FFB after it was steamed and fruitlets are removed. Study done before on the ten years application of empty fruit bunches in an oil palm plantation on soil chemical properties showed that the decomposition time of OPEFB was decreased when it is applied to soil. Within 7–15 weeks of decomposition process, empty fruit bunch (EFB) has lost it dry matter up to 50 % with the total decomposition of dry OPEFB in the soil only take less than 12 months. Other than that, the application of OPEFB in soil has increase the pH value of the soil [5]. So that, the application of OPEFB as a biomass filler in superabsorbent polymer will help in naturally for the polymer chain to decompose itself in soil besides increasing the fertility of the soil as the OPEFB help to increase the pH value of the acidic soil.

Materials and Experimental

Materials

OPEFB was chosen as biomass filler in SAP'c composites. For synthesis of SAP'c, AA was used as a monomer, MBA as a crosslinker agent while APS was used as an initiator.

Experimental

Preparation of OPEFB as a Filler

Fresh OPEFB were cut into a small pieces and dried in an oven for 5 days at 70 °C. After that, high speed grinder was used to reduce the size of dried OPEFB. Grinded OPEFB particles were than sieved to get a uniform particle size.

Synthesis of SAP'c

SAP'c was synthesized by using graft polymerization method. In an empty three neck flask, acrylic acid solution was prepared. Then, sodium hydroxide (5 M) was added to neutralize the acrylic acid solution and OPEFB was added as the filler. Mixture was continuously stirred throughout the process and crosslinking agent (MBA) was added under a nitrogen atmosphere. After 30 min of stirring, APS was added into it and temperature slowly raised up to 70 °C. SAP'c gel will formed after 30 min of reactions. All the reactions were carried out using double boil technique. The obtained gel was then washed using distilled water for several times and dried in an oven at temperatures, 70 °C for 24 h. Dried SAP'c was then milled to small particles (powder form). Similar step was repeated for pure SAP'c, without the addition of OPEFB filler.

Water Absorbency Measurement

0.53 g of SAP'c was weighed and added into a tea-bag. The tea bag was then immersed in a 1 L distilled water at room temperature. SAP'c will slowly swell until the equilibrium is reached. After 24 h, swollen gel was allowed to drain and reweighed to calculate its water absorbency by using the following equation.

$$Water\ Absorbency,\ Q(\frac{g_{H_2O}}{g_{sample}}) = \frac{m_2 - m_1}{m_1}$$

Water absorbency is expressed in grams of water stored in gel by a gram of dried gel. m_2 and m_1 represents the weight of gel after immersed in water and weight of dried gel before immersed in water. The test was repeated for three times.

Results and Discussions

Water Absorbency Test

Water absorbency test was carried out for 24 h to observe and determine the swelling capacity of the all SAP'c. Table 1 shows that the weight of all samples is increased up to 200 times from the original weight. The application of the OPEFB filler has increased the swelling capacity. However, the particle size gave effect to the swelling behavior of the SAP'c. It showed that SAP'c with a coarse OPEFB can absorbed more water with the water absorbency increment up to 0.94 % but SAP'c with fine OPEFB decreased to 5.82 %,compared to the pure OPEFB. Refer to the Sadeghi et al. studied on the swelling behavior of the SAP, dense crosslinking in SAP'c can decrease the swelling behavior of the hydrogel. It is due to the higher crosslinking density has decreased the space between the copolymer chains thus resulted in the rigid structure that cannot be expanded and hold large amount of water [6]. So in this case, fine particle size give a large surface area for the OPEFB to make contact and crosslink with the copolymer thus increased the crosslinking density of SAP'c.

Table 1 The weight of SAP'c before and after the water absorption test

Samples	Pure SAP'c	SAP'c with coarse OPEFB	SAP'c with fine OPEFB
Initial weight (g)	0.53	0.53	0.53
Final weight (g)	115.55	116.62	108.85

FTIR Analysis

FTIR analysis has been carried out to determine the functional group that exists in OPEFB and SAP'c. As shown in Fig. 1, there is a strong and broad OH-stretch with H-bonded at the band 3420.87 cm^{-1}. At the spectra with medium peak of 2920.43 cm^{-1}, C–H with asymmetric and symmetric stretch (alkanes group) formed while at the band of 1738.42, 1639.33, and 1055.21 cm^{-1}, there is C double bond O stretch, medium peak of N-H bend in amines-primary group and medium C–N stretch in aliphatic amines group for OPEFB sample.

The comparison of FTIR spectra in Fig. 2 for the three samples of SAP'c before immersed in water showed that at peak 3425.52 cm^{-1} for pure SAP'c, the peak shift

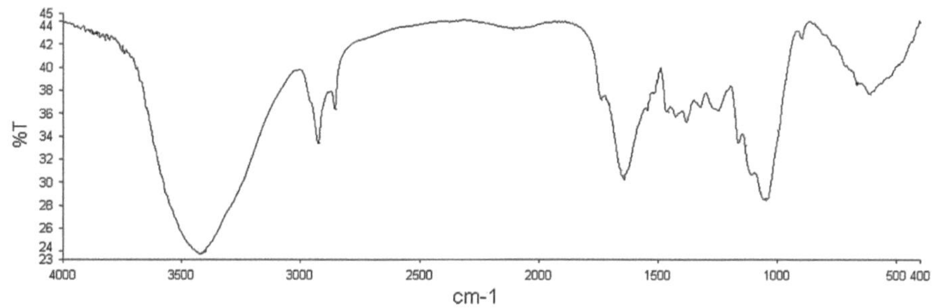

Fig. 1 FTIR spectra for OPEFB

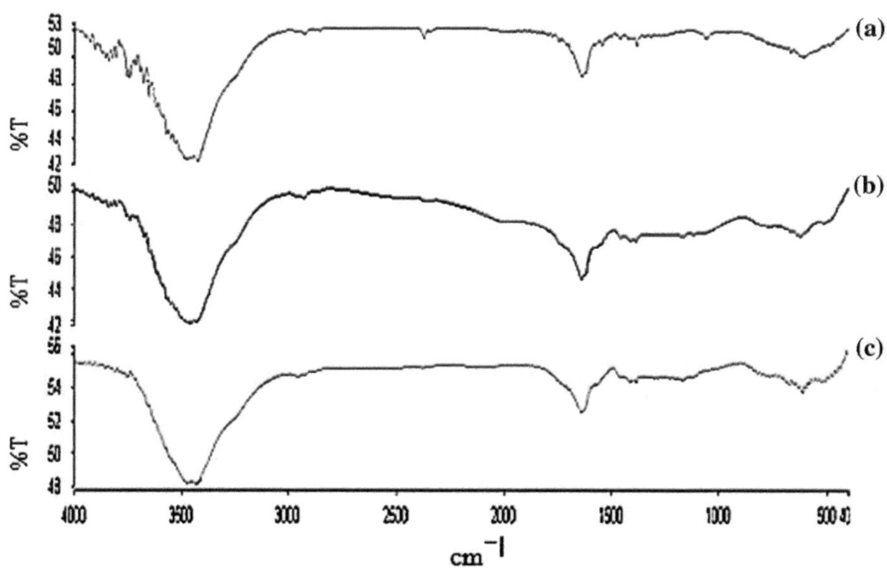

Fig. 2 FTIR spectra for **a** SAP'c with coarse OPEFB size **b** SAP'c with fine OPEFB size **c** pure SAP'c

to 3464.64 cm^{-1} for the SAP'c with the fine size of filler and for the coarse filler size, the peak shift back to 3420.61 cm^{-1}. This peak shows the existence of strong and broad O-H stretch with H-bonded. These bonds indicate the presence of –OH group in the samples. For all three samples, N–H and N = O bending at peaks 1637.38 and 615.00 cm^{-1} for pure SAP'c, 1639.19 and 621.78 cm^{-1} for SAP'c with fine filler and for SAP'c with coarse size filler was at 1639.19 and 1384.55 cm^{-1} and also –C ≡ C–H with C–H bend exist for three samples at peaks in range between 614.00 and 621.8 cm^{-1} indicates that the graft polymerization occurs in all samples and AA was grafted to the OPEFB during the graft polymerization reaction occurs [7].

Surrface Morphology Analysis Using Scanning Electron Microscopy (SEM)

From the SEM analysis, it can be seen that the addition of the OPEFB as a filler in SAP'c able to increase the space capacity in the SAP'c as the parallel alignment of the OPEFB help the water to absorb faster and thus occupied the gel space. Compared Fig. 3c, e the addition of the OPEFB give a fibrous morphology and build a surface crosslinking thus enhance the absorption ability of the SAP'c gel.

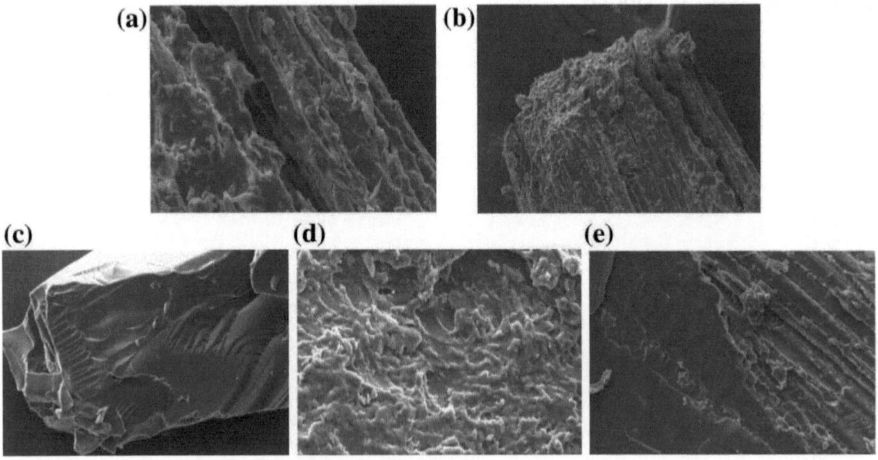

Fig. 3 Surface morphology of **a** coarse OPEFB **b** fine OPEFB **c** pure SAP'c **d** SAP'c with fine OPEFB **e** SAP'c with coarse OPEFB

Conclusion

In conclusion, superabsorbent polymer composite can be synthesized by using graft polymerization techniques. The application of well distributed OPEFB in SAP'c will enhance and optimize the absorbency and swelling properties of SAP'c. However, it is important to choose the suitable particles size to prevent SAP'c from losing their good absorbency and water retention characteristics.

References

1. Nexant Inc. (2004) Superabsorbent polymers (SAP). Nexant Chem System. PERP Program 1–6
2. Kabiri K, Omidian H, Hashemi SA, Zohuriaan-Mehr MJ (2003) Synthesis of fast-swelling superabsorbent hydrogels: effect of crosslinker type and concentration on porosity and absorption rate. Eur Polymer J 39:1341–1348
3. Doane WM, Doane SW and Savich MH (2004) Superabsorbent polymers in agriculture applications. Absorbent Technologies, Inc
4. Jamari SS, Howse JR (2012) The effect of the hydrothermal carbonization process on palm oil empty fruit bunch. Biomass Bioenergy 47:82–90
5. Abu Bakar R, Darus S, Kulaseharan S, Jamaluddin N (2010) Effects on ten year application of empty fruit bunches in oil palm plantation on soil chemical properties. Nutr Cycl Agroecosyst 89(3):341–349
6. Sadeghi M (2012) Synthesis of biocopolymer carrageenam-g-poly (AAm-co-IA)/montmorilonite superabsorbent hydrogel composite. Braz J Chem Eng 29(02):295–305
7. Zhang J, Wang Q, Wang A (2007) Synthesis and characterization of chitosan-g-poly (acrylic acid)/attapulgite superabsorbent composites. Carbohydr Polym 68(2):367–374

Exploring Non-wood Plants as Alternative Pulps: From the Physical and Chemical Perspectives

Angzzas Kassim, Ashuvila Aripin, Zainuri Hatta and Zawawi Daud

Abstract The increasing demand in wood fibre consumptions especially in pulp and paper making has pushed forward the search for alternative fibre resources. Non-woods derived fibre could be good candidates due to its abundance availability. The objective of this study is to determine the potential of non-wood plants as alternative fibres for pulp and paper-based industries based on its chemical and physical properties. The chemical properties involved in this study (cellulose, hemicellulose, lignin, hot water and 1 % NaOH solubilities and ash contents) were determined according to relevant TAPPI test, Kurscher-Hoffner and Chlorite methods. Meanwhile, the physical properties (fibre length and diameter) were determined according to the Franklin method. In order to propose the suitability of the studied non-wood plants as alternative fibre resources in pulp and paper-based industries, the obtained results are compared to properties of published wood resources. Results show that lignin content (5.67 %) and 1 % NaOH solubility (19.64 %) of cogon grass are the lowest compared to cocoa pod husk and oil palm leaf. These contents influenced the production of higher pulp yield. Although cogon grass contains short fibres length than oil palm leaf, the paper product will have higher strength due to the good inter-fibre bonding. This study conclude, based on

A. Kassim (✉) · A. Aripin
Faculty of Engineering Technology, Universiti Tun Hussein Onn Malaysia,
Batu Pahat, Malaysia
e-mail: angzzas@uthm.edu.my

A. Aripin
e-mail: ashuvila.aripin87@gmail.com

Z. Hatta · Z. Daud
Faculty of Civil and Environmental Engineering, Universiti Tun Hussein Onn Malaysia,
Batu Pahat, Malaysia
e-mail: mzainuri88@gmail.com

Z. Daud
e-mail: zawawi@uthm.edu.my

M.A. Hashim (ed.), *ICGSCE 2014*, DOI 10.1007/978-981-287-505-1_3

the chemical and physical properties, the cogon grass, cocoa pod husk, and oil palm leaf are suitable to be used as alternative fibre resources in pulp and paper-based industries with various applications such papers, boxes and craft materials.

Keywords Cellulose · Fibre · Hemicellulose · Lignin · Paper

Introduction

Perennial non-wood plant (cogon grass) and agro-wastes (cocoa pod husk and oil palm leaf) are considered problematic because they possess no economic importance for any specific purpose. The depleting natural resources, growing environmental concerns, abundant and inexpensive of non-wood plant resources are the major driving forces to utilize these annually renewable plants for various industrial applications [1]. Generally, these plants have the compositions and properties that are suitable to be used as composite, textile, pulp and paper-based manufacturing as well as in producing fuel, enzymes, chemicals and animal feed [1, 2].

In Malaysia, the plantation of cocoa was around 20,645 ha with 7019 tons production in 2010 [3]. Cocoa pod husk (CPH) represents between 70 to 75 % of the whole cocoa fruit weight [4]. That is, an average of 5000 tons of cocoa pod husk was generated in that year. Oil palm trees are also one of the most valuable crops in Malaysia with approximately 4.85 million ha of plantation area in 2010 [5]. The large plantation area has generated large biomass residues (more than 70 million tons) as 1 ha of plantation area may produce between 50 and 70 tons of wastes [6]. The residues included oil palm leaves (OPL), empty fruit bunches, shell, kernel and trunks [7]. Moreover, the cogon grass (CG) is also widely found in Malaysia as scattered patches along highways and open area. This invasive grass is problematic to plantation area and a bad plant as feedstock for animals due to the rough and sharp leaves [8].

When left on the plantation floor, these wastes create great environmental problems that constitute a barrier for easy movement and serves as habitat for pests and diseases causing organisms on the farm [9]. Inevitably, these massive wastes from CPH and OPL were dumped in the landfill for disposal.

To promote a sustainable and greener environment, the economic importance of the cogon grass, cocoa pod husks and oil palm leaves is being put forward. The first step is to evaluate the fundamental properties comprise within them. Therefore, the objective of this study is to evaluate the chemical and physical properties of cocoa pod husk, cogon grass and oil palms leaves and to make comparison of their suitability with the published literature. The information generated from this study will be utilised in exploring the potential of those perennial non woods in the applications as pulp in paper-based industries.

Experimental

Materials

For this study, samples were collected from Pusat Pembangunan Komoditi Parit Botak, Jabatan Pertanian Malaysia, Batu Pahat, Johor. Prior to air-dried, CG and OPL were cut into 2 to 5 cm length whereas CPH was chopped approximately to 1.5×1.5 cm^2. The samples were thoroughly washed to eliminate sand and other contaminants. For chemical analysis, the air-dried samples were ground to 0.40–0.45 mm and stored in air tight container for further analyses.

Methods

Chemical properties, Chemical composition of CG, CPH and OPL were performed according to Technical Association of the Pulp and Paper Industry (TAPPI) Test Method. The samples were first submitted to soxhlet extraction for 6 h according to method T 264 om-88. The evaluation of extractive substances was carried out in different liquids according to experimental parameters: hot water solubility (T 207 cm-08) and 1 % sodium hydroxide solubility (T 212 om-07) and ash content (T 211 om-07). The amount of lignin, holocellulose, hemicelluloses and cellulose were assessed by using the following respective standard methods: T 222 om-06, Chlorite and Kurschner-Hoffner methods. All experiments were conducted in triplicates.

Physical properties, To measure the fibre length of the materials, Fraklin method (Han et al. 1999) was applied. In this method, the materials were immersed into glacial acetic acid and hydrogen peroxide for a period of 24 h at 60 °C or until they were defibred. The fibre length of materials were measured under profile projector microscope (V-12, Japan) and the fibre diameters were measured by using Quata 200 Scanning Electron Microscopy.

Results and Discussions

Chemical properties, Table 1 shows the percentages of various components present in CG, CPH and OPL. Results indicated significant difference ($p < 0.05$) between different samples (of the same properties). The data shows that OPL (43.79 %) exhibited the highest content in cellulose compared to CPH (35.44 %) and CG (37.13 %). In pulp and paper-based industries, cellulose is an important constituent in papermaking. A high content of cellulose produces a high pulp yield and a good quality paper [10].

Table 1 Chemical properties of non-wood plant fibres

Chemical properties (% w/w)	Non-wood plants		
	Cocoa pod husk (CPH)	Cogon grass (CG)	Oil palm leaf (OPL)
Cellulose	35.44 ± 0.33*	37.13 ± 0.43*	43.79 ± 1.51*#
Hemicellulose	37.00 ± 0.50*#	27.36 ± 0.40*	36.41 ± 1.73*
Lignin	14.71 ± 0.35*	5.67 ± 0.26*#	19.66 ± 2.41*
Hot water solubility	17.63 ± 0.23*	3.83 ± 0.30*#	18.43 ± 0.39*
1 % NaOH solubility	20.23 ± 0.59*	19.64 ± 0.53*#	41.95 ± 2.92*
Ash	12.26 ± 0.23*	8.24 ± 0.28*	5.73 ± 0.94*#

*Significant difference ($p < 0.05$) of different plant fibres; #the most desirable content

On the other hand, hemicellulose content of CPH (37 %) is the highest compared to CG (27.36 %) and OPL (36.41 %) as shown in Table 1. High hemicellulose content is desirable for pulp and paper-based industries because it is correlated with high strength of the paper produced especially in tensile and burst indices [11, 12]. Therefore, paper products from CPH are expected to have high tensile and burst strength.

Lignin contents are at satisfactory level if the value is less than 30 % [10]. Remarkably, the lignin content of CG was found to be 5.67 %, which is the lowest compared to CPH (14.71 %) and OPL (19.66 %). Low lignin content is considered advantageous in the pulp and paper-based industries as it utilize less chemicals and energy during the pulping process [13, 14]. Low lignin will also generate whiter paper and therefore requires less bleaching.

Hot water solubility in water solvent is the measure of the extractive components of the plants [10]. The hot water solubility in CG (3.83 %) is the lowest compared to CPH (17.63 %) and OPL (18.43 %) as shown in Table 1. Low hot water solubility indicates less extractive materials in the plants and potential of generating high pulp yield after the pulping process [10].

The 1 % NaOH solubility indicates the level of degradation of the material in weak alkali. CG (19.64 %) contains the lowest content of 1 % NaOH solubility compared to CPH (20.23 %) and OPL (41.95 %). Low solubility suggests less fibre degradation during the pulping process and high pulp yield [15].

Finally, the ash content indicates the mineral component of lignocellulosic materials. The ash content was found highest in CPH (12.26 %), followed by CG (8.24 %) and OPL (5.73 %). High ash content will contribute to decrease in the strength properties of the product and reducing pulp yield [16].

Physical properties, Fibre length is an important parameter in measuring the quality of pulp to be used in paper-based industries [17], due to the strong relationship between fibre length and strength properties of the resultant pulp [18]. Table 2 shows the fibre length and diameter of CPH, CG and OPL with significance difference ($p < 0.05$) between different plants.

There are three classifications of fibre length; long, medium and short. Results in Table 2 indicates fibres obtained from CPH, CG and OPL are considered as short

Table 2 Physical properties of non-wood plants

Physical properties	Non-wood plants		
	Cocoa pod husk (CPH)	Cogon grass (CG)	Oil palm leaf (OPL)
Fibre length, mm	0.73 ± 0.09*#	0.94 ± 0.05*	1.13 ± 0.06*
Fibre diameter, μm	11.09 ± 2.40*	6.63 ± 0.40*	6.33 ± 0.06*#
Felting rate	65.83	141.78	178.52

*Significant difference ($p < 0.05$) of different plant fibres; #the most desirable content

fibre (lengths between 0.2 and 1.2 mm). Fibre length of CPH (0.73 mm) is the shortest compared to CG (0.94 mm) and OPL (1.13 mm). The shorter fibres length contributes to excellent formation and increase inter-fibre bonding as well as increases the strength properties of the produced paper [19]. In addition, short fibre length will also give smoothness of the surface morphology of produced paper [20].

Fibre diameter is also important to determine the quality of fibres in pulp and paper-based industries. The fibre diameter of OPL (6.33 μm) is slightly smaller than CPH (6.63 μm), followed by CG (11.09 μm). A small diameter of fibre indicates that less fibre flexibility.

The strength properties of paper-based products were found to be positively correlated to the felting rate (fibre length/fibre diameter). Generally, the acceptable value for felting rate is more than 33 [10]. The felting rates of all the materials are acceptable as a good fibre; CPH (65.83); CG (141.78) and OPL (178.52).

Conclusion

In conclusion, the non-wood plants (cocoa pod husk, cogon grass and oil palm leaf) could be viewed as potential resources as alternative fibres in pulp and paper-based industries. Although the chemical and physical properties are different, these plants are still appropriate to be utilised in the paper-based industries depending on the suitable applications. For example, the oil palm leaf with a greater fibre length (i.e. rougher paper surface) as compared to CPH and CG might be more suitable for craft industries or carrier bags. The chemical properties indicate cogon grass has the most favourable parameters (three out of six), indicating it might be the most suitable fibre as pulp in paper industry. In addition, cogon grass has short fibre length and considered to have good arrangement of fibre that increase the strength property in the resultant paper. Based on the chemical and physical properties, all proposed plants are suitable to be used as alternative fibre depending on its application with cogon grass being the most suitable.

Acknowledgment This research was financially supported by the Fundamental Research Grant Scheme (FRGS) and scholarship from Universiti Tun Hussein Onn Malaysia (UTHM). The authors are thankful for this financial support.

References

1. Reddy N, Yang Y (2005) Biofibers from agricultural byproducts for industrial applications. Trends Biotechnol 23(1):22–27
2. Abdul Khalil HPS, Siti Alwani M, Mohd Omar AK (2006) Chemical composition, anatomy, lignin distribution and cell wall structure of Malaysian plants waste fibers. BioResource 1 (2):220–232
3. Jabatan Perangkaan Malaysia, JPM (2011) Perangkaan Ekonomi Malaysia, Siri Masa Retrieved on 08 May 2012. www.statistics.gov.my
4. Cruz G, Pirilä M, Huuhtanen M, Carrión L, Alvarenga LE, Keiski RL (2012) Production of activated carbon from cocoa (*Theobroma cacao*) pod husk. J Civil Environ Eng 2(2):1–6
5. Malaysia Palm Oil Board Statistics, MPOB (2011). Retrieved 10 Nov 2011. http://www.mpob.gov.my
6. Shuit SH, Tan KT, Lee KT, Kamaruddin AH (2009) Oil palm biomass as a sustainable energy source: a Malaysia case study. Energy 34(9):1225–1235
7. WanRosli WD, Law KN (2011) Oil palm fibers as papermaking material: potentials and challenges. BioResources 6(1):901–917
8. Coile NC, Shilling DG (1993) Cogongrass, Imperata cylindrical (L.) Beauc: a good grass gone bad! Botany Circular 28:1–3
9. Agele SO, Agbona AI (2008) Effects of cocoa pod husk amendment on soil and leaf chemical composition and growth of cashew (*Anacardium occidentale* L).seeding in the nursery. Am-Euraasian J Sustain Agric 2(3):219–224
10. Shakhes J, Marandi MAB, Zeinaly F, Saraian A, Saghafi T (2011) Tobacco residuals as promising lignocellulosic materials for pulp and paper industry. Bioresources 6(4):4481–4493
11. Jacobs RS, Pan WL, Fuller WS, Mckean WT (1999) Genetic and environmental influences on the chemical composition of Washington state wheat straw. In: 1999 pulping conference proceedings of Atlanta, TAPPI Press. pp 839–847
12. Wathén R (2006) Studies on fiber strength and its effect on paper properties. Ph.D. Thesis, Helsinki University of Technology
13. Al-Mefarrej HA, Abdel-Aal MA, Nasser RA (2013) Chemical evaluation of some lignocellulosic residues for pulp and paper production. Am-Eurasia J Agric Environ Sci 13 (4):498–504
14. Marques, Rencoret J, Ana G, José CDR (2010) Evaluation of the chemical composition of different non-woody plants fibers used for pulp and paper manufacturing. Open Agric J 4:93–101
15. Tran AV (2006) Chemical analysis and pulping study of pineapple crown leaves. Ind Crops Prod 24:66–74
16. Ai J, Tschirner U (2010) Fiber length and pulping characteristics of switchgrass, alfalfa stems, hybrid poplar and willow biomasses. Bioresour Technol 101(1):215–221
17. Saijonkari-Pahkal K (2001) Non-wood plants as raw material for pulp and paper. Master's Thesis, University of Helsinki
18. Akpakpan AE, Akpabio UD, Ogunsile BO, Eduok UM (2011) Influence of cooking variables on the soda and soda-ethanol pulping of Nypa Fruticans petioles. Aust J Basic Appl Sci 5 (12):1202–1208
19. Chandra M (1998) Use of nonwood plant fibres for pulp and paper industry in Asia: potential in China. Master's Thesis, State University
20. Heriksson G, Elisabet B, Helena L (2009) The trees. In: Monica Ek, Göran G, Heriksson Gunnar (eds) Wood chemistry and wood biotechnology vol 1. Walter de Gruyter GmBH and Co, United State of America, pp 13–44

Separation of Pyrrole from Isododecane Using Imidazolium and Pyridinium Based Ionic Liquid at 298.15 K: Experiment and COSMO-RS Prediction

Nurshakirin Hazim Chan, Hanee Farzana Hizaddin and Ramalingam Anantharaj

Abstract The removal of aromatic nitrogen compound from diesel oil is important to produce cleaner fuel and reduce the environmental impact. Therefore, denitrification of diesel oil is carried out by liquid-liquid extraction at ambient and moderate condition. The aim of this work is to study liquid-liquid extraction (LLE) for the mixture of ionic liquids (ILs); 1-ethyl-3-methylimidazolium ethylsulfate ([EMIM][$EtSO_4$]) and 1-ethyl-3-methylpyridinium ethylsulfate ([EMPy][$EtSO_4$], nitrogen compounds; pyrrole (PYR) and model diesel compound; isododecane (ISOD). The selectivity and solute distribution ratio values were calculated, four ternary diagrams were generated and denitrification efficiency was determined to evaluate the effectiveness of the extraction. The Conductor-like Screening Model for Real Solvents (COSMO-RS) model was used to predict the composition of the ternary systems. The experimental values and COSMO-RS prediction were then compared to determine the root-mean-square deviation (RMSD) for each system. The slope of tie lines is positive for all ternary system and the average RMSD is 2.825 %. The prediction gave RMSD of 1.12 % for [EMIM][$EtSO_4$](1) + Pyrrole(2) + Isodecane (3) system; 2.51 % for [EMPy][$EtSO_4$](1) + Pyrrole(2) + Isodecane(3) system.

Keywords Ionic liquid · Pyrrole · Isododecane · Solvent extraction · COSMO-RS prediction

N. Hazim Chan · H.F. Hizaddin · R. Anantharaj (✉)
Department Chemical Engineering, Faculty of Engineering, University of Malaya, 50603 Kuala Lumpur, Malaysia
e-mail: anantharaj@um.edu.my

N. Hazim Chan
e-mail: kyara_ryn@yahoo.com.my

H.F. Hizaddin
e-mail: hanee@um.edu.my

M.A. Hashim (ed.), *ICGSCE 2014*, DOI 10.1007/978-981-287-505-1_4

Introduction

Due to the increasingly strict environmental regulations on transportation fuel, cleaner fuel has become very significant research subject. Therefore, denitrification of diesel oil is the main interest of refineries worldwide nowadays. Diesel oil is a complex mixture of saturated, unsaturated and aromatic hydrocarbon which has low amount of nitrogen-based compounds. However, the nitrogen content will increase with the boiling point of a particular oil fraction. Most of the nitrogen compound found in diesel oil is present in aromatic compound which has a low reactivity and high refractoriness compared to acidic and non-acidic sulfur compound. The nitrogen-based compounds have to be removed from the diesel oil because they are (a) main source of air pollution; (b) strong inhibitor for desulfurization process; (c) emitting NO_x during combustion which will be converted into NO_2 that is the contributor for troposphere ozone formation. The presence of nitrogen compounds as low as <30 ppm can inhibits the efficiency of hydrodesulphurization process through competitive adsorption, which make it more difficult to meet the low-sulphur requirement. Thus, it is important to reduce the nitrogen content to achieve the level of zero emission by reducing from >70 to <0.1 ppm of nitrogen compound in diesel oil [1]. The nitrogen compound can be divided into non-basic compound (5-membered ring) and basic compound (6-membered ring). The non-basic nitrogen compounds are Pyrrole (C_4H_5N), Indole (C_8H_7N), Indoline (C_8H_9N), Carbazole ($C_{12}H_9N$) and Benzocarbazole ($C_{16}H_{11}N$) while the basic nitrogen compounds are Pyridine (C_5H_5N), Quinoline (C_9H_7N) and Benzoquinoline ($C1_3H_9N$).

The aim of this work is to perform LLE by using ionic liquids to know their selectivity toward nitrogen compound for the nitrogen removal in diesel oil at 298.15 K. Thus, the nitrogen compounds used in this work is pyrrole with isododecane as the feed and two types of ionic liquids which are 1-ethyl-3-methyl-imidazolium ethylsulfate ([EMIM][$EtSO_4$]) and 1-ethyl-3-methylpyridinium ethylsulfate ([EMPy][$EtSO_4$]).

Theory

COSMO-RS Theory: COSMO-RS stand for Conductor-like Screening Model for Real Solvents which is a unique method for a prior prediction of thermodynamics phase behavior of liquid mixture. This method was developed to give a general prediction without using any experimental data. The prediction is performed by screening charge density of individual molecules to calculate their chemical potential. In the quantum chemical calculation, the activity coefficient of molecules can be calculated by dividing the molecules according to their unique area and charge and get their ratio of charge segment and area segment [2]. The activity coefficient for the component is the total of their segment activity coefficient. By using the screening charge density of molecules, COSMO-RS need no functional group parameters. The prediction and experimental values can be compared by root-mean-square deviation (RMSD), which can be represented as:

$$\mathrm{RMSD}\,(\%) = 100 \times \left[\sum_{k=1}^{a}\sum_{i=i}^{b}\sum_{j=1}^{2}\frac{\left(x_{ik}^{j} - \hat{x}_{ik}^{j}\right)^{2}}{2mc}\right]^{\frac{1}{2}} \quad (1)$$

where a is the number of tie lines, b is the number of components; and 2 is the number of phases. While x_{ik}^{j} and $\hat{x}_{ik}^{j}$ are the experimental and predicted values of composition for component i at kth tie-line in phase j, respectively.

NMR Spectroscopy Theory: NMR spectroscopy is Nuclear Magnetic Resonance spectroscopy which is one of the most information-rich techniques because it exhibits a number of important characteristics for identification and quantitative analysis. The basis of this technique is the resonance interaction between high frequency field and nuclei of a compound placed in an external magnetic field. The nuclei of isotope which is hydrogen that has a spin quantum number of 1/2 will possess a magnetic moment by the spin of the nucleus [3]. Generally, it can be said that this analysis is for the measurement for the peak areas of the hydrogen molecule (1H) of each component. The concentration of each component can be calculated using the equation below [4]:

$$\mathrm{x_i} = \frac{\mathrm{H_i}}{\sum_{\mathrm{i}=1}^{3}\mathrm{H_i}} \quad (2)$$

where x_i is the composition of component i; and H_i is the peak area of single hydrogen atom in component i.

Methodology

Chemicals: Pyrrole was supplied by Sigma Aldrich, with a purity of 98 %. 1-ethyl-3-methylimidazolium ethylsulfate ([EMIM][$EtSO_4$]) with purity of ≥95 % was supplied by Sigma Aldrich while 1-ethyl-3-methylpyridinium ethylsulfate([EMPy][$EtSO_4$]) was supplied by Merck. Isododecane with 99 % purity was supplied by Sigma Aldrich as well as deuterated chloroform ($CDCl_3$) with purity >99.8 % which was used in ^{1}H NMR analysis of extract and raffinate phase.

Procedure: 10 pieces of 20 ml culture tubes were prepared. A culture tube was put on the weighing balance and it was tare to zero. A 5 ml syringe was used to transfer nitrogen compound into the culture tube for about 0.10 g of mass. The mass of nitrogen compound was recorded and tare it to zero again to measure the mass of next compound. A 10 ml syringe was used to transfer isododecane into the same culture tube for about 1.90 g. The mass of isododecane in the culture tube was recorded and the weighing balance was tare to zero again. A 10 ml syringe was used to transfer ionic liquid for about 2.00 g into culture tube. The mass of ionic liquid was recorded. The culture tube was then covered with a screw cap and secured with parafilm tape to avoid loss due to evaporation. The screw cap was labeled with the composition of nitrogen compound in the feed (5 %). The mixture in the culture tube was mixed by shaking the tube. The same steps were repeated to prepare for the other feed compositions (10–50 %). 10 culture tubes that were prepared were put in a shaker port at which the speed and temperature of the shaker were set at 130 rpm and 298.15 K, respectively. The culture tubes were kept in the shaker port for about 6 h. After 6 h of mixing, let the mixture rest to settle for at least 12 h. For NMR analysis, a syringe was used to transfer 3–4 drops of the top layer (raffinate phase) into the NMR tube with 0.5 ml of $CDCl_3$ solvent. The NMR tube was then capped and sealed with parafilm tape and ready to proceeds with ^1H NMR analysis. The steps were repeated for bottom layer (extract phase) as well as for other compositions.

Results and Discussion

LLE Studies

The ternary diagrams are presented in Figs. 1 and 2. It can be observed that in all ternary systems, the slopes of tie-lines are all positive and they are increasing with the solute (pyrrole) concentration. This indicates that the lower the solute concentration, the higher the solvent (IL) concentration required to extracts the solute. From the diagrams, it can also be seen that as the tie-lines length increases, the higher the immiscibility. In LLE studies, there two type of ternary system. Type I system consist of a system with one pair of compound that partially miscible and one immiscible region. Type II system is a system with two pair of compound that partially miscible and one immiscible region. In this case, the tie-lines for all systems are exhibiting both type of system, where the nitrogen compound and hydrocarbon is partially miscible in ionic liquid while ionic liquid is immiscible in hydrocarbon. Therefore, the raffinate phase is rich with hydrocarbon while ionic liquid is rich in extract phase.

Solvent Selection Parameters

The parameters of solvent selection are solute distribution ratio (β) and selectivity (S). It is significant to know whether the ILs used have high selectivity towards nitrogen compound or not. Therefore, the ability of IL in extracting nitrogen compound is measured by obtaining both of the parameters which can be defined by following equations:

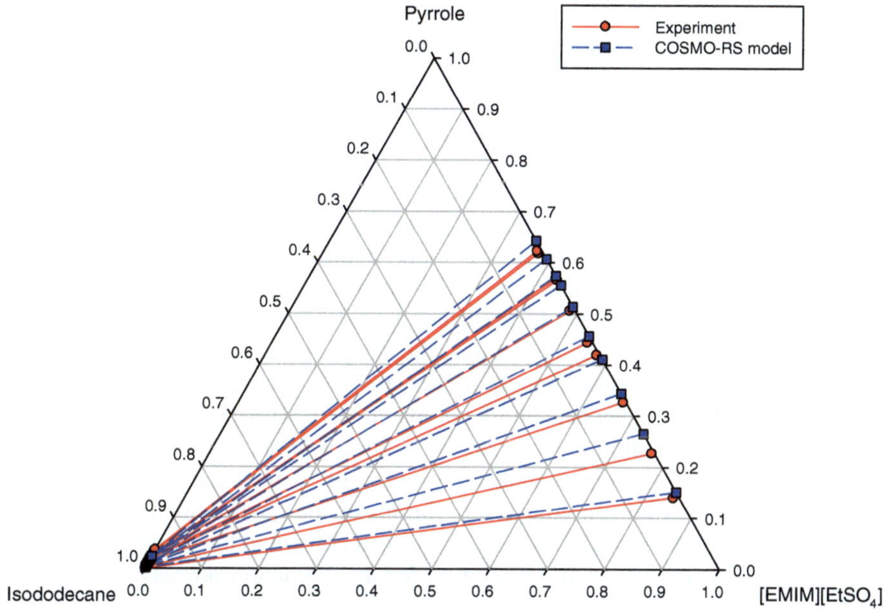

Fig. 1 Experimental tie-lines and COSMO-RS prediction for LLE of ternary diagram system {[EMIM][$EtSO_4$] (1) + Pyrrole (2) + Isododecane (3)}

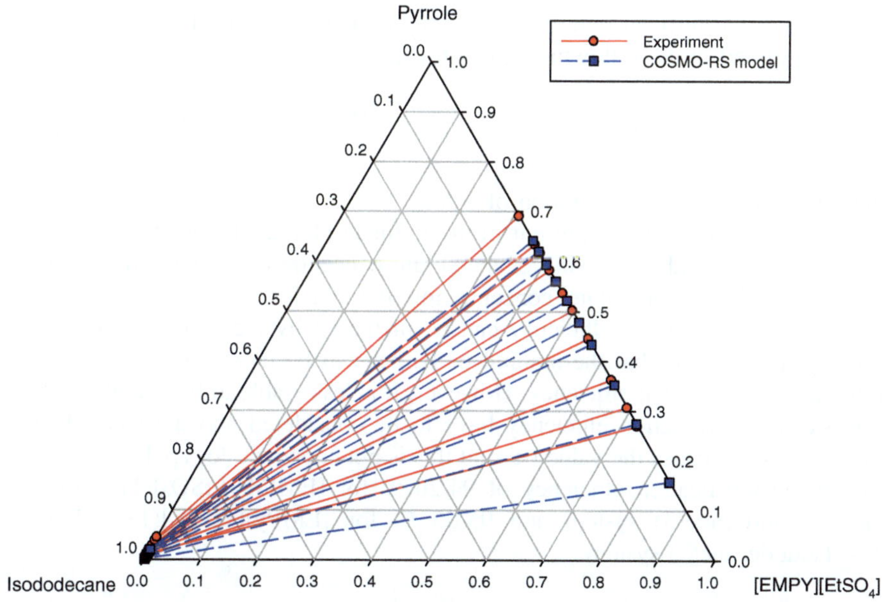

Fig. 2 Experimental tie-lines and COSMO-RS prediction for LLE of ternary diagram system {[EMPy][$EtSO_4$] (1) + Pyrrole (2) + Isodecane (3)}

$$\text{Solute distribution ratio, } \beta = \frac{x_2^{II}}{x_2^{I}} \quad (3)$$

$$\text{Selectivity, } S = \frac{x_2^{II} x_1^{I}}{x_2^{I} x_1^{II}} \quad (4)$$

where x refers to the concentration or mole fraction; subscript 1 refers to the hydrocarbon; subscript 2 refers to the ionic liquid; superscript I refers to the hydrocarbon rich phase (raffinate layer); and superscript II refers to the ionic liquid rich phase (extract layer). It can be seen that the selectivity at low concentration is very high. As the concentration increases, the selectivity also increases until some point where the concentration is high, it decreases. By comparing both of the ILs, [EMPy][$EtSO_4$] have higher selectivity compared to [EMIM][$EtSO_4$]. On the other hand, the solute distribution ratio for all the system is greater than unity (>1). This indicates that the nitrogen compounds are successfully extracted by ILs. The higher the concentration of nitrogen compound, the lower the quantity of IL required to extracts the nitrogen compound. Since the nitrogen content in the commercial diesel is very low, great amount of IL is required for the extraction process.

Denitrification Efficiency

The tendency of nitrogen removal can be represented by denitrification efficiency which is defined in the following equation [5]:

$$\textit{Denitrification efficiency}\,(\%) = \frac{x_{initial} - x_{final}}{x_{initial}} \times 100 \quad (5)$$

where x refers to the concentration of solute.

In this case, IL does not present in the raffinate phase. The denitrification efficiency decreases with increasing of the solute concentration. Therefore, it can be said that at the lower solute concentration the efficiency is very high. Some researcher found that the efficiency can be increased by increasing IL: Oil mass ratio [5]. This means that the amount of IL have to be increased in order to get higher nitrogen removal efficiency. However, it is less viable to be operated. Thus, multi-steps denitrification efficiency have to be considered and the ideal IL: Oil mass ratio is 1 because the value beyond this is not economic for application in the industry. Denitrification efficiency of 94.26 % for [EMIM][$EtSO_4$](1) + Pyrrole (2) + Isododecane(3) system and 95.06 % for [EMPy][$EtSO_4$](1) + Pyrrole (2) + Isododecane(3) system.

COSMO-RS Prediction

Figures 1 and 2 are illustrated to compare the experimental LLE data with COSMO-RS prediction of the ternary systems. At the immiscible region, the shape and length of immiscible region are slightly deviated, particularly in extract phase. The deviation may due to interactions between pyrrole with IL in the mixture where the hydrogen bonding plays a main role in the interaction. From Figs. 1 and 2, it can be observed that the experimental value for pyrrole is nearly matched to the COSMO-RS prediction at the raffinate phase. Pyrrole is 5-membered ring nitrogen compound while pyridine is 6-membered ring nitrogen compound. Therefore, pyrrole shows higher selectivity because it has higher electronegativity of heteroatom. The nitrogen atom in pyrrole molecule contains lone-pair electrons which are not tied up in the π–electron cloud of the ring [6]. The nitrogen atom has hydrogen atom attached to it in which it can interact with the electronegative atom in IL. Thus, experimental LLE data for pyrrole is nearly matched to the prediction at the raffinate phase. The deviation of experimental LLE data from COSMO-RS prediction is measured by root-mean-square deviation (RMSD). The RMSD is 1.12 % for [EMIM][$EtSO_4$](1) + Pyrrole(2) + Isododecane(3) system; 2.51 % for [EMPy] [$EtSO_4$](1) + Pyrrole(2) + Isododecane(3) system. The average RMSD for all ternary system is 2.825 %.

Conclusion

In this work, the following ternary systems, which are [EMIM][$EtSO_4$](1) + Pyrrole (2) + Isododecane(3), [EMPy][$EtSO_4$](1) + Pyrrole(2) + Isododecane(3) system were studied. The experimental LLE data were generated in ternary diagram corresponding with COSMO-RS prediction. Generally, the tie-lines for all the ternary system have positive slope. The selectivity of ionic liquid is higher at low solute concentration. Ionic liquid [EMPy][$EtSO_4$] have higher selectivity compared to [EMIM][$EtSO_4$]. The solute distribution ratio is greater than unity for all the system. The denitrification efficiency for system containing pyrrole is higher by using [EMPy][$EtSO_4$] with denitrification efficiency of 94.26 and 95.06 % respectively. Lastly, the experimental data were predicted with COSMO-RS model which give an average RMSD of 2.825 % for all ternary system.

Acknowledgment The modeling and experimental work reported in this article was financially supported by a research grant number **RP006E-13SUS** under University Malaya Research Grant (UMRG) from University of Malaya, Malaysia.

References

1. Anantharaj R, Banerjee T (2010) COSMO-RS-based screening of ionic liquids as green solvents in denitrification studies. Ind Eng Chem Res 49:8705–8725
2. Anantharaj R, Banerjee T (2013) Liquid-liquid equilibrium studies on the removal of thiophene and pyridine from pentane using imidazolium-based ionic liquids. J Chem Eng Data 58: 829–837
3. Perry RH, Green DW, Maloney JO (1984) Perry's chemical engineers' handbook, 6th edn. McGraw-Hill Book Company, New York, pp 4–17
4. Ravilla UK, Banerjee T (2012) Liquid liquid equilibria of imidazolium based ionic liquid + pyridine + hydrocarbon at 298.15 K: experiments and correlations. Fluid Phase Equilib 324:17–27
5. Wilfred CD, Chong FK, Man Zakaria, Azmi Bustam M, Ibrahim MMM, Chan ZP (2012) Extraction of dibenzothiophene from dodecane using ionic liquids. Fuel Process Technol 93:85–89
6. Anantharaj R, Banerjee T (2011) Fast solvent screening for the simultaneous hydrodesulfurization and hydrodenitrification of diesel oil using ionic liquids. J Chem Eng Data 56: 2720–2785

The Development of Amine-Based Potential Kenaf Sorbent for Carbon Dioxide (CO_2) Capture

Nabilah Zaini and Khairul Sozana Nor Kamarudin

Abstract Adsorption process becomes a promising, cost-effective and efficient separation technique for CO_2 capture. Inspired by the commercialized technique for capturing and sequestering CO_2, the development of solid amine sorbents relies on support material has become a great interest amongst researchers worldwide. Based on this new invention, a research on kenaf as cost-potential agricultural source was conducted to study the adsorptive performance towards carbon dioxide (CO_2). The research work was initiated by impregnating a series of amine groups such as MEA, DEA and MDEA on kenaf via wet impregnation method. The CO_2 adsorption study was carried out by flowing 300 cm^3/min of purified CO_2 to the PSA column and compressed up to 1.5 bar. Result obtained indicates that MEA has achieved the highest amount of CO_2 adsorbed amongst the other types of amines. Therefore, MEA was selected for further study by varying the ratio of kenaf to MEA as 1:0.5, 1:0.7, 1:1, 1:2, 1:5, 1:7 and 1:10. It is shown that the ratio of kenaf to MEA of 1:1 have recorded the highest amount of CO_2 capture capacity (2.07 mmol CO_2/g sorbent) as compared to 1:0.5, 1:0.7, 1:2, 1:5, 1:7 and 1:10 with 0.78, 1.71, 1.77, 1.20, 0.96 and 0.89 mmol CO_2/g sorbent, respectively. This study inferred that the introduction of MEA has enhanced the adsorptive capability of kenaf towards CO_2.

Keywords Kenaf · Amine · Impregnation · Adsorption · CO_2 · Pressure swing adsorption (PSA)

N. Zaini (✉) · K.S.N. Kamarudin
Faculty of Petroleum and Renewable Energy Engineering, Universiti Teknologi Malaysia, UTM Skudai, 81310 Johor Bahru, Johor, Malaysia
e-mail: nabilahzaini88@gmail.com

K.S.N. Kamarudin
e-mail: sozana@petroleum.utm.my

M.A. Hashim (ed.), *ICGSCE 2014*, DOI 10.1007/978-981-287-505-1_5

Introduction

For many years, there has been a growing concern about global warming and climate change that prompted the development towards more-efficient and improved processes for carbon dioxide (CO_2) capture from various sources. The greenhouse gases that mainly consist of CO_2 have increased by 30 % over the past 200 years [1]. CO_2 capture from large point sources was identified as a major option to address the problem and known as CO_2 capture and storage [2]. CO_2 capture and storage (CCS) technology was implemented via amine-based absorption, cryogenic techniques, membranes diffusion, adsorption by using solid sorbent and oxygen recovery from O_2/CO_2 recycle combustion [3]. Among of these options, amine-based regenerative chemical absorption processes have been widely practiced for capturing CO_2. However, this technology exists several drawbacks such as low CO_2 loading capacity, high energy consumption, high equipment corrosion rate, amine degradation by SO_2, NO_2 and O_2, large equipment size required, solvent leakage, fouling of the process equipment, have limited amine concentration in aqueous phase because of the viscosity and foaming issues and possible environmental problems due to high volatility of amine [4–9]. Therefore, the cost-effective and efficient process for CO_2 capture should be replaced the existing technology in order to mitigate the global warming issues.

The economic crisis has led the researchers to turn their interest and expertise into adsorption field. Adsorption is a promising method since it is cost-effective and efficient CO_2 separation technique. It requires less regeneration energy than aqueous amine-based process, higher adsorption capacity and selectivity towards CO_2, low heat capacity than aqueous amine solvent and ease of handling [2]. Inspired by the commercialized technique for capturing and sequestering CO_2, the development of solid amine sorbents had become a great interest amongst researchers worldwide. The introduction of green material as a sorbent that originated from agricultural sources makes them very competitive since it is cost-effective as compared to the existing amine-based support. Therefore, a research on kenaf as agricultural source was conducted to study the adsorptive performance towards CO_2. To the best of the authors' knowledge, there are no literatures on the modification of kenaf by using amine functional group for CO_2 capture. Thus, this paper presents the modification of kenaf by impregnating different types of alkanolamines onto kenaf surface in order to determine the CO_2 adsorption capacity.

Experimental

Material and Chemicals, In this study, kenaf core without outer periphery bark with particle size larger than 6 mm was supplied by National Kenaf and Tobacco Board (NKTB), Kelantan. The sample was then grinded into smaller particle size (300 μm). The sample was stored in air-tight container to prevent moisture build up and fungi infections and was used for impregnation procedure without having pre-treatment process. All chemicals used in this study were of highly concentrated reagent grade; MEA, DEA and MDEA. Methanol (CH_3OH, Merck) was selected as solvent in amine impregnation procedure since it has lower boiling point (64.7 °C) than water (100 °C). This initial adsorption study involves the use of purified CO_2 (99.999 %, Mega Mount Industrial Gases Sdn Bhd) and purified N_2 (99.999 %, Mega Mount Industrial Gases Sdn Bhd) for regeneration process.

Amine-Impregnation Procedure, The synthesis of amine-impregnated kenaf was carried out via conventional method known as wet impregnation method. Initially, kenaf sample was wetted in methanol solution according to the ratio solid to liquid of 1:20 (weight percent) and the mixture was maintained in contact under mechanical agitation of 600 rpm for 20 min in two stages prior to be filtered and air-dried overnight. Alcoholic amine solution was prepared by stirring MEA with methanol for 20 min. Then, the dried kenaf sample was added into the alcoholic amine solution for 15 min prior to be agitated for 5 h impregnation. Finally, the product of MEA-impregnated kenaf was filtered and air-dried before being stored in desiccators for further analysis. The similar procedure was repeated for DEA and MDEA.

CO_2 Adsorption and Regeneration Study, In this study, the CO_2 adsorption process was carried out by using a single-column pressure swing adsorption system (PSA). The column used has a dimension of 15 cm height and 1 cm diameter. The PSA operation system involves four phases; pressurization, adsorption, desorption and purging. The adsorption process was conducted at 1.5 bar with a feed gas of 300 cm^3/min for 5 min. Whilst, the regeneration process was implemented by changing the pressure to 1 bar. The schematic diagram of single-column PSA operation system is shown in Fig. 1. In this study, the adsorption process was

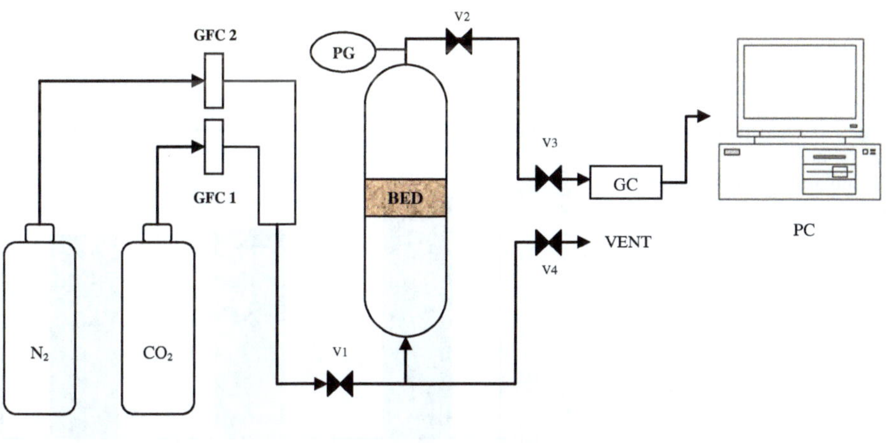

Fig. 1 Schematic diagram of single-column PSA system

executed in three (3) consecutive cycles of operation. Gas chromatography (7820A) from Agilent Technologies was used to measure the amount of gas leaving the PSA system in each cyclic operation.

Result and Discussion

CO_2 Adsorption Study, In this study, a series of amine groups containing N-functional groups (MEA, DEA and MDEA) have been studied for the adsorption of CO_2. Initially, a sample containing 50 wt% amine-impregnated kenaf was pressurized with purified CO_2 at 1.5 bar in order to determine the highest types of amine for capturing CO_2. The results were presented in Fig. 2.

Figure 2 presents a comparison of CO_2 capture capacity for MEA, DEA and MDEA impregnated onto kenaf sorbent. It is demonstrated that kenaf that has been impregnated by MEA has achieved the highest CO_2 capture capacity (0.78 mmol CO_2/g) compared to DEA and MDEA with 0.68 and 0.66 mmol CO_2/g, respectively. Generally, the structures of alkanolamines containing at least one hydroxyl group (–OH) and one amine group (–NH_2) that reacts directly with CO_2 to form carbamate ions for primary and secondary amines and bicarbonate or carbonate for tertiary amine. The reaction of primary and secondary amines with CO_2 can be represented in Eq. (1). Whilst, the reaction of tertiary amine with CO_2 is indicated in Eq. (2) [9].

$$2RR'NH + CO_2 \leftrightarrow RR'NCOO^-(\text{carbamate}) + RR'NH_2^+ \quad (1)$$

$$RR'NCOO^- + H_2O \leftrightarrow RR'NH + HCO_3^-(\text{bicarbonate}) \quad (2)$$

where $R = C_2H_4OH$. The ability to form carbamate ions result from a direct reaction of primary and secondary amine with CO_2 could produces faster CO_2 capture kinetics [10].

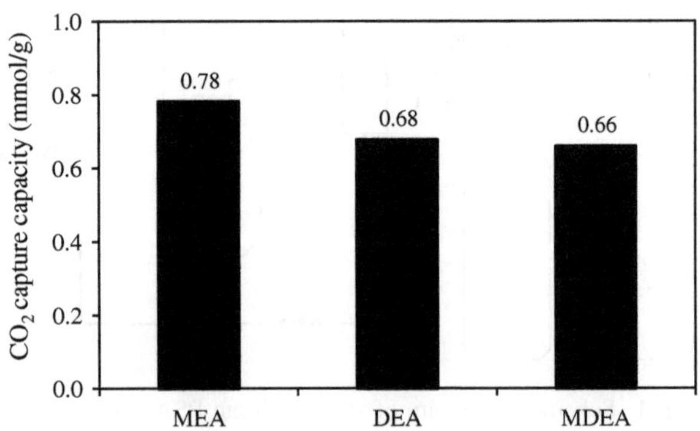

Fig. 2 CO_2 capture capacity for different types of alkanolamines on kenaf at 1.5 bar

Although the formation of carbamate ions proceed rapidly, the CO_2 capture capacity for MEA and DEA is reduced according to the stoichiometry requirement of two amine molecules for each CO_2 molecule reacted (Eq. 1). Since the adsorption time is only 5 min, the tertiary amine did not transform into carbamates because of slow reaction rate. Therefore, the adsorbent has low affinity towards CO_2 which is shown in Fig. 2. The impregnation of amine has introduced nitrogen functional groups on kenaf that provides the additional sites for CO_2 adsorption; hence improve the capture capacity towards CO_2. Moreover, this result could also be attributed to the effect of molecular size since MEA has the lowest size of molecule (61.08 g/mol) compared to DEA (105.14 g/mol) and MDEA (119.16 g/mol). The intrusion of MEA could enhance the basicity of kenaf support through the numbers of amine impregnated. In the other perspective, it can be said that the smaller size of amine could increase the numbers of amine impregnated on kenaf support and vice versa. This argument can be proved through the pH value of decanted amine solution gained during experimental work. The pH values for each prepared sample were recorded accordingly as tabulated in Table 1. It is clear that basicity of the sample increases by ascending order of MDEA < DEA < MEA. The higher basicity of prepared sample will increase the tendency as well as affinity towards acidic CO_2, thereby enhancing the capability of capturing CO_2. To further study the adsorption characteristics of CO_2 on amine-impregnated kenaf, the amount of MEA was increased up to 10 times of their weight and was presented in Fig. 3.

Table 1 Parameters for amine-impregnated kenaf sample

Sample	pH values	
	Alcoholic amine solution	Decanted alcoholic amine solution
50 wt% MEA	11.23	10.60
50 wt% DEA	10.62	10.24
50 wt% MDEA	10.21	10.18

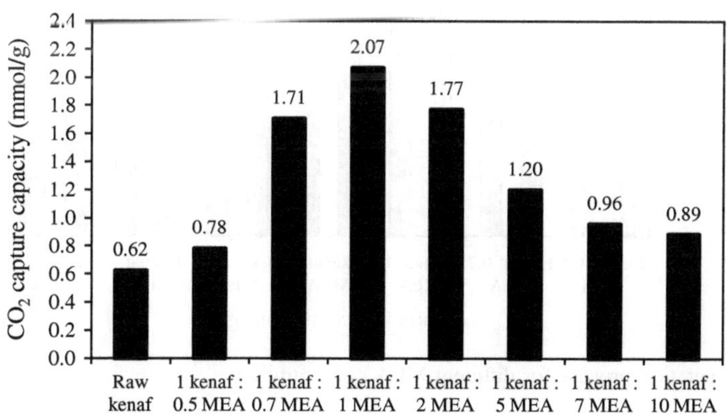

Fig. 3 CO_2 capture capacity for different MEA concentration on kenaf at 1.5 bar

From Fig. 3, it can be inferred that the performance of kenaf after MEA modification increased significantly as compared to the raw kenaf. This is due to the additional basic adsorption sites exist on the surface that was discussed previously. Then, the MEA concentration was varied to 0.5, 0.7, 1, 2, 5, 7 and 10 times of sorbent weight. Results gained show that the ratio of kenaf to MEA as 1:1 indicates higher CO_2 capture capacity (2.07 mmol CO_2/g) compared to other ratios (1:0.5, 1:0.7, 1:2, 1:5, 1:7 and 1:10) with the capture capacity of 0.78, 1.71, 1.77, 1.20, 0.96 and 0.89 mmol CO_2/g, respectively. It might be due to the effect of the number of adsorption sites on sorbent. The increment of adsorption sites is directly proportional to the MEA concentration up to 1:1. Further increased the MEA concentration might block the pore channel of kenaf, less adsorption due to the effect of steric hindrance. This finding reveals that the ratio of kenaf to MEA of 1:1 is the best concentration for MEA impregnation on kenaf sorbent. Further increase of MEA concentration could reduce the CO_2 capture capacity of MEA-modified kenaf. It can be conveyed that the additional concentration of MEA could not give any increment to the number of basic adsorption sites.

Regeneration Study, Regeneration study is a crucial and important step in order to ensure the adsorbent can be used for the next cycle operation. In spite of having high adsorption capacity towards adsorbate, the adsorbent must also display a stable adsorption performance for cyclic operation. In this study, the regeneration process was implemented by reducing the adsorption pressure of 1.5 bar to 1 bar for three (3) consecutive cycles of operation. The regeneration performance for each sample is shown in Fig. 4. It can be inferred that the adsorption capacity was almost constant for three (3) consecutive cycles of adsorption and desorption. The results indicate that the adsorption performance was stable under those conditions.

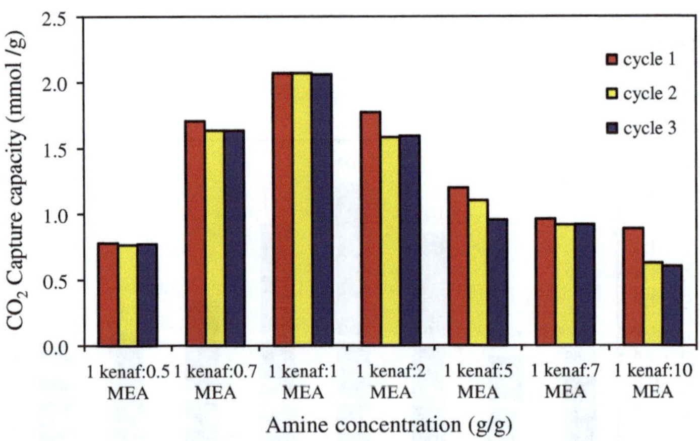

Fig. 4 Regeneration process for different MEA concentration

Conclusion

This finding revealed that the functionalized of kenaf by MEA could improve the adsorption of CO_2 as compared to the raw kenaf. The optimum CO_2 uptakes can be achieved as high as 2.07 mmol CO_2/g for a ratio of kenaf to MEA of 1:1 at 1.5 bar and 298 K. The high adsorption of CO_2 on MEA-modified kenaf can be ascribed by the presence of nitrogen-functionalized group as well as the chemical adsorbate-adsorbent interaction. However, further increase the amount of MEA loadings on kenaf could reduce the CO_2 capture capacity. This study also shows that MEA-modified kenaf have a stable cyclic adsorption/desorption performance up to three (3) cycles of operation. Indeed, kenaf with large pore size could be a good and promising host for the development of novel CO_2 adsorbent.

Acknowledgments The authors gratefully acknowledge the financial support given by Ministry of Education Malaysia (MOE), Ministry of Science, Technology and Innovation (MOSTI) Vote no. 4S047 and Universiti Teknologi Malaysia (UTM). An appreciation also dedicated to National Kenaf and Tobacco Board (NKTB), Kelantan for providing the kenaf sample.

References

1. IPCC (2001) Chapters 1 and 2. Climate change: the scientific basis. Contribution of working group to the third assessment report of the intergovernmental panel on climate change. Cambridge University Press, Cambridge
2. Samanta A, Zhao A, Shimizu GKH, Sarkar P, Gupta R (2012) Post-combustion CO_2 capture using solid sorbents: a review. Ind Eng Chem Res 51:1438–1463
3. Gray ML, Soong Y, Champagne KJ, Pennline H, Baltrus JP, Steven RW Jr, Khatri R, Chuang SSC, Filburn T (2005) Improved immobilized carbon dioxide capture sorbents. Fuel Process Technol 86:1449–1455
4. Màmun S, Nilsen R, Svendsen HF, Juliussen O (2005) Solubility of carbon in 30 Mass % monoethanolamine and 50 Mass % methyldiethanolamine solutions. J Chem Eng Data 50:630–634
5. Mandal BP, Bandyopadhyay SS (2005) Simultaneous absorption of carbon dioxide and hydrogen sulphide into aqueous blends of 2–Amino–2–Methyl–1–propanol and diethanolamine. Chem Eng Sci 60:6438–6451
6. Mandal BP, Bandyopadhyay SS (2006) Absorption of carbon dioxide into aqueous blends of 2–amino–2–methyl–1–propanol and monoethanolamine. Chem Eng Sci 61:5440–5447
7. Yue MB, Sun LB, Cao Y, Wang ZJ, Wang Y, Yu Q, Zhu JH (2008) Promoting the CO_2 adsorption in the amine–containing SBA–15 by hydroxyl group. Microporous Mesoporous Mater 114:74–81
8. Cao C, Zhang K, He C, Zhao Y, Guo Q (2011) Investigation into a gas–solid–solid three phase fluidized–bed carbonator to capture CO_2 from combustion flue gas. Chem Eng Sci 66:375–383
9. Yu CH, Huang CH, Tan CS (2012) A review of CO_2 capture by absorption and adsorption. Aerosol Air Qual Res 12:745–769
10. Filburn T, Helble JJ, Weiss RA (2005) Development of supported ethanolamines and modified ethanolamines for CO_2 capture. Ind Eng Chem Res 44:1542–1546

Potential Effect of Palm Oil Fuel Ash as Micro-Filler of Polymer Concrete

A.K. Nur Hafizah, M.W. Hussin, M. Ismail, M.A.R. Bhutta, M. Azman, P.J. Ramadhansyah, A. Nur Farhayu and A.S.L. Nor Hasanah

Abstract This paper discusses the feasibility of using an agricultural waste called palm oil fuel ash (POFA) as micro-filler in polymer concrete. As a cellulose waste material, such potential of POFA is often neglected since it tends to uptake excessive resin during the mixing process. Hence, the filler characterization was involved in this study to investigate the potential of POFA become as micro-filler in polymer concrete. The filler characteristics of unground and ground POFA (UPOFA and GPOFA) were examined under microstructural examination, which included particle size analyzer (PSA), morphology image, and termogravimetric

A.K. Nur Hafizah · M.W. Hussin (✉) · A. Nur Farhayu · A.S.L. Nor Hasanah
Institute of Smart Infrastructure and Innovative Construction (ISIIC), Faculty of Civil Engineering, Universiti Teknologi Malaysia, 81310 Johor, Malaysia
e-mail: warid@utm.my

A.K. Nur Hafizah
e-mail: fiza_johor2003@yahoo.com.sg

A. Nur Farhayu
e-mail: farhayu_rui@yahoo.com

A.S.L. Nor Hasanah
e-mail: amoi_1464@yahoo.com

M. Ismail
Departments of Structure and Materials, Faculty of Civil Engineering, Universiti Teknologi Malaysia, 81310 Johor, Malaysia
e-mail: mohammad@utm.my

M.A.R. Bhutta
Department of Civil Engineering, Faculty of Applied Science, University of British Columbia, Vancouver, BC V5P 1X6, Canada
e-mail: aamer.bhutta@yahoo.com

M. Azman · P.J. Ramadhansyah
Department of Geotechnics and Transportation, Faculty of Civil Engineering, Universiti Teknologi Malaysia, International Campus, 81310 Johor, Malaysia
e-mail: azmanmohamed.kl@utm.my

P.J. Ramadhansyah
e-mail: ramadhansyah@utm.my

M.A. Hashim (ed.), *ICGSCE 2014*, DOI 10.1007/978-981-287-505-1_6

and differential thermal analyses (TGA and DTA). A total of ten mix designs of polymer concrete were employed for flowability and compression test. Then, the workability and strength of polymer concrete were investigated after incorporating two types of filler with different filler content. Two major outcomes had been achieved. Firstly, the filler had their physical features significantly changed after physical surface modification. Secondly, finer filler had given superior workability and compressive strength. In conclusion, the agricultural waste of POFA can be incorporated as micro-filler in polymer concrete after being physically modified and mixed with the appropriate mix design to obtain superior mechanical properties.

Keywords Agricultural waste · Palm oil fuel ash · Filler · Polymer concrete

Introduction

Polymer concrete (PC) is produced solely using polymer resin as concrete binder, dry inert granular of aggregate, and occasionally filler. Such PC mix does not contain cement materials and water; its hardening takes the form of the polymerization process when additives, catalysts, or accelerators are added. Generally, any PC with filler tends to have denser packing of the solids concrete which enhances the mechanical properties. The type of filling materials, i.e., natural, granulated, or synthetic, affects the composite characteristics [1].

Currently, the incorporation of natural filler originating from calcium carbonate and sand in PC production is still being debated. In spite of this, the incorporation of green material with the waste in concrete that has PC enhancing characteristics has recently come into the lime light as a more attractive alternative to researchers. The characteristics of PC incorporated with filler are governed by several factors such as the filler type, its size and shape, and the type as well as amount of binder used [2]. The filler exists as small particles that fit into the gaps between larger particles to enhance the mechanical properties of PC, and such has been proven by several researchers. For instance, Atzeni et al. [3] have shown that the addition of fly ash as filler in epoxy mortar can enhance their specimens' mechanical properties better than conventional quartz flour filler mainly because of the great fineness of fly ash filler. Other than the aforementioned fly ash, palm oil fuel ash (POFA) is also attractive alternative filler. POFA can be found in abundance as agricultural waste that is often uncontrollably dumped in landfills. Nevertheless, not all waste ashes have the potential to become PC filler. The wrong selection of filler material may lead to worse PC quality. Not only that, it affects the PC's workability and process ability [4]. Most natural waste source from agricultural plant is cellulose [5, 6] which has a structure that attracts liquid into PC. This can lead to excessive resin consumption, which is not cost effective, and thus jeopardize the production of PC even when filler is used. Until today, no study has so far been done on the incorporation of POFA as PC filler because of the aforesaid potential setbacks.

The objective of this paper is to give an insight into the potential incorporation of palm oil fuel ash (POFA) as micro-filler into PC, gauged through microstructure and strength examination. The first outcomes of microstructure experiments on raw and ground POFA (UPOFA and GPOFA) were compared with previous findings. The workability and compressive strength of PC with appropriate mix design was investigated to support the outcomes. The outcomes of this work will be beneficial to researchers and engineers working in the field of cellulose filler in PC.

Material and Method

Binder, The polymer binders used in this study was the isophthalic unsaturated polyester resin. Its general properties are shown in Table 1. Polyester resin has been chosen in this study since it is widely used in the composites industry and is the cheapest among thermoset resins. Additionally, the resin can be easily handled, pigmented, filled and fiber reinforced in liquid form [7].

Aggregates and Filler, Crushed coarse aggregate and natural river fine aggregate were used as the inert granular material. The aggregate was oven-dried at 100 ± 5 °C. The moisture content of the inert granular material was limited below 0.1 %. The size of coarse aggregate of between 10 and 12 mm was used in this study.

Micro-filler used in this study was the palm oil fuel ash (POFA) since it was cheap and readily available. The POFA was collected from the southern part of Peninsular Malaysia. It was oven-dried at 100 ± 5 °C and the moisture content was kept constant below 0.1 % before being further divided into two portions, i.e., UPOFA and GPOFA. The UPOFA was obtained directly by passing the POFA through the 300 μm sieve to separate their natural substances and homogenize the particles size. Meanwhile, the GPOFA was obtained directly by passing the POFA through the 45 μm sieve and then grinding them. After that, it was grounded using a laboratory grinding mill with rod bar to obtain finer filler.

Microstructure Examination, Particle size distribution was also analyzed to observe the changes in size before and after the particle grinding process. For both GPOFA and UPOFA, this was done using the particle size analyzer that employed the wetting method. The particle was dispersed using distilled water to avoid agglomerate condition before the test was run to get convincingly accurate results.

Morphology test was conducted to examine the surface of GPOFA and UPOFA under field emission scanning electron microscopy (FESEM) with 250 times

Table 1 Properties of isophthalic unsaturated polyester resin

Properties	Isophthalic unsaturated polyester resin
Density (g/cm^3)	1.1
Viscosity (Brookfield), 25 °C, 60 rpm (cps)	538

Table 2 Mix proportion of polymer concrete

Mix proportion				
Polymer concrete type	Polyester binder (kg/m^3)	POFA filler (kg/m^3)	Fine aggregate (kg/m^3)	Coarse aggregate (kg/m^3)
PC-GPOFA PC-UPOFA	132[a]	78	1240	750
		97	1190	
		116	1141	
		136	1091	
		155	1042	

[a]Polyester binder was mixed with 1 % MEKP and 0.5 % CoNp

magnifications. Before that, both GPOFA and UPOFA specimens were coated so that significant morphology image could be captured.

Termogravimetric and differential thermal analysis (TGA and DTA) was employed where automatic thermal analyzer was used to investigate the thermal behaviour of POFA. The measurement was carried out between 30 and 1000 °C with heating rate set at 10 °C/min in nitrogenic atmosphere. The finding was similar since both GPOFA and UPOFA exhibited identical thermal behaviour due to both POFA from similar origin.

Mix Proportion, Ten mix proportions of PC incorporating POFA as filler are given in Table 2. Since resin is expensive, the low content of resins (binder) was fixed at 12 % by weight of resin. Then, they were separately mixed with methyl ethyl keton peroxide (MEKP) as catalysis (harderner) and cobalt naphthenate (CoNp) as initiator. The unground and ground POFA micro-filler that occupied the PC spaces varied from 8 to 16 % by weight of POFA. To achieve the strength of PC inert granular material, fine and coarse aggregates were used. The coarse aggregate content was fixed at 30 % by weight of coarse aggregate. The PC without any filler was designated as the control specimen.

Specimen, The dry PC mix was prepared before the wet mix, and consisted of inert granular aggregates and filler. The wet mix had additional polyester binder mixed together with its initiator and promoter. Then, the wet mix was poured into the dry mix and immediately mixed together using concrete mixer. Releasing agent was applied onto the mould surface for easy demoulding. The fresh concrete was thereafter poured into the mould and compacted with a vibrating table for 20 s. The specimens were subjected to post-curing by oven-heating temperature at 70 °C for 4 h. The specimens were prepared strictly according to JIS A 1181 [8].

Flowability Test, Flowability test was conducted as per JIS R 5201 [9] after mixing the PC to measure its workability, which was governed by the flowing ability of actively-mobilized materials with respect to the flow spread. In this study, the actively-mobilized materials were the polymer binder and filler, which were used to fill in the gaps of inert materials. Both materials were blended at various filler content and no compaction was applied during the test. The blended polymer was poured into v-funnel slump cone (inside diameter of v-funnel con is 70 and 100 mm while the height is 50 mm) and immediately uplifted once the cone was full of blended polymer. The flow spread diameter was measured five times in every test after the flow spread ceased. This method was employed since PC is sticky and has zero slumps. Similar test had been conducted to examine the flowing ability of self-compacting mortar using v-funnel slump cone by Safiuddin et al. [10], which proposed that the flow spread could be indicative of the workability of PC.

Strength Test, The cured specimens were tested for compressive strength according to JIS A 1181 [8]. A total of 66 cubical specimens including control specimens (without any filler) with 100 mm by 100 mm in cross-section were prepared for the compression test. Compression test was carried out by using compression machine with capacity of 200 kN. The appropriate mix design at different filler content was investigated at this point.

Result and Discussion

Particle Size Distribution, The particle size distribution curve of UPOFA and GPOFA are shown in Fig. 1. The distribution showed that both UPOFA and GPOFA had well-graded particles present as both coarse and fine micro-particle, but more than 90 % had passed through 300 and 45 μm sieve, respectively. The mean particle size of GPOFA was 14.21 μm and that of UPOFA was 86.00 μm. Smaller particle size was obtained by grinding. These were confirmed clearly in the FESEM image as shown in Fig. 4a. The UPOFA was considered as having coarse micro-particle since the original mean particle size in the study of Bakar et al. [11] was not more than 18 μm; the UPOFA was comparatively coarser in this sense.

Particle Morphology, Fig. 2a, b shows the morphology of GPOFA and UPOFA, respectively. The morphology results indicated that the UPOFA had a combination of spherical and irregular shaped particles with a sizeable fraction and porous cellular structures. Meanwhile, GPOFA had more uniform spherical shaped particles as compared to UPOFA. The morphology of UPOFA was similar to that found by Noorvand et al. [12] using SEM. The opening size of pores for UPOFA ranged between 8.3 and 17 μm; this opening size may have affected the workability

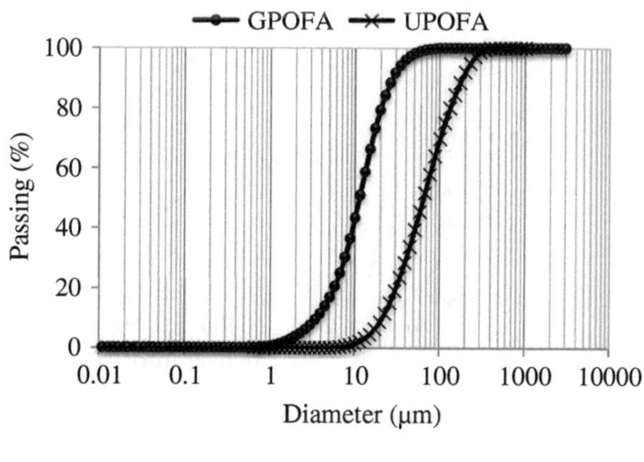

Fig. 1 Particle size distribution

Fig. 2 FESEM image of ground POFA (**a**) and unground POFA (**b**)

of fresh PC since it has higher tendency of absorbing liquid resin. This is in agreement with Haidar et al. [13]. Other than that, it may have increased the number of voids and capillaries in the structure, thus resulting in the reduction of densely pack structure [12]. The existing UPOFA opening pore size also showed its original particle features as cellulose based agricultural waste, which can be physically modified by grinding to make it finer and reduce the particle pores (see Fig. 2a). Additionally, a known fact, the fineness of particle give the highest surface area and this was supported by Bakar et al. [11]. This also demonstrated the potential of GPOFA as filler in PC which is higher than UPOFA.

Termogravimetric and differential thermal analysis (TGA and DTA), TGA and DTA were obtained in concomitant in a single test as shown in Fig. 3. The Figure has also depicted that the first DTA intensity peak of POFA occurs at the boiling point of H_2O, i.e., 100 °C. Major POFA weight loss was observed between 500 and 780 °C, which happened at the second DTA intensity peak. The significant weight loss of POFA was interpreted to be due to the production of silica. In PC any inherent moisture should be eliminated to enhance the bonding between materials,

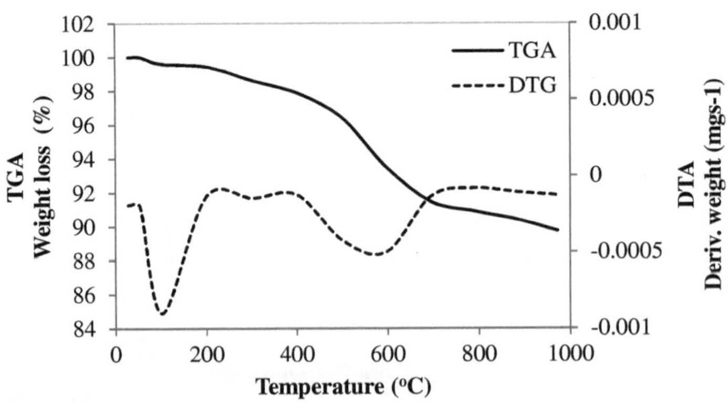

Fig. 3 TGA and DTA analysis for POFA

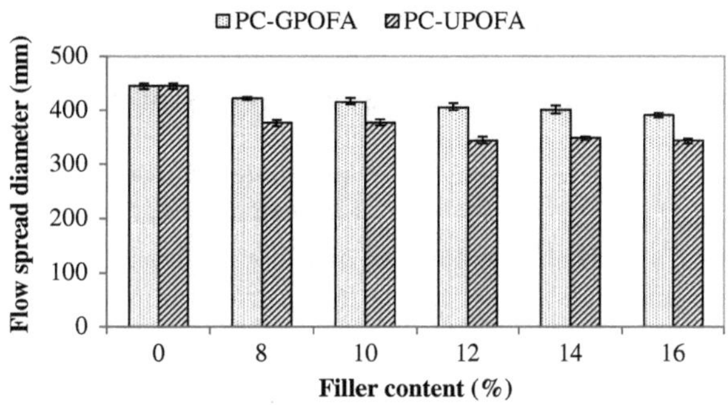

Fig. 4 Flow spread diameter of blended polymer with ground and unground POFA

and the moisture content in this regard is usually controlled below than 0.1 % at 100 ± 5 °C [14–16]. These conditions were kept in this study. The TGA and DTA examinations are very important to gauge the POFA's performance as PC filler at the post-curing process and under high pressure processing [17, 18]. Additionally, the results indicated that POFA is able to withstand thermal degradation up to 590 °C.

Workability, Fig. 4 presents the flow spread diameter of blended polymer (polymer binder and POFA filler) as obtained in the flowability test. Higher flow spread diameter indicates higher flowability and thus higher workability of PC and vice versa. From the results, the overall flowability of various blended polymer had gradually decreased when the filler content was increased. The decrease in flowability of blended polymer reduced about 12 and 23 %, respectively for PC-GPOFA and PC-UPOFA, respectively, as compared to neat polymer. Nevertheless, generally, the blended polymer with GPOFA had shown superior flowability to that of blended polymer with UPOFA. This showed that the grinding of POFA had significantly improved the flowability of blended polymer. The worst flowabilty exhibited by unground POFA (UPOFA) was predominantly caused by the open cellulose structure of original POFA, as seen in Fig. 4b. Such structure has encouraged excessive uptake of polyester resin and this has been observed by Bignozzi et al. [4] as well.

Compressive Strength, The compressive strength of different PCs with various filler content is shown in Fig. 5. Generally, the filler had increased the specimens' compressive strength approximately up to 57 and 18% for both PC-GPOFA and

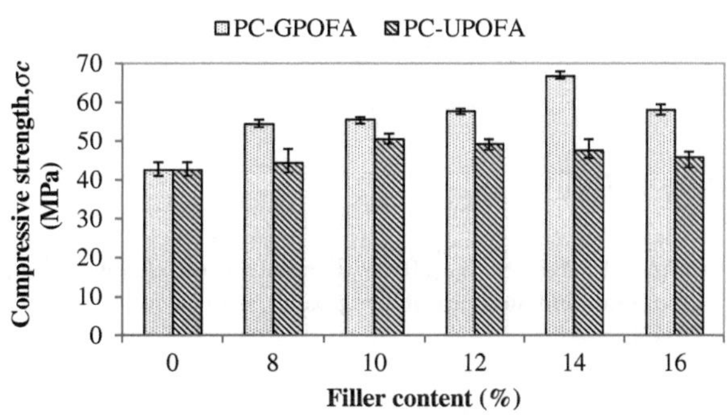

Fig. 5 Compressive strength of polymer concrete with ground and unground POFA

PC-UPOFA, respectively, as compared to neat PC. uced about 12 and 23 %, respectively for PC-GPOFA and PC-UPOFA, respectively, as compared to neat polymer. Moreover, the results demonstrated that an increment in ground POFA added could give superior compressive strength for all POFA. Lower compressive strength was obtained with unground POFA because of the earlier mentioned original cellulose-pores structure of POFA. This had caused a decrease in the concrete mix's workability. Results had also indicated that the cellulose-pores structure in original POFA might have taken up the liquid resin; such is in agreement with the findings of Bignozzi et al. [4], Haidar et al. [13]. Additionally, the compressive strength had been influenced by the workability of the PC, which had increased when GPOFA was added. However, the optimum filler content necessary for Iso-GPOFA and Iso-UPOFA was 14 and 10 %, respectively. This demonstrated that GPOFA had satisfactorily filled in the gaps between inert materials. In conclusion, the agricultural waste of POFA has high potentiality to be used as filling material in PC after being physically modified.

Conclusion

This paper had described the potential of agricultural POFA waste as filler for PC with the following conclusions drawn:

1. The grinding process would give satisfactory POFA fineness.
2. The original cellulose-pores structure of POFA had significantly diminished after grinding.
3. POFA is able to withstand thermal degradation up to 590 °C.
4. Ground POFA (GPOFA) had increased the flowability of blended polymer and enhanced the workability and strength of PC.
5. The agricultural waste of POFA has the potential to be used as micro-filler in PC after being subjected to physical surface modification.

Acknowledgment The authors wish to thank the Ministry of Education, Malaysia (MOE) and Research Management Centre (RMC), Universiti Teknologi Malaysia (UTM) for sponsoring this project under the research university grants (RUG) No. Q.J130000.7122.03H35 and No. Q.J130000.2509.06H56. The authors are also thankful to the staff of Structures and Materials Laboratory, Faculty of Civil Engineering for the facilities and support for experimental work.

References

1. Ateş E, Barnes S (2012) The effect of elevated temperature curing treatment on the compression strength of composites with polyester resin matrix and quartz filler. Mater Des 34:435–443

2. Roger NR, Hancock M (2003) General principles guiding selection and use of particulate materials. In: Rothon RN (ed) Particulate-filled polymer composites, 2nd edn. Rapra Technology Limited, UK, pp 5–51
3. Atzeni C, Massidda L, Sanna U (1990) Mechanical properties of epoxy mortars with fly ash as filler. Cement Concr Compos 12(1):3–8
4. Bignozzi M, Saccani A, Sandrolini F (2000) New polymer mortars containing polymeric wastes. Part 1. Microstructure and mechanical properties. Compos A Appl Sci Manuf 31(2): 97–106
5. Raveendran K, Ganesh A, Khilar KC (1996) Pyrolysis characteristics of biomass and biomass components. Sci Technol Fuel Energy 75(8):987–998
6. Kaddami H, Dufresne A, Khelifi B, Bendahou A, Taourirte M, Raihane M, Sami N (2006) Short palm tree fibers—thermoset matrices composites. Compos A Appl Sci Manuf 37(9): 1413–1422
7. Yang H, Lee LJ (2001) Comparison of unsaturated polyester and vinylester resins in low temperature polymerization. J Appl Polym Sci (January 2000) 1230–1242
8. Japanese Industrial Standard (2005) Test methods for polymer concrete. JIS A 1181, Japan
9. Japanese Industrial Standard (1997) Physical testing method for cement. JIS R 5201, Japan
10. Safiuddin M, West JS, Soudki KA (2011) Flowing ability of the mortars formulated from self-compacting concretes incorporating rice husk ash. Constr Build Mater 25(2):973–978
11. Bakar BHA, Ramadhansyah PJ, Azmi MJM (2011) Effect of rice husk ash fineness on the chemical and physical properties of concrete. Mag Concr Res 63(5):313–320
12. Noorvand H, Abdullah A, Ali A, Demirboga R, Farzadnia N (2013) Physical and chemical characteristics of unground palm oil fuel ash cement mortars with nanosilica. Constr Build Mater 48:1104–1113
13. Haidar M, Ghorbel E, Toutanji H (2011) Optimization of the formulation of micro-polymer concretes. Constr Build Mater 25(4):1632–1644
14. Mun KJ, Choi NW (2008) Properties of poly methyl methacrylate mortars with unsaturated polyester resin as a crosslinking agent. Constr Build Mater 22(10):2147–2152
15. Jo B, Park S, Kim C (2006) Mechanical properties of polyester polymer concrete using recycled polyethylene terephthalate. Constr Build Mater 103:219–226
16. Jo B-W, Park S-K, Park J-C (2008) Mechanical properties of polymer concrete made with recycled PET and recycled concrete aggregates. Constr Build Mater 22(12):2281–2291
17. Kubota H, Motors G (1975) Curing of highly reactive polyester resin under pressure: kinetic studies by differential scanning calorimetry. J Appl Polym Sci 19:2279–2297
18. Waigaonkar S, Babu BJC, Rajput A (2011) Curing studies of unsaturated polyester resin used in FRP products. Indian J Eng Mater Sci 18(February):31–39

Liquid Membrane Formulation for Succinic Acid Extraction from Simulated Aqueous Waste Solution

Norela Jusoh, Norasikin Othman and Nur Alina Nasruddin

Abstract Succinic acid is developing as one of the most competitive new bio-based chemicals. However, the biological production of succinic acid results in difficult product recovery and not economically competitive. Thus, the development of economical method to extract and purify succinic acid is really essential. One of the promising methods for succinic acid extraction is emulsion liquid membrane (ELM). In this study, the ELM process will be develop to recover succinic acid from simulated aqueous waste solution through screening of ELM components; the carrier, diluent and stripping agent using liquid-liquid extraction method. Several parameters such as the possibility of using palm oil as diluent, types of carrier and their concentration, and stripping agents were examined. Outcomes indicate that 21.39 g/l of succinic acid and 0.41 g/l of acetic acid was extracted using Amberlite LA2 in palm oil. Besides, 16.34 g/l succinic acid was recovered selectively over acetic acid by Na_2CO_3 in stripping process. Therefore, it can be concluded that Amberlite LA2, palm oil, and Na_2CO_3 has high possibility to be used in liquid membrane formulation for succinic acid extraction.

Keywords Liquid membrane formulation · Amberlite LA2 · Succinic acid · Aqueous waste solution

N. Jusoh · N. Othman (✉) · N.A. Nasruddin
Centre of Lipid Engineering and Applied Research (CLEAR), Department of Chemical Engineering, Faculty of Chemical Engineering, Universiti Teknologi Malaysia, 81310 Johor Bahru, Johor, Malaysia
e-mail: norasikin@cheme.utm.my

N. Jusoh
e-mail: norela289@gmail.com

N.A. Nasruddin
e-mail: nuralina88@gmail.com

M.A. Hashim (ed.), *ICGSCE 2014*, DOI 10.1007/978-981-287-505-1_7

Introduction

Succinic acid is developing as one of the most competitive of the new platform chemicals. The suitability of succinic acid as a platform chemical is based on the ease of its biotechnological production and the huge variety of chemicals that can be produced from it by chemical conversion. Some of the molecules that can be derived from succinic acid are 1,4-butanediol, maleic anhydride, 2-pyrrolidinone, and tetrahydrofuran. These molecules can be applied in a wide range of application, such as in polymers [1], industrial solvents, and specialty chemicals [2]. Moreover, succinic acid also can be applied in the food industry as flavoring agents [3] and as the counterion in many pharmaceutical applications [4].

Nevertheless, the production of succinic acid through biological process is not economically competitive because of low product concentration. In addition, the existence of acetic acid as major by-product complicates the product recovery. In order to increase the competitiveness of the biotechnological process over petrochemical process, the minimization of the production cost is very crucial. Various methods have been proposed for succinic acid recovery, including crystallization [5], liquid-liquid extraction [6], and membrane filtration [7], but these methods are constrained by high energy or high chemical inputs.

One of the promising separation methods for succinic acid is emulsion liquid membrane (ELM). ELM provides the advantage of combination of extraction and stripping process [8]. This combination can remove the equilibrium limitation between the aqueous and organic phases. Besides, with the use of appropriate carrier for transport mechanism, specific molecular recognition can be achieved, where succinic will be highly selective over acetic acid.

Therefore, in this study, an ELM process will be developed for selective separation of succinic acid from aqueous waste solution by screening the ELM components which are carrier, diluent and stripping agent. The screening process was done by liquid-liquid extraction process, changing various experimental variables, including different type of carrier, diluents, and stripping agents with different concentration.

Materials and Methods

Materials, Amberlite LA2 as carrier was obtained from Merck Company. Amberlite LA2 used, was a mixture of straight-chain secondary amine mixture (M = 374 g/mol). Trioctylamine (TOA) (>93 % assay) and Tridodecylamine (TDA) (>95 % assay) were purchased from Merck. Kerosene as diluent was purchased from Sigma Aldrich. Palm oil, also as diluent used in this study, was ordinary cooking oil (BURUH) acquired from a supermarket. Sodium Hydroxide (NaOH) (98 % assay) was purchased from J. T. Baker, while Sodium Carbonate, Na_2CO_3 (99.5 % assay) was purchased from Merck. In addition, succinic acid (SA) (99.0 %

assay) and acetic acid (AA) (99.7 % assay) were purchased from Sigma Aldrich and J. T. Baker, respectively. All these reagents and solutions were used directly as received without further purification.

Experimental, The experiment on the carrier screening was carried out by mixing an equal volume (10 ml) of organic solution (carrier in kerosene) with a simulated aqueous waste solution (40 g/l succinic acid, 10 g/l acetic acid) at 320 rpm using a mechanical shaker (IKA-KS 130 Basic, Germany). The solution was then carefully poured into a separating funnel for phase separation for about 15 min. Sample of aqueous phase at the bottom of the separating funnel was taken for concentration measurement of succinic and acetic acid using High Performance Liquid Chromatography (HPLC). The same procedures were repeated for other parameters such as diluents and carrier concentrations.

After that, stripping process was carried out and it is the reverse extraction step. Therefore, organic phase (succinic acid loaded phase) was taken from the extraction process in section Materials and Methods. An equal volume of succinic acid loaded organic phase and aqueous stripping solution were mixed together at 320 rpm using mechanical shaker for 24 h. The mixture was then carefully poured in separating funnel and was left for about 15 min for the phase separation. Aqueous phase at the bottom of the separating funnel was taken for succinic and acetic acid concentration measurements using HPLC. The stripping procedures were repeated for different stripping agent concentration. All experiments were carried out at ambient temperature (±26 °C). The extraction and stripping performance of succinic and acetic acid was evaluated using Eqs. (1) and (2).

$$Amount\,extract\,(g/l) = \frac{[A]_{i(aq)} - [A]_{f(aq)}}{[A]_{i(aq)}} \tag{1}$$

$$Amount\,strip\,(g/l) = [A]_{fs(aq)} \tag{2}$$

where $[A]_{i(aq)}$ is the initial acid concentration in aqueous phase (g/l); $[A]_{f(aq)}$ is the acid concentration in aqueous phase after extraction (g/l); $[A]_{fS(aq)}$ is the acid concentration in aqueous phase after stripping (g/l).

Besides that, distribution ratio (D) and separation factor (α) for the extraction of succinic and acetic acid were determined using the following equation;

$$D_{SA} = \frac{[SA]_{org}}{[SA]_{aq}} \tag{3}$$

$$D_{AA} = \frac{[AA]_{org}}{[AA]_{aq}} \tag{4}$$

$$\alpha_{\frac{SA}{AA}} = \frac{D_{SA}}{D_{AA}} \tag{5}$$

where D_{SA} is the distribution ratio of succinic acid; D_{AA} is the distribution ratio of acetic acid; $\alpha_{SA/AA}$ is the separation factor of succinic over acetic acid.

Result and Discussion

Effect of carrier on succinic acid extraction, The selection of suitable carrier is very crucial for the ELM system to perform well and selectively form complex with succinic acid. Generally, secondary or tertiary amine carriers are typical carriers of carboxylic acid [9]. In this study, different types of amine carrier were used to extract succinic acid, which is TOA, TDA, and Amberlite LA2 as shown in Fig. 1. The current study clearly shows that Amberlite LA2 gives the best performance for succinic acid extraction (19.98 g/l) compared to TOA (10.98 g/l) and TDA (3.19 g/l). This is due to the Amberlite LA2 is one of the secondary amine, while TOA and TDA are both tertiary amine. Secondary amine can easily form complex with succinic acid and form ammonium salt. The process of complex formation involved the ion pair association between the alkylammonium cation and the acid radical. Since Amberlite LA2 is a secondary amine, it can easily react with acid radical by losing the H bonding. This finding in line with the results found by Ricker [10] who reported that the extractability of secondary amine was higher than that of tertiary amine.

Besides, the selectivity of succinic acid and acetic acid extraction was also performed and tabulated in Table 1. The highest value of separation factor was obtained using Amberlite LA2 as the carrier. This means that Amberlite LA2 not only can extract high amount of succinic acid, but it also selectively forms complex with succinic acid over acetic acid. Therefore, the use of Amberlite LA2 was

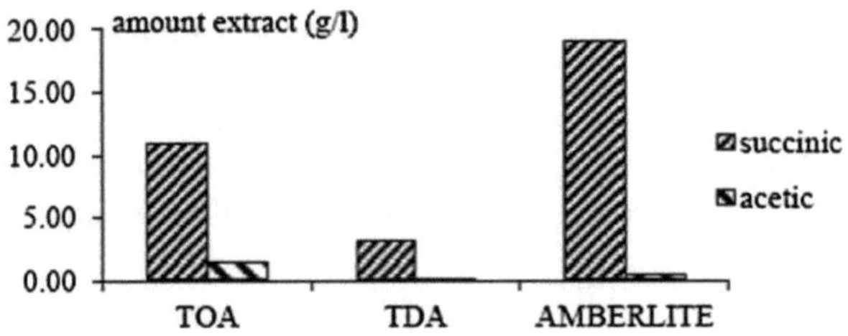

Fig. 1 Effect of carrier types for succinic acid extraction (diluent: kerosene; [succinic acid]: 40 g/l; [acetic acid]: 10 g/l; [carrier]: 0.5 M; agitation speed: 320 rpm; extraction time: 24 h)

Table 1 Selectivity of succinic from succinic and acetic acid mixtures by various types of carrier

Carrier	D_{SA}	D_{AA}	$\alpha_{SA/AA}$
TOA	0.54	0.25	2.15
TDA	0.11	0.01	8.48
Amberlite LA2	1.51	0.06	23.81

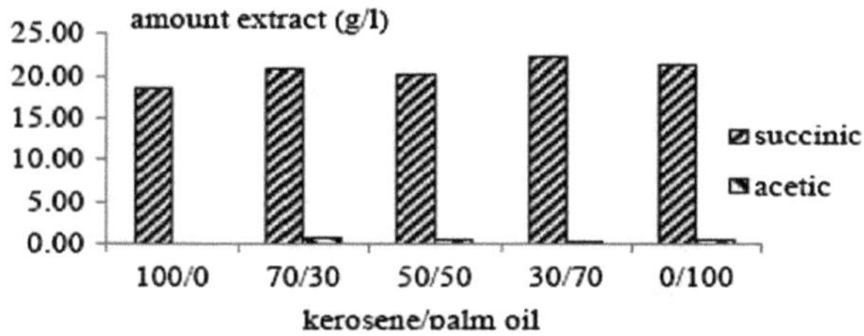

Fig. 2 Effect of diluent composition on extraction of succinic acid ([Amberlite LA2]: 0.5 M; [succinic acid]: 40 g/l; [acetic acid]: 10 g/l; agitation speed: 320 rpm; extraction time: 24 h)

Table 2 Selectivity of succinic from succinic and acetic acid mixtures by various types of carrier

Kerosene/palm oil %	D_{SA}	D_{AA}	$\alpha_{SA/AA}$
100/0	1.41	0.01	134.73
70/30	1.97	0.11	17.79
50/50	1.76	0.06	27.73
30/70	2.39	0.04	57.82
0/100	2.11	0.06	34.10

preferred for the ELM formulation to separate succinic and acetic acid and used in the following section.

Effect of diluent on succinic acid extraction, In this study, kerosene and palm oil and was used as diluents in different composition. The effect of these diluents on separation of succinic and acetic acid were illustrated in Fig. 2. Besides, the selectivity of succinic acid was shown in Table 2. The present study shows that similar amount of succinic acid was extracted using different composition of kerosene and palm oil between 18–22 g/l of succinic acid. The results indicate that the composition of kerosene and palm oil did not give much effect on succinic acid extraction. Nevertheless, there was a really large different on the selectivity of succinic acid as shown in Table 2, where kerosene gives the best selection of succinic acid with 134.73 separation factor value. However, kerosene is not environmentally friendly and the toxicity effect of kerosene will be harmful to human. Hence, it is highly desirable to replace it with another material, such as palm oil. According to Table 2, the combination of 30/70 kerosene to palm oil also leads to high separation factor (57.82) followed by a comparatively separation factor using pure palm oil (34.10). The 30/70 kerosene to palm oil provide larger separation factor, but the combination can complicate the extraction process and kerosene is still being used although in a small proportion. Therefore pure palm oil is selected as possible diluent in this study. This is because the valuable property in palm oil over kerosene seen as renewable, nontoxic and readily available. In addition, it works well for the extraction of heavy metal [11] and phenol [12]. Hence, pure palm oil will be considered as diluent for the next study.

Effect of stripping agent types, In the ELM system, driving force for the succinic acid extraction is pH difference between the feed and receiving phase.

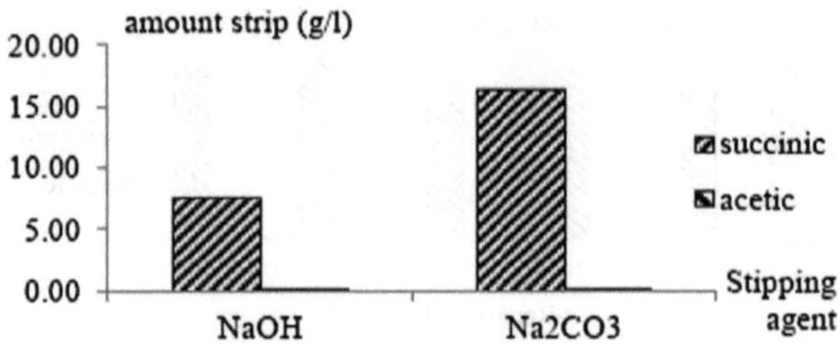

Fig. 3 Effect of stripping agent types on succinic acid extraction ([stripping agent]: 1 M; agitation speed: 320 rpm; stripping time: 24 h)

Therefore, alkaline solution such as NaOH and Na_2CO_3 were screened in this study as stripping agent for succinic acid extraction. Figure 3 shows the effect of stripping agent type of solute from succinic loaded organic phase. The results indicate there is almost no acetic acid stripped in the process due to very small amount of acetic acid in extracted into loaded organic phase. The amount of succinic acid stripped using Na_2CO_3 is 16.34 g/l compared to NaOH which is only 7.59 g/l. This is because Na_2CO_3 dissociates, and forms 2 mol of sodium ion in comparative to NaOH that gives only 1 mol. Therefore, more succinic acid will be stripped using Na_2CO_3. The result of this study seems to be consistent with other research, who reported that Na_2CO_3 show better performance than NaOH in stripping process of succinic acid [13].

Effect of Amberlite LA2 concentration on extraction of succinic acid, The effect of concentration of Amberlite LA2 on the amount of succinic acid extracted is shown in Fig. 4. The amount of succinic acid extracted was increased by increasing the concentration of Amberlite LA2. This indicates that Amberlite LA2 plays an important role in the extraction of succinic acid and it shows that the extraction of succinic acid was very selective toward acetic acid. Besides that, this result shows that smaller amount of Amberlite LA2 was required to extract succinic acid and can be proved by examining the stoichiometric reaction of Amberlite LA2 and succinic acid.

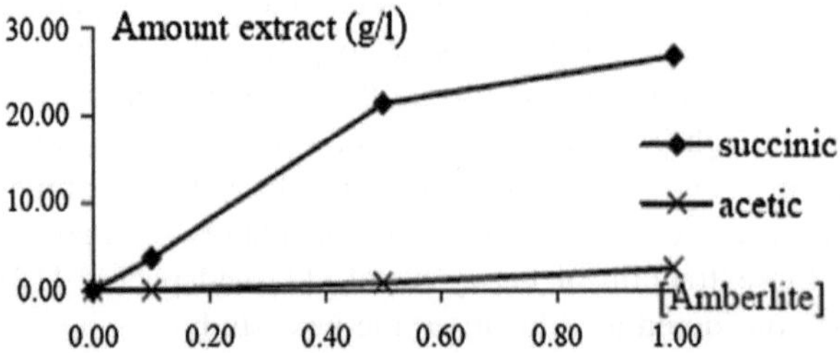

Fig. 4 Effect of Amberlite LA2 concentration on succinic acid extraction (diluent: palm oil; [succinic acid]: 40 g/l; [acetic acid]: 10 g/l; agitation speed: 320 rpm; extraction time: 24 h)

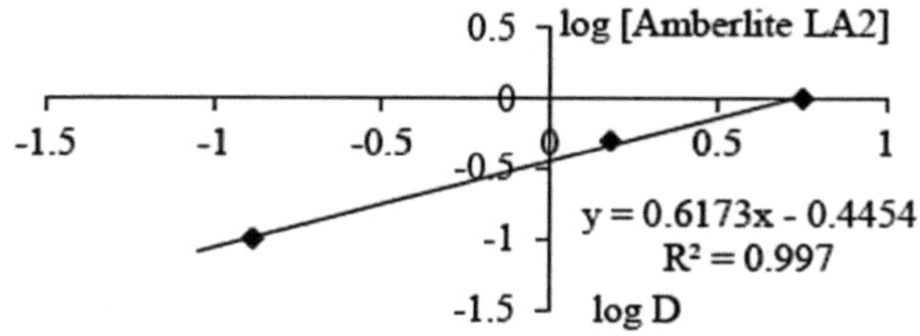

Fig. 5 Stoichiometric plot of equilibrium extraction of succinic acid using Amberlite LA2

The reaction of Amberlite LA2 with undissociated succinic acid can be described by the reaction:

$$m(H_2A)(aq) + nB(org) \rightarrow (H_2A)_m B_n(org) \quad (6)$$

where m and n is the number of moles,(H_2A)(aq) is undissociated succinic acid in aqueous phase, $B(org)$ is Amberlite LA2 in organic phase, and $(H_2A)_mB_n(org)$ is the complex formed.

Based on the extraction reaction and result from Fig. 4, the nature of extracted solute was examined, by plotting a graph of log D versus log [Amberlite LA2] as shown in Fig. 5. The slope obtained was 0.617, which implies that the Amberlite LA2 form 1:1 complex with succinic acid and presented in Eq. 7.

$$(H_2A)(aq) + B(org) \rightarrow (H_2A)B(org) \quad (7)$$

Effect of Na_2CO_3 concentration, Effect of concentration Na_2CO_3 on stripping of succinic acid was shown in Fig. 6. According to the result, there is approximately no acetic acid because it was stripped from organic loaded phased contained a very low amount of acetic acid. On the other hand, the amount of succinic acid stripped was increased when high concentrations of Na_2CO_3 were used. The result indicates that the amount of succinic acid strip was dependent on concentration of Na_2CO_3. The reaction of Na_2CO_3 and acid-amine complex is expressed by:

$$m(H_2A)B + nNa^+ \rightarrow (H_2A)_2Na + B \quad (8)$$

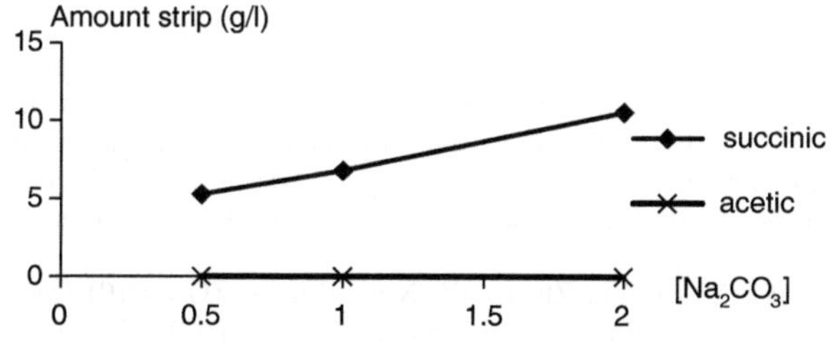

Fig. 6 Effect of Na_2CO_3 concentration on succinic acid extraction (Agitation speed: 320 rpm; time: 24 h)

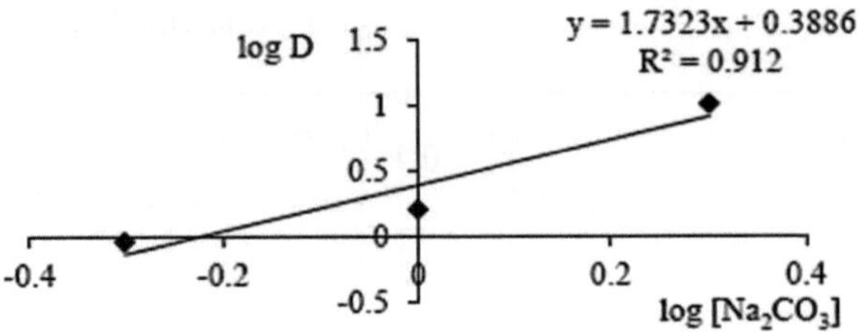

Fig. 7 Stoichiometric plot of equilibrium stripping of succinic acid by Na_2CO_3 stripping agent

To evaluate the nature of the stripped solute, a graph of log D versus log [Na_2CO_3] was plotted as shown in Fig. 7. From the graph, slope obtained was 1.723 which is nearly to 2.0. Hence, molar ratio of succinic-amine complex to sodium ion is found to be 1:2. Thus, it shows that one mole of complex reacts with 2 mol of sodium ion in stripping reaction as shown in Eq. 9.

$$(H_2A)B + 2Na^+ \rightarrow (H_2A)_2Na + B \tag{9}$$

Conclusion

The finding of this study provides an approach for selective extraction of succinic acid toward acetic acid. It can be concluded the best liquid membrane was formulated using Amberlite LA2 as carrier, palm oil as diluent, and Na_2CO_3 as stripping agent.

Acknowledgments The authors would like to acknowledge the Ministry of Higher Education (MOHE), Universiti Teknologi Malaysia (RU Research Grant; GUP:Q.J130000.2544.04H47), and Centre of Lipid Engineering and Applied Research (CLEAR) for financial support to make this research possible.

References

1. Xu J, Guo BH (2010) Poly(butylene succinate) and its copolymers: research, development and industrialization. Biotechnol J 5:1149–1163
2. Zeikus JG, Jain MK, Elankovan P (1999) Biotechnology of succinic acid production and markets for derived industrial products. Appl Environ Microbiol 51:545–552
3. Tsukatani T, Matsumoto K (2000) Flow-injection fluorometric quantification of succinate in food stuffs based on the use of an immobilized enzyme reactor. Anal Chim Acta 416:197–203
4. Thakker C, Martinez I, San KY, Bennet GN (2012) Succinate production in *Escherichia coli*. Biotechnol J 7:213–224
5. Li Q, Wang D, Wu Y, Li WL, Zhang YJ, Xing JM, Su ZG (2010) One step recovery of succinic acid from fermentation broths by crystallization. Sep Purif Technol 72:294–300

6. Kurzrock T, Weuster-Botz D (2011) New reactive extraction systems for separation of bio-succinic acid. Bioproc Biosyst Eng 34:779–787
7. Wang C, Li Q, Tang H, Yan DJ, Zhou W, Xing JM, Wan YH (2012) Membrane fouling mechanism in ultrafiltration of succinic acid fermentation broth. Bioresour Technol 116: 366–371
8. Othman N, Chan KH, Goto M, Mat H (2006) Emulsion liquid membrane extraction of silver from photographic waste using CYANEX 302 as the mobile carrier. Solvent Extr Res Dev 13:191–202
9. Hong YK, Hong WH, Han DH (2001) Application of reactive extraction to recovery of carboxylic acids. Biotechnol Bioprocess Eng 6:386–394
10. Ricker NL (1978) Recovery of carboxylic acids and related organic chemicals from wastewater by solvent extraction. PhD. Dissertation, University of California, Berkeley
11. Sanna B, Rose FK (2011) A study of the heavy metal extraction process using emulsion liquid membranes. Master's thesis, Chalmers University of Technology, Sweden
12. Venkateswaran P, Palanivelu K (2006) Recovery of phenol from aqueous solution by supported liquid membrane using vegetable oils as liquid membrane. J Hazard Mater 131: 146–152
13. Lee SC (2011) Extraction of succinic acid from simulated media by emulsion liquid membranes. J Membr Sci 391:237–243

Preservation of Natural Textile Colorant from *Xylocarpus Moluccensis* by Encapsulating into PVA Alginate

Norul Azilah Abdul Rahman, Ruziyati Tajuddin and Siti Marsinah Tumin

Abstract Natural dye/colorant extracts used for dyeing source are incapable to be stored for a substantial period of time due to their unstable property in aqueous medium. Therefore, this study explores a suitable method to preserve the extracts for storage and produce natural textiles dyes with longer shelf life. PVA Alginate was used as a pure/blank beads to encapsulate the colorant from *Xylocarpus moluccensis* using ultrasonic cleaner approach at different ultrasonic volume and time. Results show the highest encapsulation efficiency was achieved using ultrasonic cleaner at optimum condition of high sonic volume in 80 min time. Fourier-transform infrared spectrosopic (FTIR), scanning electron microscope (SEM) and x-ray diffraction (XRD) were used to characterize the encapsulated colorant. The characterization via these methods indicates the colorant was successfully encapsulated into the PVA Alginate beads. Thus, the encapsulation approach studied in this work showed a promising process for preservation of natural textiles dyes.

Keywords Xylocarpus moluccensis · PVA alginate · Ultrasonic cleaner · Encapsulation

N.A. Abdul Rahman · R. Tajuddin (✉) · S.M. Tumin
Faculty of Applied Sciences, Universiti Teknologi MARA (UiTM), 40450 Shah Alam, Selangor, Malaysia
e-mail: ruziy039@salam.uitm.edu.my

N.A. Abdul Rahman
e-mail: norul22453@yahoo.com

S.M. Tumin
e-mail: sitimarsinah@salam.uitm.edu.my

M.A. Hashim (ed.), *ICGSCE 2014*, DOI 10.1007/978-981-287-505-1_8

Introduction

There are many challenges in developing and preserving natural dye/colorant. Natural textiles dyes extracted from plant materials are commonly in liquid aqueous form which must be used immediately due to its low stability. It is suggested that the liquid extracts should be converted to a solid form so that they can be preserved. To date there is no reported research on any method to preserve natural textile dyes. Nevertheless, preservation method based on encapsulation approach is a common practice in food industry in order to entrap colorant or flavor within solid matrix as for the purpose of converting some product into a more conveniently handled solid form. β-cyclodextrin [1], Polyethyleneglycol [2], AIF (acetone insoluble fraction) phospholipids [3], alginate [4] and PVA Alginate [1] are among the substances commonly used as encapsulation agents.

PVA-alginate is widely used in enzyme activity [5] and bacteria growth (microbiology) [6] studies as well as to remove hazardous waste (chemical, textile and pharmaceutical industries) [7–9]. It has porous structure on the surface and inner to entrap and immobilize compound. According to Dave and Madamwar [5], PVA contributes to the strength and high crosslinking capacity whereas alginate reduces the agglomeration and increases the surface properties. These claim strongly support the rationale of choosing PVA-alginate as the encapsulation agent in this study. PVA-alginate was exploited as the pure/blank beads to trap and immobilize natural colorant extracted from the heartwood of *Xylocarpus moluccensis* plant.

Experimental

Materials, Wood and bark wastes of *Xylocarpus moluccensis* plant were obtained from craft small industries located in Pulau Carrey, Klang Selangor, Malaysia. The sample was ground and stored at room temperature before use. Sodium hydroxide and preparation beads, calcium chloride were purchased from Merck, Germany. Boric acid was purchased from R & M Marketing, U.K. Poly(vinyl alcohol) and Alginic acid sodium salt from brown algae (sodium alginate) were supplied by Sigma-aldrich, UK.

Extraction Methods, Ultrasonic extractions were carried out as described previously in our reported study [10]. The optimum parameters were used.

Preparation of beads for encapsulation, Approximately 10 g of PVA and 1 g of Sodium alginate were weighed and placed in a sample bottle. 100 mL of distilled water was added and stirred using glass rod. The bottle was then placed into the ultrasonic cleaning system (Powersonic 420, 100-420-230, AC 230 V, 50 Hz, 700 W, Hwashin Technology Seoul Korea) for 60 min at temperature of 0 °C and medium sonic volume. The PVA Alginate mixture was then dropped into a mixed solution of 100 mL boric acid (1 %) and calcium chloride (1 %) using peristaltic pump (Master Flex®4s®, USA) and silicone tubing to form beads. The PVA–alginate beads were then stored at 4 °C for 24 h. For the encapsulation process, the beads were dried using oven (Memmert, Germany) and added with colorant extracts, followed by ultrasonic process. The ultrasonic condition was optimized by varying the ultrasonic volume (low, medium and high) and encapsulation time (20, 30, 40, 60, and 80 min) while the temperature was read after the encapsulation process. The encapsulated beads were then dried using oven. To determine encapsulation efficiency (EE), the volume of colorant loaded into PVA Alginate beads was calculated based on Eq. (1) below:

$$\%\mathrm{EE} = \frac{\text{Volume of loaded colorant}}{\text{Volume of initial colorant}} \times 100\,\% \tag{1}$$

Characterization Methods

Fourier-transform infrared (FTIR) spectroscopic analysis, The Fourier transform IR spectra (FT-IR) of samples were recorded with a Perkin Elmer in the range of 400 and 4000 cm^{-1} by averaging 16 scans number at 4 cm^{-1} resolution as KBr pellets.

Scanning electron microscope (SEM), The surfaces of the samples were imaged using a Carl Zeiss electron microscope (Supra 40VP, Germany) at 10 kV with 3 k magnification. The samples were deposited on a copper stub and coated with gold, using vapor deposition techniques.

X-ray diffraction (XRD), X-ray diffractograms of the samples were recorded over a 2 theta range of 2–50° by X-ray diffractometer (Rikagu Ultima IV, Japan) with a step angle of 0.02°/min. The voltage and current were utilized at 40 kV and 20 mA respectively.

Result and Discussion

Encapsulation efficiency, The encapsulation efficiency determined by the percentage of *Xylocarpus moluccensis* colorant adsorbed into PVA Alginate beads was measured based on ultrasonic volume and encapsulation time as shown in a plot in Fig. 1. As can be observed, the encapsulation efficiency increased with increasing of ultrasonic volume and encapsulation time, and it reached maximum at 80 min time with percentage adsorption of 98 %. Further increased in ultrasonic encapsulation time at high sonic volume, altered the structure of PVA Alginate beads which caused a slight reduction of percentage adsorption. This indicates that the equilibrium of PVA alginate and colorant solution was completely achieved within 80 min.

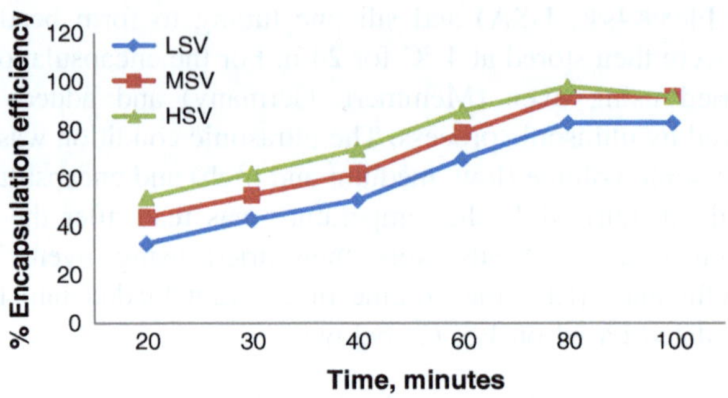

Fig. 1 Percentage of encapsulation efficiency based on ultrasonic volume and encapsulation time. LSV is lower sonic volume, MSV is medium sonic volume and HSV is high sonic volume

Characterization

Fourier-Transform Infrared Spectroscopic Analysis

FTIR spectra of PVA Alginate (pure/blank beads), XM (colorant extract from Xylocarpus *moluccensis*) and HSV80 (encapsulated XM colorant at high sonic volume in 80 min) are shown in Fig. 2 and the assignment of bands is presented in Table 1. General characteristics of PVA Alginate were viewed at 3405.79 cm^{-1} (–OH stretching), 2927.53 cm^{-1} (–CH2 asymmetric stretching), 1627.85 cm^{-1} and 1429.13 cm^{-1} (C = O asymmetric stretching and C = O symmetric stretching

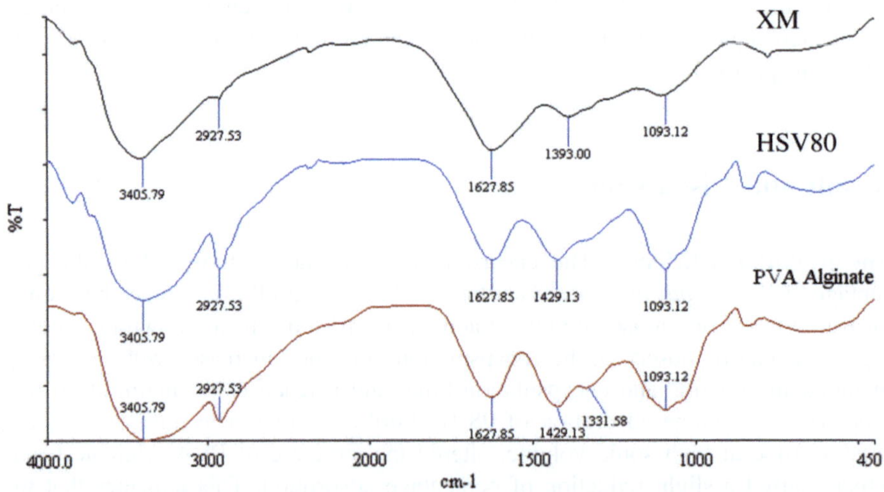

Fig. 2 FTIR spectra of PVA Alginate beads (pure bead/blank beads), XM (colorant extract of *Xylocarpus moluccensis*) and HSV80 (encapsulated XM colorant at high sonic volume with 80 min encapsulation time)

Table 1 Assignment of FTIR bands for PVA alginate beads (pure bead/blank beads), XM (dye extract *Xylocarpus moluccensis*) and HSV80 (encapsulated bead with XM at high sonic volume with 80 min encapsulation time)

Samples	OH (cm^{-1})	CH_2 (cm^{-1})	C = O (cm^{-1})	C = O (cm^{-1})	CH_3 (cm^{-1})	C–O (cm^{-1})	Aromatic (cm^{-1})
(a) PVA A	3405.79	2927.53	1627.85	1429.13	1331.58	1093.12	
(b) XM	3405.79	2927.53	1627.85	1393.00		1093.12	1093.12–1393.00
(c) HSV80	3405.79	2927.53	1627.85	1429.13		1093.12	1093.12–1429.13

respectively), 1331.58 cm^{-1} (–CH_3 symmetric bending) and 1093.12 cm^{-1} (C–O bending). This result was similar to those reported by Ravindrachary et al. [11] and De Queiroz [12].

For XM sample, the same characteristics were observed at 3405.79 cm^{-1} (–OH stretching), 2927.53 cm^{-1} (–CH_2 stretching) and 1627.85 cm^{-1} (C = O asymmetric stretching), however the C = O symmetric stretching band was shifted to 1393.00 cm^{-1}. A weak and broad absorption band attributed to benzene ring was observed from 1093.12 to 1393.00 cm^{-1} and this overlapped with the absorption band of C–O bending at 1093.12 cm^{-1}. This result is in agreement with that reported by Rosu et al. [13] for alizarin (9,10-anthraquinone) compound. They suggested that the presence of absorption band at 1627/1629 cm^{-1} was due to the conjugate carbonyl group vibrations (by hydrogen bonds with adjacent O–H groups) that represented the characteristic absorption band of 9,10-anthraquinone compound.

XM loaded into pure/blank beads (HSV80) resulted in a markedly increased intensity of the absorption bands as shown at 3405.79 cm^{-1} (–OH stretching), 2927.153 cm^{-1}(–CH_2 stretching), 1093.12 cm^{-1} (C–O bending) and 1093.12–1429.13 cm^{-1} (benzene ring), indicating an increase in the content of alizarin, which might come from colorant extract of *Xylocarpus moluccensis*. The disappearance of absorption band for –CH_3 symmetric bending at 1331.58 cm^{-1} (previously shown in PVA-Alginate spectrum) also indicates the successful encapsulation of XM colorant into the PVA Alginate.

Scanning Electron Microscope (SEM) Analysis

The morphology obtained for PVA Alginate and HSV80 samples were then imaged by SEM as shown in Fig. 3. The aggregation and pores of PVA Alginate beads were obviously observed in Fig. 3a, whereas in Fig. 3b, both aggregation and pores of beads were almost invisible owing to complete adsorption of colorant after encapsulation process. These aggregates seemed to be melting, combining with each other and the outer surface was smooth due to the possibility of some XM colorant content surrounding the particle surface.

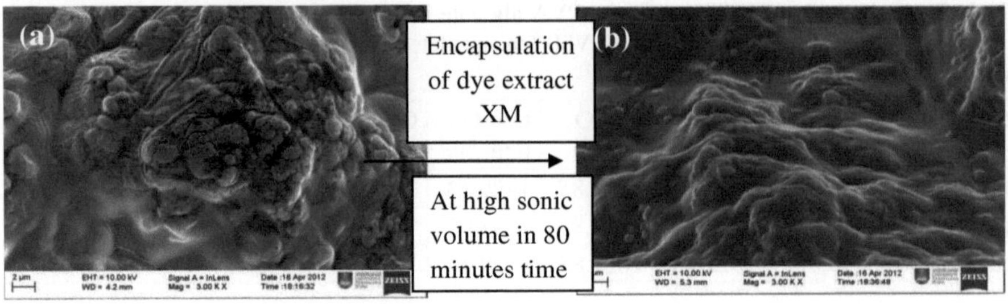

Fig. 3 SEM picture of **a** PVA Alginate (pure bead/blank beads), and **b** HSV80 (encapsulated XM colorant at high sonic volume with 80 min encapsulation time). **a** PVA alginate (pure/blank beads), **b** HSV80 (encapsulated colorant)

X-Ray Diffraction (XRD) Analysis

Figure 4 illustrates the X-ray diffractograms of PVA Alginate, XM and HSV80 samples. Peaks with significant intensity at 20° and a shoulder at 30° were present in PVA Alginate diffractogram which were similar to a result reported by Udaseen et al. [14]. The intensities of both peaks were increased after encapsulation process as observed in HSV80 diffractogram. This implies that the remarkably higher intensity of peak at 20° was the result of inclusion complexes between XM and the PVA Alginate beads. In contrast, diffractogram of XM did not show any characteristic peak for the colorant, indicating that the colorant extract was in an amorphous state.

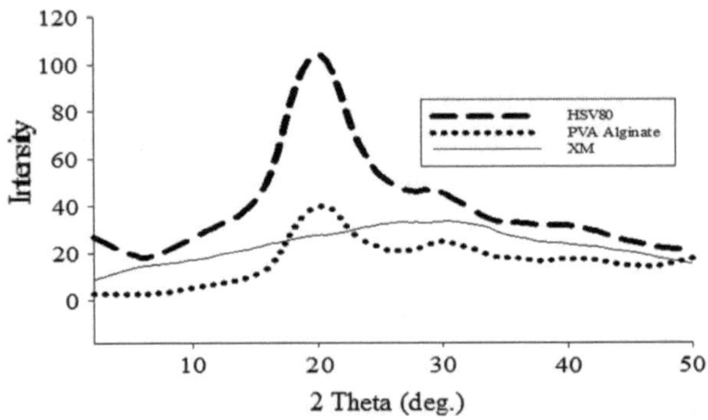

Fig. 4 X-Ray diffractograms for PVA alginate (pure bead/blank beads), XM (colorant extract *Xylocarpus moluccensis*) and HSV80 (encapsulated XM colorant at high sonic volume with 80 min encapsulation time)

Conclusion

In this study, colorant extract from *Xylocarpus moluccensis* was successfully encapsulated into PVA Alginate using ultrasonic method. PVA Alginate was found to be an efficient encapsulation matrix for colorant encapsulation. The encapsulation efficiency of 98 % colorant adsorption was achieved at optimum condition of high sonic volume in 80 min time. Characteristics of encapsulated colorant were successfully determined using FTIR, SEM and XRD analysis which indicate the potential achievement of this encapsulation process for preservation of natural textile dyes.

Acknowledgments We wish to express our gratitude to the Faculty of Applied Sciences, Universiti Teknologi MARA Shah Alam for the support in this research project

References

1. Norasiha H, Mimi Sakinah AM, Rohaida CM (2011) Characterization of β-cyclodextrin complexes with natural dye. In: National conference on postgraduate research (NCON-PGR), pp. 98–106
2. Santos DT, Albarelli JQ, Beppu MM, Meireles MAA (2011) Stabilization of anthocyanin extract from jabuticaba skins by encapsulation using supercritical CO_2 as solvent. Food Res Int 1–8
3. El-Zawahry MM, El-Shami S, El-Mallah MH (2007) Optimizing a wool dyeing process with reactive dye by liposome microencapsulation. Dyes Pigm 74:684–691
4. Sanjay S, Kumar A, Suneetha V, Mishra B, Gopinath R, Yadav S et al (2012) Synthesis and activation of Immobilized beads by natural dye extracts. Int J Drug Dev Res 4:304–310
5. Dave R, Madamwar D (2006) Esterification in organic solvents by lipase immobilized in polymer of PVA–alginate–boric acid. Process Biochem 41:951–955
6. Yujian W, Xiaojuan Y, Hongyu L, Wei T (2006) Immobilization of *Acidithiobacillus ferrooxidans* with complex of PVA and sodium alginate. Polym Degrad Stab 91:2408–2414
7. Zhang Y, Kogelnig D, Morgenbesser C, Stojanovic A, Jirsa F, Lichtscheidl-Schultz I, Krachler R, Li Y, Keppler BK (2011) Preparation and characterization of immobilized [A336] [MTBA] in PVA–alginate gel beads as novel solid-phase extractants for an efficient recovery of Hg(II) from aqueous solutions. J Hazard Mater 196:201–209
8. Idris A, Misran E, Hassan N, Abd Jalil A, Seng CE (2012) Modified PVA-alginate encapsulated photocatalyst ferro photo gels for Cr(VI) reduction. J Hazard Mater 227–228:309–316
9. Kumar SS, Kumar MS, Siddavattam D, Karegoudar TB (2012) Generation of continuous packed bed reactor with PVA–alginate blend immobilized *Ochrobactrum* sp. DGVK1 cells for effective removal of N, N-dimethylformamide from industrial effluents. J Hazard Mater 199–200:58–63
10. Abdul Rahman NA, Tumin SM, Tajuddin R (2013) Optimization of ultrasonic extraction method of natural dyes from *Xylocarpus moluccensis*. Int J Biosci Biochem Bioinf 3:53–55
11. Ravindrachary V, Ismayil SP Nayak, Dutta D, Pujari PK (2011) Free volume related fluorescent behavior in electron beam irradiated chalcone doped PVA. Polym Degrad Stab 96:1676–1686
12. De Queiroz AAA, Passos ED, Alves SDB, Silva GS, Fernandez LRV, Higa OZ, Vítolo M (2006) Alginate-poly(vinyl alcohol) core-shell microspheres for lipase immobilization. J Apppl Polym Sci 102:1553–1560

13. Rosu M-C, Suciu R-C, Lazar MD, Bratu I (2013) The influence of alizarin and fluorescein on the photoactivity of Ni, Pt and Ru-doped TiO_2 layers. Mater Sci Eng 178:383–390
14. Udaseen S, Asthana S, Raveendran NT, Kumar K, Samal A, Pal K, Pramanik K, Ray SS (2014) Optimization of process parameters for nozzle—free electrospinning of poly (vinyl alcohol) and alginate blend nano-fibrous scaffolds. Int J Enhanced Res Sci Technol Eng 3 (1):405–411

Coagulant from Chemically Modified Fish Scale for Textile Wastewater Treatment

Mohibah Musa, Ku Halim Ku Hamid, Hanafiah Zainal Abidin and Miradatul Najwa Muhd Rodhi

Abstract Textile manufacturing is one of the largest industrial users of process water. This production gives rise to the major pollutant in this industry where the discharging residue from dyeing process contains organic and inorganic compounds and some potentially toxic solvents. This effluent is the environmental challenge for textile industry where the wastewater often rich in color, high pH, high Chemical Oxygen Demand (COD) and Biological Oxygen Demand (BOD) also low biodegradability. Therefore, complete dye removing technology of users and environmentally friendly is very necessary. In this research, biodegradable material for coagulation-flocculation process tailored for textile wastewater treatment was developed based on polymeric material from fish scales. This coagulant with a particle size of 75 μm was tested for real wastewater from industry via jar tests to determine the optimum operating condition such as the dosage of coagulant and pH. The treated textile wastewater was found to be complying with the Parameter Limits of Effluents of Standards A and B from Environmental Quality (Sewage and Industrial Effluents) Regulations 1978, from the result, the optimum pH and dosage is 5.0 mg/l of coagulant dosage at pH 5.0. COD reduction is about 98 % meanwhile the reduction of colour achieves more than 90 %.

Keywords Fish scale · Textile wastewater · Polymeric material · Coagulant · Biodegradable

---

M. Musa (✉) · K.H. Ku Hamid · H. Zainal Abidin · M.N. Muhd Rodhi
Faculty of Chemical Engineering, Universiti Teknologi MARA, 40450 Shah Alam, Selangor, Malaysia
e-mail: mohibah@salam.uitm.edu.my

K.H. Ku Hamid
e-mail: kuhalim@salam.uitm.edu.my

H. Zainal Abidin
e-mail: hanafiah@salam.uitm.edu.my

M.N. Muhd Rodhi
e-mail: miradatul@salam.uitm.edu.my

M.A. Hashim (ed.), *ICGSCE 2014*, DOI 10.1007/978-981-287-505-1_9

Introduction

The population of the world is increasing year by year and its effect considerable amount of stress on the environment. Human needs water to survive and move on with their activities and nowadays water will continue to become a major factor for survival. Currently, water had been used by household approximately 50 % and the rest had been used by industrial and agricultural activities. Textile manufacturing is one of the largest industrial users of process water. Water is used extensively throughout batik processing operations. Dyeing among the most water intensive processes in batik production, the environmental challenge for the textile industry is associated with liquid waste, which tends to dominate over air-emissions and solid wastes in term of environmental effects [1].

The effluent from dyeing and finishing processes often rich in colour, containing residual of dyes and chemical, and needs proper treatment before releasing into the environment [2]. It's also been characterized by high pH, high temperature, high COD and low biodegradability [3].

The effluent from textile wet processing requires several treatment options before it can be discharged. Conventional activated sludge is one of the technology that's been applied for the treatment. In this type of treatment, it has the capability to remove large fractions of COD. Dye removal is most often achieved by sorption processes, rather than by biodegradation. For effluents with COD over 5000 mg/l, anaerobic treatment becomes more significant. A combined anaerobic and aerobic treatment can also be effective for removal of azodyes [7]. However, this system removes poorly most of the dyes used, and is clearly ineffective in decolourizing textile effluent, even when mixed and treated together with sewage. Removal of dyes is possible by activated carbon treatment. But, it is expensive and the level of colour removal depends on the type of dye and complete colour removal is rarely achieved [1, 4].

Ozone (O_3) is one of the strongest oxidizing agents that is readily available. It is used to reduce colour, eliminate organic waste, reduce odour and reduce total organic in the water. Ozone is one of the strongest commercially available oxidizers, making it popular for primary disinfection of potable water as well as for colour and organic removal in wastewater applications. But, ozone cannot completely mineralize the organic dye to carbon dioxide and water even at high dose usage [4]. This due to the decolonization rate decreasing with increasing initial dye colour [5]. Up to date there is no economic and effective method to treat textile wastewater. Therefore, a serious effort has to be carried out to solve this problem.

Colour from textile wastewater is hard to remove due to resistance of dyes to biological degradation where they are not readily degraded under the aerobic conditions prevailing in biological treatment plant [6]. The colour of textile effluent is unacceptable under Malaysian Environmental Regulation besides the other parameter such as COD, BOD, total iron, etc. [7]. Regulation for effluent and air emissions, however, are becoming increasingly stringent and more rigorously enforced, which is prompting the industry to develop alternative methods for the management of process wastes. Due to the factors stated above, balancing the environmental protection and its economic viability and sustainable development is the biggest challenges in the textile industry. In order to keep the economic growth, new way of treatment is vital to preserve the environment. To face the current challenges, textile effluents need an efficient treatment system. Coagulation is one of the techniques used to remove the pollution base on suspended particles. This method has been applied in removal of colour textile by using miscellaneous type of coagulant.

Coagulation method using inorganic coagulants with the addition of polymer flocculant gives almost 100 % of colour removal [4]. Removal of some reactive dyes from synthetic wastewater by combining Al(III) coagulation /carbon adsorption process, achieved complete decolorisation and almost total elimination (99.9 %) of reactive dyes from wastewater [8]. In this work, the use of fish scale as the coagulant was studied. Amount of fish scale generated as waste allows an abundant source of biomaterial. Fish scale found to be effectively used as an organic wastewater coagulant [8].

The proteins which present in the organic fraction in fish scale seem to be the major factor of adsorption ability. Other biomaterial such as Chitosan, had proven its ability in dye removal where the charge neutralization in main mechanism for dye coagulation with chitosan is at acidic pH [9]. For this work, fish scales were dried and ground before being added as a coagulant in powder form. An experimental study were carried out in a local textile wastewater to assess the removal efficiency of a coagulation process. Up to date there is no serious effort to utilize fish scale as a coagulant for wastewater treatment in Malaysia.

Experimental

Raw material preparation, The raw material (scales) were collected at the local market. Firstly, this material were washed and thermally treated with distilled water at temperature of 800 °C for 30 min. Then, this pre-treated material were dried a convection oven at 80–90 °C for 18 h. Then, the scales were treated with 5 % HCL at 40 °C for 6 h in order to produce organic-rich material and to make the scales amber and very fragile. After 6 h, the solution will filter off and the scales will be washed with distilled water until the pH of the washes becomes neutral. Then, it will be dry in a convection oven at 80 °C for 18 h. The treated fish scale will be ground into a powder form with particle size of 75 μm.

Coagulation-Flocculation Process, The process were carried out via Jar Test at room temperature. For each test coagulant dose were varied from 1.0 to 5.0 g and pH from 5 to 8 in order to determine the optimum operating conditions for colour removal. Wastewater were transferred into the jar and stirred at 50 rpm for 5 min. The pH of the wastewater were adjusted by adding 0.1 M HCl or 0.1 M NaOH. The sample will be stirred at 180 rpm for 10 min prior to additional of organic coagulant from fish scale were added. The rapid mixing will be continued for another 2 min. The speed of the mixing will be reduced to 50 rpm for 15 min before leaving the sample for sedimentation process about 4 h.

Sample analysis, At each condition applied (different dosage and pH), COD, BOD, colour, metallic ion level and colour Removal Efficiency (CRE) will be determined. For metallic ion level, the test were run using ICP and amount of metallic ion present in the sample before and after jar test treatment were measured. The Colour Removal Efficiency (CRE) was calculated from the difference between initial and final absorbance value. The observance was measured using UV-VIS Spectrophotometer. The CRE will be calculated as follows:

$$\mathrm{CRE} = \frac{(1 - \mathrm{A1})}{\mathrm{A0}} \times 100\,\%$$

where A0 and A1 are the absorbance of the samples before and after jar test treatment. The final result were compared to the Parameter Limits of Effluents of Standards A and B from Environmental Quality (Sewage And Industrial Effluents) Regulations 1978, DOE.

Result and Discussion

To investigate the reason of high efficiency and the textile wastewater treatment mechanism of fish scale based coagulant, the variation of zeta potential hydrolysis products as a function of pH was measured. In this series of experiments, the pH value was varied by the addition of HCl or NaOH solution. It needs to be pointed out that the pH values in zeta experiments were the same as the pH of coagulant solution (0.5 g/L). The working solution used in this study was distilled water. The data obtained are summarized in Fig. 1. Zeta potential is a controlling parameter of charge neutralization of coagulant can usually used to interpret the trend of

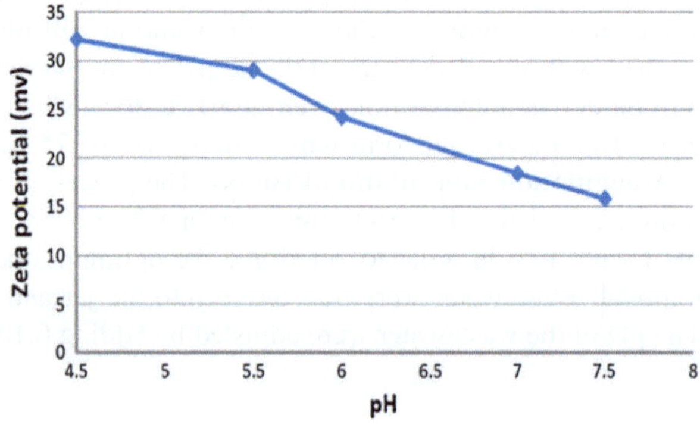

Fig. 1 The variation of zeta potential with respect to pH for fish scale based coagulant

Table 1 Wastewater quality for industrial dye wastewater

Properties	Unit	Value
Chemical oxygen demand (COD)	ppm	4700.00
Colour	m^{-1}	2120.00
Cadmium (Cd)	ppm	0.31
Chromium (Cr)	ppm	2.23
Copper (Cu)	ppm	6.78
Iron (Fe)	ppm	2.62
Mercury (Hg)	ppm	3.33
Plumbum (Pb)	ppm	16.13

coagulation efficiency. The zeta potential of fish scale based coagulant decreased from 32.2 to 15.8 mV as the pH was increased from 4.5 to 7.5. The value of the zeta potential is practically constant in a wide range of pH (from 4 to 8). The textile in water has a negative charge on its surface, so the higher the positive charge on the coagulant surface, the higher the latex concentrate removal efficiency.

Table 1 shows the wastewater quality of industrial dye wastewater, the sample was collected from batik industries from Kuala Terengganu. The main parameter that was measured is mentioned in the table below. From the study the COD value of the wastewater is around 4700 ppm. Value for colour is around 2120 m^{-1}. In addition, heavy metal ion concentration been also measured in this study where Cd, Cr, Cu, Fe, Hg and Pb concentration approximately 0.31, 2.23, 6.78, 2.26, 3.33 and 16.13 ppm respectively.

Figure 2 shows the result of COD removal by difference coagulant dosing i.e. 1.0, 2.0, 3.0, 4.0 and 5.0 g/ml. The lowest value of COD was found in sample with a pH value around 5.0 ppm at coagulant dosing about 5.0 g/ml with concentration of 130 ppm. The highest value for COD is found in sample at pH 8.0, the value reaches until 4650 ppm compare to other.

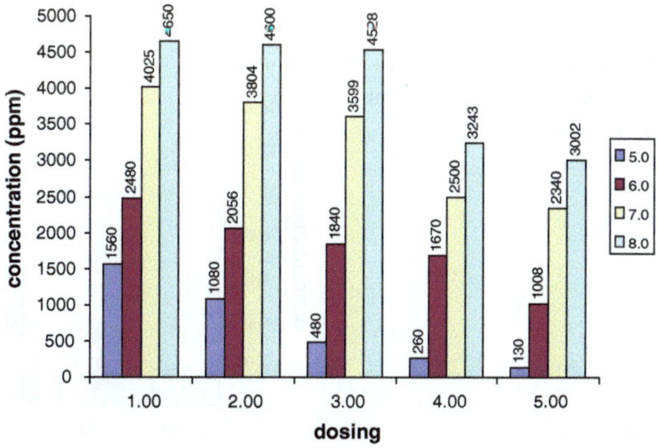

Fig. 2 Result for COD removal by difference coagulant dosing

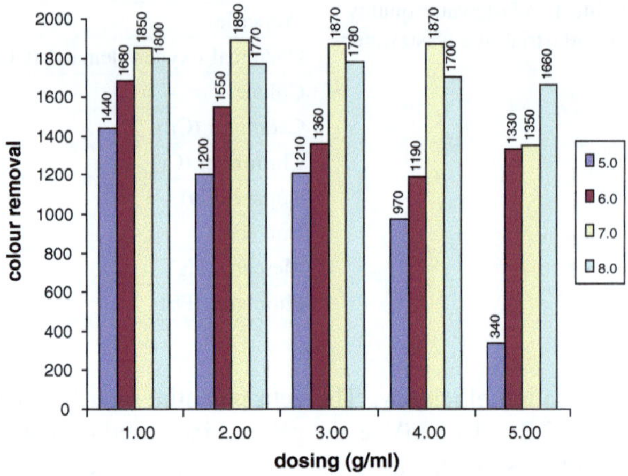

Fig. 3 Result for colour removal by difference coagulant dosing

Figure 3 shows the result of colour removal by difference coagulant dosing. The result show that the highest value of colour at 1890 m^{-1} was detected at doses of coagulant around 2.0 g/ml the value is around. Meanwhile, the lowest value of colour was detected at pH 5.0 and coagulant dosing about 5.0 g/ml, the value reaches to 340 m^{-1}.

Figures 4, 5, 6, 7 and 8 shows the value for removing heavy metal ions by difference in coagulant dosing i.e. 1.0, 2.0, 3.0, 4.0 and 5.0 g/ml. Several common heavy metal ions were measured to indicate the performance of the coagulant. For the cadmium ion the lowest value is about 0.01 ppm the value was indicated by the experiment using pH 5.0 at coagulant dosing about 4.0 and 5.0 g/ml. For chromium ion the lowest value is 0.39 ppm. The value found from experiments using pH 5.0 and dosing value is around 5.0 g/ml. The pattern of lowest value also show for the other remaining metal ion. The lowest value was shown at pH 5.0 and at doses of coagulant around 5.0 g/ml.

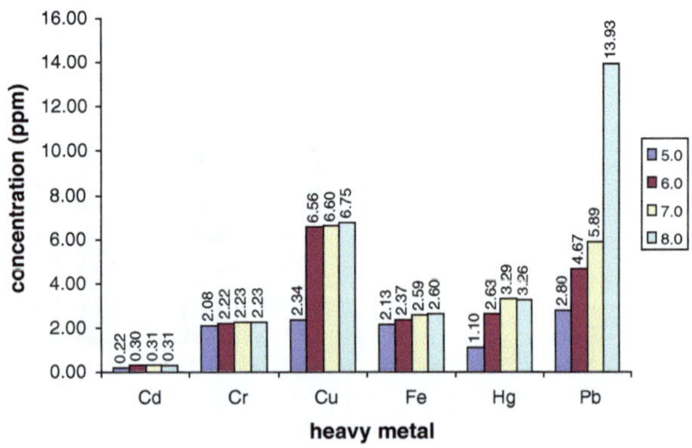

Fig. 4 Result for removal of heavy metal ion by 1.0 g/ml of coagulant dosing

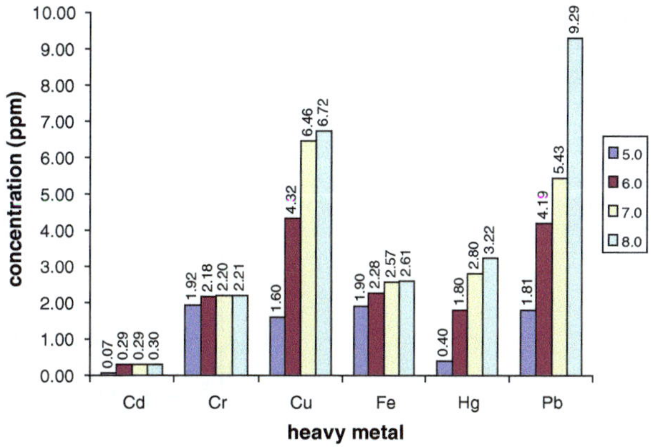

Fig. 5 Result for removal of heavy metal ion by 2.0 g/ml of coagulant dosing

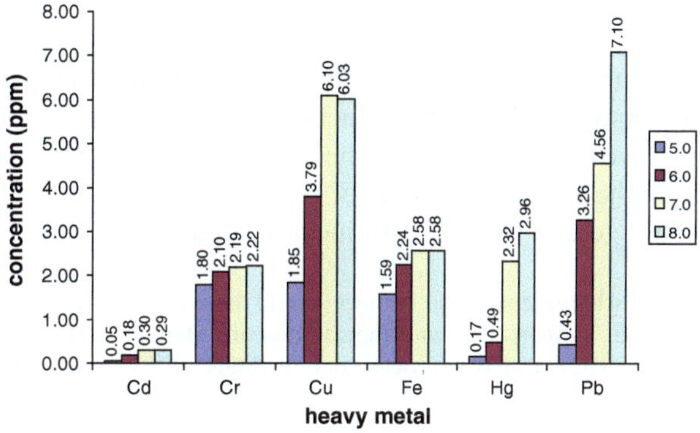

Fig. 6 Result for removal of heavy metal ion by 3.0 g/ml of coagulant dosing

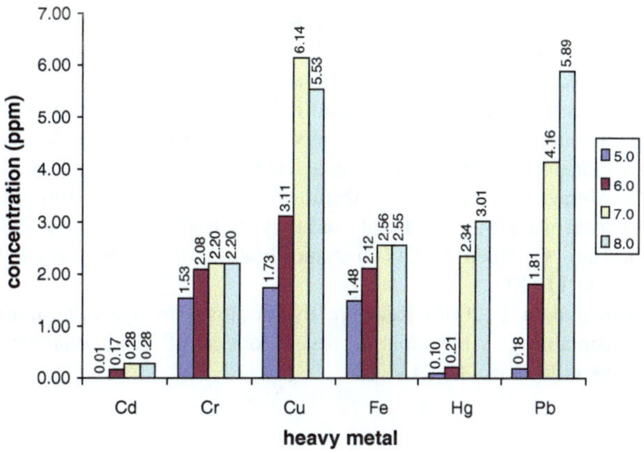

Fig. 7 Result for removal of heavy metal ion by 4.0 g/ml of coagulant dosing

Fig. 8 Result for removal of heavy metal ion by 5.0 g/ml of coagulant dosing

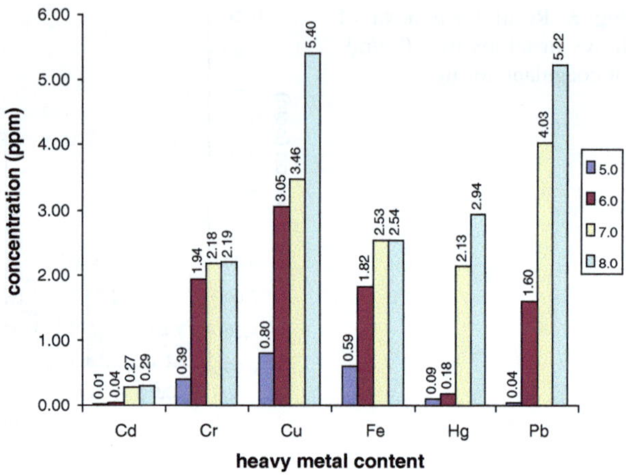

Conclusion

From the result, the optimum dosage and pH is 5.0 mg/l of coagulant dosage at pH 5.0 respectively. COD reduction is about 98 % meanwhile the reduction of color achieve more than 90 %. The treated textile wastewater was found to be complying with the Parameter Limits of Effluents of Standards B from Environmental Quality (Sewage and Industrial Effluents) Regulations 2013.

Acknowledgments The authors would like to thank the Ministry of Education Malaysia for financial support via Fundamental Research Grant Scheme (FRGS). The authors also grateful acknowledge use of services and facilities of Chemical Engineering Faculty and Universiti Teknologi MARA.

References

1. Ahmad AL, Harris WA, Syafiie Ooi BS (2002) Removal of dye from wastewater of textile industry using membrane technology. Journal Teknologi 36(F): 31–44
2. Chakraborty S, Purkait MK, DasGupta S, De S, Basu JK (2003) Nanofiltration of textile plant effluent for color removal and reduction in COD. Sep Purif Technol 31:141–151
3. Shyam SundarP, KarthikeyanN, Prabhu KH (2002) Wastewater and its treatment in textile industry. Institute of Chemical Technology, University of Mumbai
4. Ahmad AL, Puasa SW, Abiding S (2005) Crossflow ultrafiltration for removing direct-15 dye from wastewater of textile industry. School of Chemical Engineering, USM, Penang
5. Tang C, Chen V (2002) Nanofiltration of textile wastewater for water reuse. Desalination 143:11–20
6. Koyuncu I (2002) Reactive dye removal in dye/salt mixtures by nanofiltration membranes containing vinylsulphone dyes: effect of feed concentration and cross flow velocity. Desalination 143:243–253

7. Tan BH, Teng TT, Omar AKM (2000) Removal of dyes and industrial dye wastes by magnesium chloride. Wat Res 34(2):597–601
8. Hood LF, Zall RR (1980) Recovery, utilization and treatment of seafood processing wastes. In: Conell JJ (ed) Advances in fish science and technology. Fishing News Books, Ltd., Surrey, England
9. Guibal E, Roussy J (2007) Coagulation and flocculation of dye-containing solutions using a biopolymer (chitosan). React Funct Polym 67:33–42

Part II
Chemical and Biotechnology Engineering

An Investigation on N-Containing Organic Compounds as Corrosion Inhibitors for Carbon Steel in Saline Environment

Md Amin Hashim, Izni Mariah Ibrahim and Sxureha Yunus

Abstract The adsorption types of corrosion inhibitors are typically organic compounds made-up of element comprising of N, O, P and S. In this study, the relative inhibition effect of synthesized fatty amides mixture, pyridine and pyrrole as corrosion inhibitors on carbon steels in saline water had been compared using linear polarization resistance method (LPRM). It is still unknown which of these organic compounds performed best than another. The experimental results had shown that compounds containing elemental N performed as corrosion inhibitor in protecting carbon steels when immersed in saline water. The synthesized fatty amides mixture is found to be better than pyridine and pyrrole, indicating the presence of various chain lengths of fatty amides molecules provide a better surface on carbon steels.

Keywords Organic corrosion inhibitor · Electron lone pair · Physisorption · Chemisorption · Carbon steel · Static saline environment

Introduction

Corrosion inhibitors are admixture of organic and inorganic chemicals with the effect that delays the corrosion rate of carbon steels when added in small quantity in an aqueous environment. The presence of stringent environmental regulations prohibited inorganic corrosion inhibitors such as chromates, nitrites or other oxides to be used for protection of carbon steels. Previous researchers suggested that they

M.A. Hashim (✉) · I.M. Ibrahim · S. Yunus
Faculty of Chemical Engineering, Universiti Teknologi MARA,
40450 Shah Alam, Selangor, Malaysia
e-mail: aminhashim@salam.uitm.edu.my

I.M. Ibrahim
e-mail: iznimariah@yahoo.com

S. Yunus
e-mail: sxureha@yahoo.com

M.A. Hashim (ed.), *ICGSCE 2014*, DOI 10.1007/978-981-287-505-1_10

could be replaced by organic compounds such as carboxylic acid derived organics [1]. Further, carboxylic acid derived organics are well established industrial position, whilst their applications are widespread and increasing. The main reason for using carboxylic acid derived organics is due to its availability and relatively inexpensive. They are derived from non-edible animal and vegetables fats.

In this research, as-synthesized fatty amides, pyridine and pyrrole were used as the working corrosion inhibitors. These organics performed with relatively high efficiency as adsorption type of corrosion inhibitors, inhibiting carbon steels when immersed in saline water at room temperature and under static environment [2]. When the concentrations of organic inhibitors are low, the adsorbed inhibitor will be parallel alignment to the steel surfaces. When a further increase in the inhibitor concentration was added, this causes the adsorbed inhibitor molecules to align in parallel with each other and perpendicular to the steel surface. In this manner, the steel surface becomes hydrophobic. Hence, the adsorbed inhibitors molecules can act as waterproof barrier on the steel surface [3].

Experimental

Preparation of fatty amides. Mixtures of fatty amides were synthesized from as-received palm fatty acid distillates (PFAD) with analytical grade isoproylamine, added in excess, and at reaction temperature of 80 °C. The selected reaction path followed closely as suggested by Cossy and Pale-Grosdemange [4]. A selected amount (in ppm) of the as-synthesized fatty amides was used in the tests to determine the efficiency of inhibitor.

Pyridine and pyrrole. As-received, analytical grades (AR) of the chemicals were used in a prescribed concentration in 3.5 wt% of saline water at room temperature and at static condition.

Test specimen and solution. Commercially available carbon steel was used as test specimen, and prepared for polarization resistance measurements as described by Hashim et al. [5]. The carbon steel was analyzed for chemical composition using Inductively Coupled Plasma (ICP) with Optical Emission Spectrometer (OES) and CHNS analyzers. The chemical composition is as shown in Table 1. The test medium used was a 3.5 wt% sodium chloride solution.

A 3-electrode system was used to determine the R_p values of the steel specimen immersed in the test solution in static condition and at ambient temperature of 25 °C. The VoltaLab, Model PGP 201 potentiostat was used in the Linear Polarization

Table 1 Chemical composition of a typical carbon steel test specimen

Element	wt%	Element	wt%
Carbon	0.04	Copper	0.059
Aluminum	0.23	Chromium	0.060
Manganese	0.74	Zinc	0.54
Nickel	0.056	Lead	0.02
		Iron	Balance

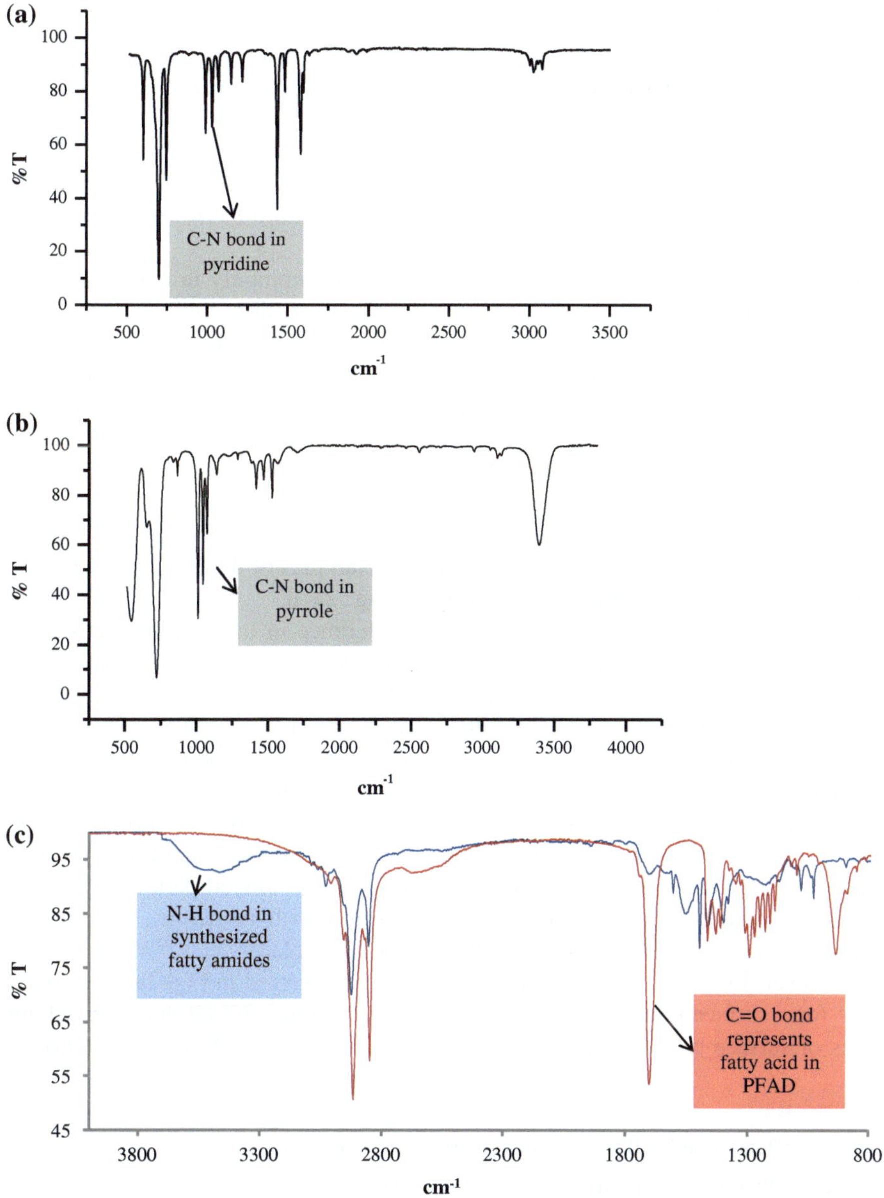

Fig. 1 Observed FTIR spectra of **a** pyrrole, **b** pyridine, and **c** synthesized fatty amides used as corrosion inhibitors [2, 7]

Resistance Method (LPRM) mode. The inhibition effects of the other organics were similarly tested with increasing volumetric quantities. Surface morphology of specimen was examined using SEM before and after conducting polarization tests.

Result and Discussion

Steel specimen chemical composition. Table 1 shows chemical composition of a typical carbon steel specimen.

Characterization of synthesized fatty amides, pyridine and pyrrole organics. Typical FTIR spectrums for these organics are as shown in Fig. 1.

Table 2 shows differences in the observed and literature FTIR spectrum for the organics used as corrosion inhibitors.

From Table 2, it is observed that elemental N is commonly in existence in the selected organic molecules which made-up the corrosion inhibitors organics. Elemental nitrogen is known to have a pair of lone electron from the sp^2 electron configuration which promotes surface adsorption [8]. Surface adsorption, either physisorption or chemisorptions enhances delay in the diffusion time of oxygen molecules toward the metallic surfaces to enhance cathodic reaction.

Liner Polarization Resistance Method (LPRM) measurement, R_p values obtained were measured by using the available potentiostat and galvanostat. Through measured R_p values, the corrosion rate is determined to be inversely proportional to the R_p of the test specimen [9]. Examples of the R_p-values obtained from the measurements done are as shown in Fig. 2a–d.

Observations from Fig. 2a, b show large difference in the measures R_p-values between fatty amides mixture added in various concentrations to the pyrrole and pyridine organics added in the same concentration. It is also observed that the inhibitive mechanism of the added fatty amides mixture is spontaneous in nature and it is an effective inhibitor for first hour after addition of the organics.

Surface analysis, Fig. 3a shows surface morphology of the steel specimen immersed in the saline solution, for a test period of 1 h, without any form of surface protection. It is observed to be covered with surface hydroxides and for a specimen which had been immersed in the saline solution for a prolonged period of time and dried, the surface layer is observed to be covered with numerous, short length surface cracks indicating corrosion activity had taken place.

However, Fig. 3b, c do not exhibit phenomenon observed in Fig. 3a. There were some form surface protections shown in Fig. 3b, c. Figure 3c did not revealed any forms of surface cracks indicating a better surface protection from the addition of

Table 2 Difference in FTIR wavelength: observed versus literature

Radical	Wavelength (cm^{-1})		Inferences
	Observed value	Literature value [7]	
C–H	838–699	680–860	Presence of C–H bond in the organic
N–H	3500–3180	3500–3300	Presence of N–H bond in the organics
C–N	1200–1025	1200–1029	Presence of C–N bond in the organics
C=O	1710–1558	1800–1678	Presence of C=O in the synthesized inhibitor

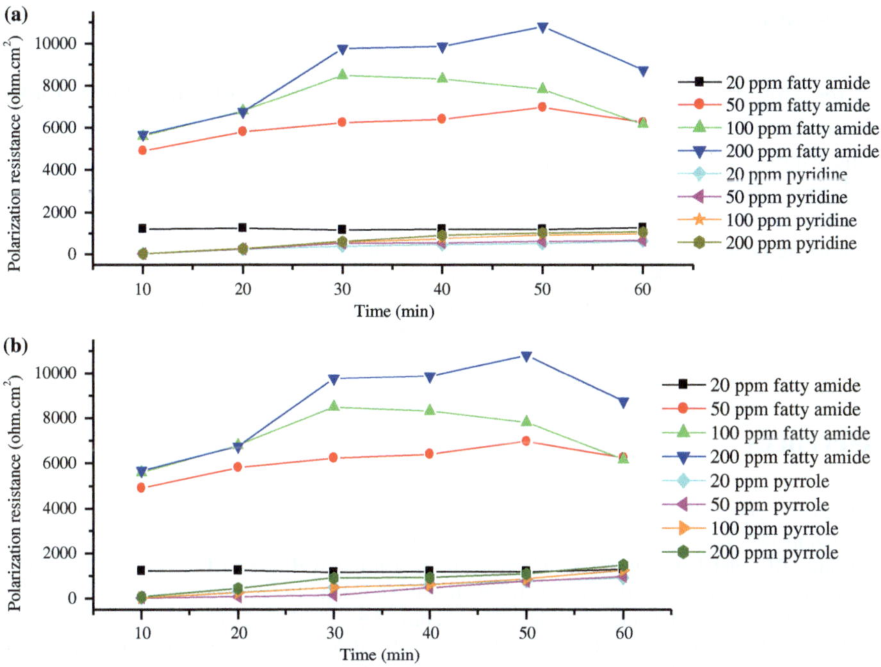

Fig. 2 Differences in the measured R_p-values in the test aqueous solution containing **a** fatty amides mixture and pyridine and, **b** fatty amides mixture and pyrrole. All tests were conducted at room temperature and in static condition

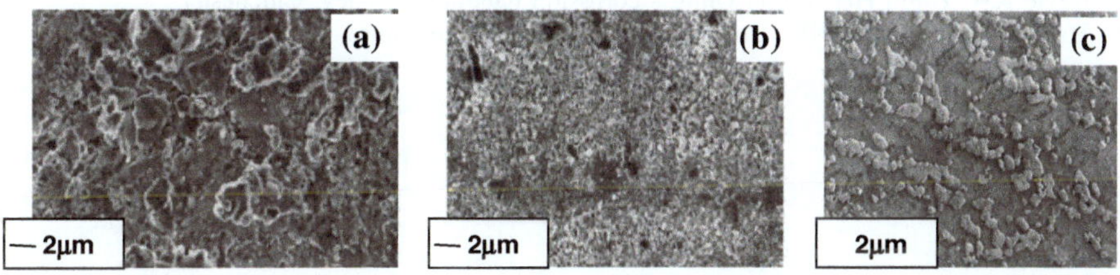

Fig. 3 Comparative steel specimen surface morphology by scanning electron microscopy. **a** Surface morphology without inhibitor, **b** with addition of 20 ppm pyrrole, and **c** with addition of 200 ppm of fatty amides mixture [5, 6]

the fatty amides mixture. The surface protection offered by fatty amides mixture is due to elemental nitrogen adsorption on the steel surface which is either physisorption or chemicsorption in nature. Previous work by Hashim et al. [10] indicated chemisorption is the type of surface adsorption preferred due to the presence of an electron lone- pair on elemental nitrogen in fatty amide molecule.

Conclusion

The research demonstrated that pyrrole, pyridine and the synthesized fatty amides mixture are corrosion inhibitors for protection of carbon steels when exposed to saline water at room temperature and under static fluid environment. Under the test conditions, as-synthesized fatty amides mixture out-performed other organics.

Acknowledgment The authors would like to express their thanks and gratitude to the Research Management Institute (RMI), the Academic and International Affairs (HEA) and the Faculty of Chemical Engineering of the Universiti Teknologi MARA for financial and other advisory supports in completing and presenting results of this research project.

References

1. Kern P, Landolt D (2001) Electrochim Acta 47:589–598
2. Ibrahim IM, Yunus S, Hashim MA (2013) Int J Sci Eng Res 4:1–12
3. Blinks BP, Fletcher PDI, Hicks JT, Horsup DI, Durnie WH (2005) Proceedings of NACE, Corrosion Paper No.05307. NACE International, Houston, Texas
4. Cossy J, Pale-Grosdemange C (1989) Tetrahedron Lett 30:2771–2774
5. Hashim MA, Shamsudin S, Lim CG, Morat A, Toff MRM (2001) Investigation into inhibitive effect of synthesized fatty amides on commercial-grade low-carbon steel, Presented at the 2nd Asian Science and Technology Week Conference. Universiti Brunie Darussalam
6. Ibrahim IM, Jai J, Hashim MA (2014) Appl Mech Mater 575:206–209
7. Coates J (2000) In: Meyers RA (ed) Interpretation of infrared spectra, a practical approach. Wiley, Chichester, pp 10815–10837
8. Al A, Hamzi H, Zarrok A, Zarrouk R, Salghi B, Hammouti S, Al-Deyab M, Bouachrine A, Amine F (2013) Guenoun. Int J Electrochem Sci 8:2586–2605
9. Revie RW, Uhlig HH (2011) Uhlig's corrosion handbook. Wiley, Chichester
10. Hashim MA, Salleh SNM, Toff MRM (2009) Mater Res Innov 13:291–294

Evaluation on the Effect of Inhibitor Ratio to the Wax Deposition in Malaysian Crude

N. Ridzuan, Z. Yaacob and F. Adam

Abstract It is important to understand the fundamental nature of petroleum wax formation since this problem is still a challenging issue in the crude oils industry. The application of a wax inhibitor was introduced in this study to reduce the wax. Poly (ethylene-co-vinyl acetate) (EVA) and cocoamide-diethanolmines (DEA) with different ratios were used to evaluate their performance as an inhibitor using cold finger equipment. A total of 10 ml inhibitor at different ratio of EVA and DEA was injected into crude oil sample at constant cold finger and water bath temperature of 10 and 50 °C, respectively. The Percent Inhibition Efficiency (PIE) was increased as the amount of EVA in DEA was increased. The use of pure EVA had successfully showed the highest PIE value which is 23.08 % compare to other ratios. The GC-FID analysis found that the $C_{28}H_{58}$ (n-octacosane) wax solid deposit was reduced almost 90 % if pure EVA was used. The addition of DEA in the EVA inhibitor does not give a significant effect to the prevention of wax solid formation. It is highly recommended to use pure EVA as a single wax inhibitor in controlling wax solid deposition.

Keywords Cold finger · Wax deposition · Ethylene-co-vinyl acetate (EVA) · Cocoamide-diethanolamine (DEA)

N. Ridzuan (✉) · Z. Yaacob · F. Adam
Faculty of Chemical and Natural Resources Engineering, Universiti Malaysia Pahang, Lebuhraya Tun Razak, 26300 Kuantan, Pahang, Malaysia
e-mail: norida@ump.edu.my

Z. Yaacob
e-mail: zyaacob@gmail.com

F. Adam
e-mail: fatmawati@ump.edu.my

M.A. Hashim (ed.), *ICGSCE 2014*, DOI 10.1007/978-981-287-505-1_11

Introduction

An increase in world crude oil demand has driven oil production industry to tap into the deeper ocean to increase the offshore production. However, wax tends to deposit from crude oil has become a very critical concern in the seabed due to the low temperatures condition. This problem can cause a major problem in oil exploration and transportations facilities [1–5]. Crude oil is a complex mixture of hydrocarbons, consisting of paraffins, asphaltenes, resins, aromatics and naphthenes [6]. Wax deposits are commonly contributed by asphaltenes and paraffins. For paraffins, the deposit commonly originated from alkanes with high molecular weight range of C_{18}–C_{70}. Paraffin can precipitate out from crude when the equilibrium condition is changed resulting the loss of solubility of the paraffin in crude or when the temperature of the crude drops below the cloud point temperature [7].

Many techniques including removal and prevention approach have been introduced to minimize the wax problems such as chemical, mechanical and thermal methods. Among them chemical method or also known as 'paraffin inhibitors' was widely applied. Paraffin inhibitor can be divided into three types which are solvents, dispersants and crystal modifiers. The effectiveness of certain paraffin inhibitor was varied from one crude oil well to another well. It is required to find the best inhibitor for the crude oil from specific well [8].

The paraffin inhibitors are commonly exist as polymeric compounds which constituted of hydrocarbon chain and polar section. The hydrocarbon section will connect the interaction between the additive and paraffin. While polar section content in the inhibitor is responsible for the wax crystals morphology modification which is necessary to inhibit the aggregation stage. It is strongly believed that both segments can reduce or eliminate the build-up of wax. An effective wax inhibitors can create a weaker deposit which subject to easy wax removal process if shears force is applied the field flow. Several polymers have been evaluated as wax crystal modifiers. EVA co-polymers is widely used due to the good efficiency as crude oil flow modifiers and wax deposition inhibitor [5, 9]. EVA acts as Pour Point Dispersant (PPD) where its capabilities of building into wax crystals through the altering growth and surface characteristics of the crystal. Thereby, it is reduces the tendency of the crystals to stick to metal surfaces such as pipe walls. In addition, it has a the tendency to reduce viscosity and influence rheology behaviour by lowering down the pour point and wax appearance temperature [10]. Lucas and co worker [9] reported that the different amount of vinyl acetate functional group content in EVA will give different results in wax deposition efficiency.

On the other hand, DEA had been reported as one of good oil-in-water emulsifier which can prevents the gelling problem in crude oil and can act as an inhibitor. Although the individual mechanism of each PPD is well understood, but there is very limited studies have been conducted on the combined effect of PPD with emulsifier such as DEA in reducing the wax deposition rate. Hence, the main objective of this study is to evaluate the performance of the EVA:DEA ratio using cold finger method by determining the best formulation of EVA:DEA ratio which can demonstrate the best paraffin inhibition efficiency (PIE).

Experimental

Materials

Chemicals, Poly (ethylene-co-vinyl acetate) (vinyl acetate 25 wt.%, melt index 19 g/10 min), cocoamide-diethenolamine (DEA) (purity 99.5 %), cyclohexane (purity 99.5 %), acetone (purity 99.5 %), heptane (purity 99.5 %) and petroleum ether were obtained from Sigma Aldrich. Raw crude oil was kindly supplied by Petronas Refinery from Kertih, Terengganu, Malaysia.

Preparation of EVA-solvent, VA-solvent solution (400 ppm) was prepared by dissolving appropriate amount of EVA in cyclohexane under continuously stirring at 300 rpm and 60 °C for one hour.

Measurement of Sample

Physical Properties

Table 1 showed the physical properties of the light crude oil sample used. Wax appearance temperature (WAT) was determined using differential scanning calorimetry (DSC). The viscosity was determined by Brookfield DV-III using spindle type 31. Pour point measurement using cloud and pour point bath was supplied by Koehler Company. The density has been was measured by pycnometer apparatus Model Micromeritis AccuPyc II 1340. The wax content was determined by Gas Chromatography with Flame Ionization Detector (FID). A capillary column ZB1, 30 m long and with 0.32 mm internal diameter and 0.25 μm film thickness, was used. The samples were injected into the column using an auto-sampler at injector temperature of 250 °C and a split/splitless ratio of 1:50. The temperature of the detector was set up at 300 °C. Analysis was performed under a helium flow rate of 1.2 ml/min and a temperature program of 40–280 °C at 12 °C/min and held for 5 min at the highest temperature.

Table 1 Physical properties of the crude oil sample

Sample name	Viscosity mPa.s at 40 °C	Pour point (°C)	Density (abs) g/cm^3	SG sample	°API	WAT, °C
Light	5.18	3	0.8505	0.8505	34.87	33

Cold Finger Experimental Set up

The cold finger apparatus as shown in Fig. 1 was fabricated to evaluate the rate of wax deposition in crude oils. This devices is suitable to correlate wax deposition data to the temperature differences between the bulk oil and wall [4, 11, 12]. The cold finger was placed within the center of stainless steel jar filled with 300 ml of crude oil. The cold finger and jar were then placed inside a heating water bath to control the crude oil temperature. An impeller was also attached at the middle of the cold finger to make the field flow influences both the shear stress field and the rate of heat transfer at the cold finger surface. Crude oils with and without wax inhibitor were tested in the cold finger experiment. The experiment was carried out for two hours to presume the aging effect upon the wax deposited onto the cold finger.

Evaluation of Paraffin-Inhibition Efficiency

Paraffin inhibition efficiency (PIE) was calculated using Eq. 1:

$$PIE(\%) = \frac{W_f - W_t}{W_f} \times 100 \tag{1}$$

where W_f is the reference amount of wax deposition without chemical treatment (g) and W_f is the amount of paraffin deposition with chemical treatment (g) [12]. Five different EVA:DEA ratios were studied.

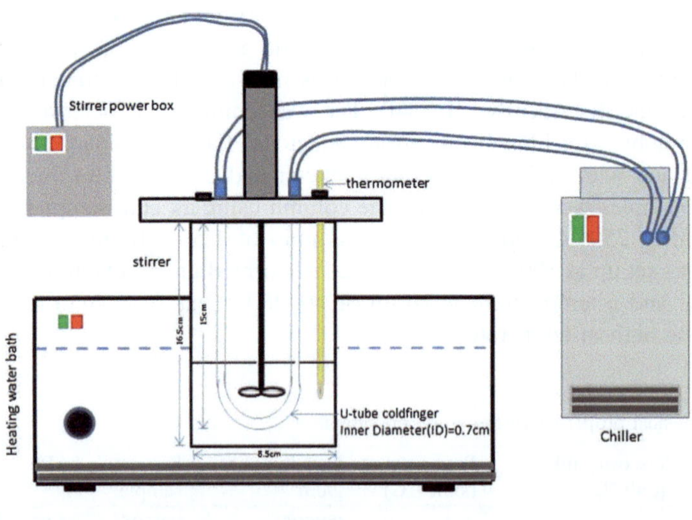

Fig. 1 Schematic diagram of cold finger apparatus

Results and Discussion

Comparative Study on the Effects of Different Wax Inhibitor Ratios

Five different ratios of EVA:DEA were examined based on the capability to decrease the gross amount of paraffin deposit. Figure 2 shows the effect of inhibitor ratio to the wax deposit and PIE at 0 rpm. The solid line represents the amount of wax deposit in gram and the dotted line represents the PIE (%). The wax deposit or crystal solid formed without the presence of the inhibitor (for control purpose) is about 1.30 g. However, under the same running condition, the amount of wax deposit reduced to 1.00 g if pure EVA was used. This reduction is equivalent to the 23.08 % in PIE value. The physical state of the deposited wax was observed visually be a soft amorphous gel. The amount of wax deposit was slightly reduced if the amount of EVA in DEA solution was increased as shown in Fig. 2. This graphical trend describes the presence of more amount of EVA in the DEA solution will reduce the formation of wax on the cold finger.

If the crude oils contain of pure DEA (10 ml of DEA), the PIE value was 6.15 %. Nevertheless, DEA can helps to suppress the wax formation is some extent when compared to the control experiment using pure crude oil without any inhibitor. It might suggest that the activity coefficient of DEA is lower compared to EVA inhibitor. The waxes solid formed was the highest at 100 % DEA concentration. This result shows that pure DEA performance exhibit 50 % lesser than EVA alone. The solid state of the deposited wax was found to be hard amorphous gel. The solid state of the deposit at 5 ml:5 ml ratios of EVA:DEA was observed to be slightly softer than the wax deposited from the pure DEA. The solid deposit was easily

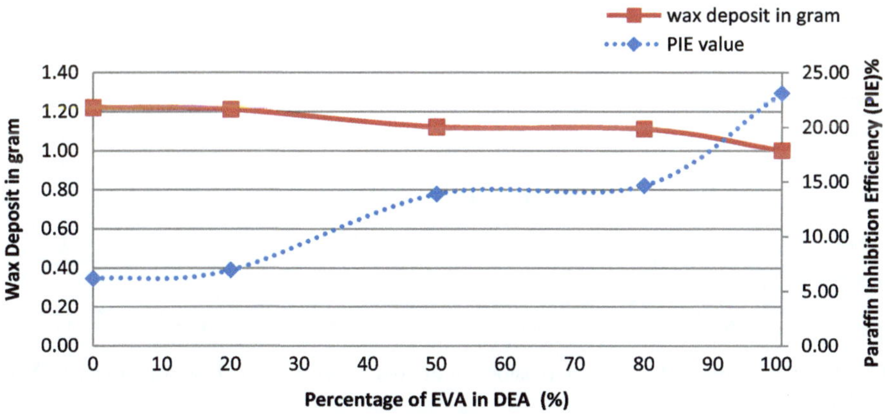

Fig. 2 The effect of inhibitor ratio to the wax deposit and PIE at 0 rpm

removed from the cold finger compared to pure DEA. This concludes that the presence of more EVA in DEA helps the solid structure softer with less amount of solid wax.

The study reported that EVA will aggregate to the structure of the paraffin wax by burying itself inside the wax crystals and absorbed itself on the surface of the wax crystals. Therefore, it can coalesce with wax crystals and interfere with the growth of crystals. Thus, the ability of crystals to interconnect and form networks decreases. Therefore, more crystals with smaller size can be produced [5].

Figure 3 shows an example of a wax solid deposited which was analyzed by GC-FID. This graph represents the solid formed during an experiment run at equal ratio of EVA:DEA. The major compound n-octacosane ($C_{28}H_{58}$) was used to represent wax solid because the highest composition amount presence in crude oil. Table 2 shows the details analysis of $C_{28}H_{58}$ percentage in wax solid and its percent inhibition compared to the wax deposit (blank sample). This result gives an indication of the inhibitor performance. The amount of $C_{28}H_{58}$ wax solid decreased as the amount of EVA in DEA was increased. The percentage of $C_{28}H_{58}$ in wax solid (wt.%) also found a similar trend to total amount (weight) of wax solid as described in Fig. 2. The inhibition percentage of $C_{28}H_{58}$ wax solid also increased significantly when the 100 % of EVA used. It can be concluded that EVA performed a significant chemical inhibitor compared to DEA in particular to inhibit the major Malaysian crude compound n-paraffin $C_{28}H_{58}$ to deposit or precipitate.

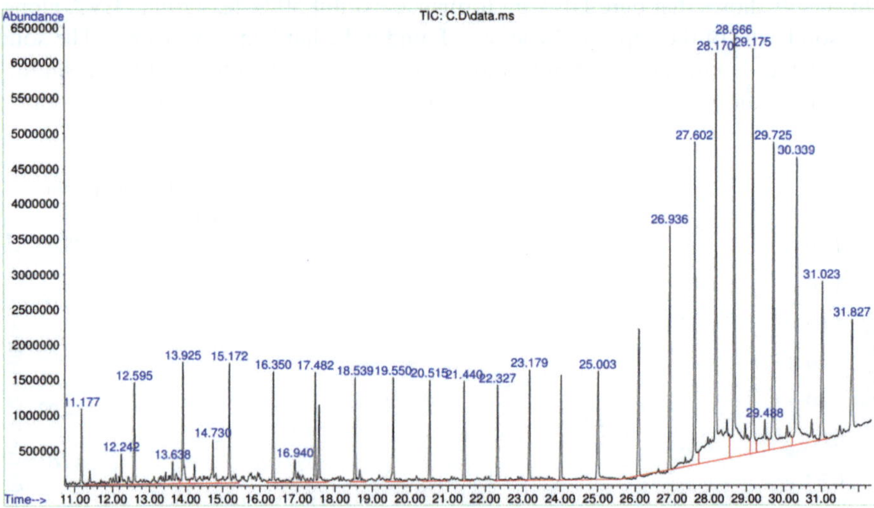

Fig. 3 An Example of GC-FID chromatogram of cold finger wax solid deposit at equal EVA: DEA ratio

Table 2 Percentage of $C_{28}H_{58}$ wax compound in total wax and it inhibition percentage

Sample	Ratio of EVA to DEA	% of $C_{28}H_{58}$ in the total wax (wt.%)	Percent inhibition of $C_{28}H_{58}$ in wax total (%)
Vial A	0:10	1.52	5.50
Vial B	2:8	1.56	3.42
Vial C	5:5	1.20	25.93
Vial D	8:2	0.24	85.41
Vial E	10:0	0.19	88.32
Vial F	Without inhibitor	1.61	NA

Conclusion

The study found that EVA inhibits the formation of waxes solid more significantly compared to DEA. The increasing of DEA amount in EVA does not significantly reduce the wax solid formation because of the less activity coefficient. GC-FID analyzed that the amount $C_{28}H_{58}$ wax reduced when 100 % of EVA was used as inhibitor. This study was successfully evaluated the suitable inhibitor ratio to inhibit the wax solid deposit for Malaysian crude oils. The details understanding on the behavior of a chemical inhibitor to wax solid should be further studied to understand the role of both chemical inhibitors at molecular level perspective.

Acknowledgment The authors express high appreciation to Universiti Malaysia Pahang and the Ministry of Higher Education, Malaysia, through *Research Acculturation Grant Scheme* (*RAGS-RDU 121414*) for having this research project successfully conducted.

References

1. Misra S, Baruah S, Singh K (1995) Paraffin problems in crude oil production and transportation: a review. SPE Pro Facil 10:50–54
2. Jang YH, Blanco M, Creek J, Tang Y, Goddard WA (2007) Wax inhibition by comb-like polymers: support of the incorporation-perturbation mechanism from molecular dynamics simulations. J Phys Chem B 111:13173–13179
3. Lee HS (2008) Computational and rheological study of wax deposition and gelation in subsea pipelines. University of Michigan, Michigan
4. dos Santos JDST, Fernandes AC, Giulietti M (2004) Study of the paraffin deposit formation using the cold finger methodology for Brazilian crude oils. J Pet Sci Eng 45:47–60
5. Jafari Ansaroudi HR, Vafaie-Sefti M, Masoudi S, Behbahani TJ, Jafari H (2013) Study of the morphology of wax crystals in the presence of ethylene-co-vinyl acetate copolymer. Pet Sci Technol 31:643–651
6. Singh P, Venkatesan R, Fogler HS, Nagarajan N (2000) Formation and aging of incipient thin film wax-oil gels. AIChE J 46:1059–1074
7. Abdallah EM, El-khatib N (2011) Modeling wax deposition in crude oil presented at National Postgraduate Conference (NPC). Kuala Lumpur, 19–20 Sept 2011, pp 1–7

8. Lashkarbolooki M, Esmaeilzadeh F, Mowla D (2011) Mitigation of wax deposition by wax-crystal modifier for Kermanshah crude oil. J Dispers Sci Technol 32:975–985
9. Lucas EF, de Machado ALC (1999) Poly (ethylene-co-vinyl acetate)(EVA) copolymers as modifiers of oil wax. Pet Sci Technol 17:1029–1041
10. Pedersen KS, Rønningsen HP (2003) Influence of wax inhibitors on wax appearance temperature, pour point, and viscosity of waxy crude oils. Energy Fuels 17:321–328
11. Jennings DW, Weispfennig K (2006) Effect of shear on the performance of paraffin inhibitors: coldfinger investigation with Gulf of Mexico crude oils. Energy Fuels 20:2457–2464
12. Bello O, Fasesan S, Teodoriu C, Reinicke K (2006) An evaluation of the performance of selected wax inhibitors on paraffin deposition of Nigerian crude oils. Pet Sci Technol 24:195–206

Solvent Extraction of Castor Beans Oil: Experimental Optimization via Response Surface Methodology

Jibrin Mohammed Danlami, Agus Arsad and Muhammad Abbas Ahmad Zaini

Abstract Optimization of percentage of oil yield extracted from castor seed was carried out using response surface methodology and Box-Behnken design. Castor oil was extracted using solvent extraction with three different solvents; hexane, petroleum-ether and ethanol. The linear and quadratic effects of the two variables studied were most significant in affecting the oil yield. The maximum yield of oil extracted was 50.9, 49.0 and 59.5 % using hexane, petroleum ether and ethanol respectively as solvents and the optimum conditions were at a time of 3 h and an average particle size of 1 mm which were within the experimental domain. The model equations gave a good fit with the coefficient of determination of 0.9216, 0.9587 and 0.9005 for hexane, petroleum ether and ethanol, respectively, and hence the suitability of the response surface methodology for the optimization of percentage oil yields.

Keywords Solvent extraction · Response surface methodology · Optimization · Castor · Ricinoleic acid

Introduction

Castor oil is viscous, pale yellow and non-drying oil with a taste and in most times used as purgative [1]. Castor oil can either be applied in its refined, hydrogenated form or crude form. Typically, 28 % is refined, 65 % of it is processed, 12 % is

J. Mohammed Danlami · M.A. Ahmad Zaini
Centre of Lipid Engineering and Applied Research (CLEAR),
Ibnu Sina Institute for Scientific and Industrial Research,
Universiti Teknologi Malaysia, 81310 Johor Bahru, Malaysia
e-mail: jibrin349@yahoo.com

M.A. Ahmad Zaini
e-mail: abbas@cheme.utm.my

A. Arsad (✉)
Enhanced Polymer Research Group (ENPRO), Department of Polymer Engineering,
Faculty of Chemical Engineering, Universiti Teknologi Malaysia,
81310 Johor Bahru, Malaysia
e-mail: agus@cheme.utm.my

M.A. Hashim (ed.), *ICGSCE 2014*, DOI 10.1007/978-981-287-505-1_12

hydrogenated, 5 % is processed to manufacture other derivatives and 20 % is dehydrated. Castor oil is extracted from castor beans by either aqueous traditional technique, mechanical expression such as hydraulic pressing or screw pressing or through solvent extraction [2, 3]. The preferred technique largely depends on some socioeconomic factors and also the technology available within the area of use, [4] mechanical pressing is often used because it ensures the extraction of non contaminated oil at a comparatively low cost [5, 6]. The production and quality of oil extracted using mechanical expression depends on the pre-pressing process conditions, such as the purity, dehulling, particle size and moisture content of the seed. Since processing conditions modifies the chemical properties of vegetable oils, appropriate combination of the processing conditions is needed to extract crude oil with properties close to that of the refined oil. However, studies have shown that moisture content, heating temperature and heating time had pronounced effects on both the quality and the quantity of oil expressed from some oil seeds [4, 7–9].

Generally, extracted castor oil contains small amounts of free fatty acids, phospholipids, mono- and diacylglycerols, protein degradation products, waxes, hydrocarbons, moisture, dirt, and other contaminants, a refining process has to be applied prior to derivatization of the oil. A typical refining process consists of the following steps: degumming, neutralization, bleaching, deodorization, and sometimes winterization [10–12]. The degumming step is employed to reduce the phospholipids and the metal content of the crude oil by combining it with water and acid, such as citric or phosphoric acid [10–12]. The efficiency of solvent extraction is affected by extraction time and particle size. To study the effect of these variables on the output, the response surface methodology (RSM) is an effective technique [13]. The principles and applications of RSM have been well described [14, 15]. The aim of the present research is to get a better understanding of the relation between the variables (time and particle size) and determine the optimum conditions for the extraction of castor oil. RSM is an efficient statistical technique for modelling and analyzing the influence of several variables on the response of a system and to optimize this response with minimum number of experiments. RSM is a faster and more economical method for gathering research results than classic one-variable-at-a-time or full-factors experimentation.

Materials and Methods

Materials

The castor bean seed was obtained from Ancient Greenfield PL, India. All Chemicals used for seed oil extraction were hexane, petroleum ether and ethanol and were purchased from Sigma Aldrich, Malaysia.

Determination of Moisture Content of the Seeds

200 g of the cleaned and ground sample was weighed and dried in an oven at 90 °C for 24 h and the weight was taken after every 2 h. The procedure was repeated until a constant weight obtained. After each 1 h, the sample was removed from the oven and placed in the desiccator for 30 min to cool. It was then removed and re-weighed. The percentage moisture in the seed was calculated from the formula:

$$\text{Moisture} = \frac{100(W_1 - W_2)}{W_1} \tag{1}$$

where W_1 = Original weight of the sample before drying; W_2 = Weight of the sample after drying.

Soxhlet Extractor Operation and Percentage Oil Determination

150 mL n-hexane solvent was measured and poured into a round bottom flask of 1000 mL. 10 g of the measured seed sample was inserted in a thimble placed in the extractor. The round bottom flask was heated to 60 °C equivalent to the boiling point of hexane. As the solvent boils, there was a rise of the solvent vapour into the condenser to the top through a vertical tube. The condensate solvent dropped into the thimble containing the castor seed sample. The extract, then fills the siphon tube by seeping through the pores of the thimble and refluxed into the flask. The circle continues for l, 2 and 3 h depending on the operating condition. The waste was then dried, cooled in a desiccator and weighed to quantify the extracted oil. The entire process was repeated in duplicate by placing another 10 g of the sample into the thimble for the next batch of extraction. The ensuing mixture (miscella) containing the oil at the end of the extraction was heated to recover the solvent from the oil by using a rotary evaporator. The same process was repeated for ethanol and petroleum ether as an extracting solvent and recovery.

Castor Oil Recovery

The % yield of castor oil was calculated using Eq. (2)

$$\%\text{Yield} = \frac{Y_1 - Y_2}{Y_2}(100) \tag{2}$$

where Y_1 and Y_2 are the weights of the beans after extraction, in grams.

Table 1 Experimental design for each of the independent variables used in developing the experimental data for optimizing the solvent extraction of oil from castor seeds

Independent variables	Symbol		Coded levels		
	Coded	Uncoded	−1	0	1
Time, (min)	X_1	T	1	2	3
Particle size, (mm)	X_2	P	1	2.36	3.35

Statistical Analysis

Design of Experiment

To achieve the optimum oil yield by employing solvent extraction, the variables considered (time and particle size) were optimized using RSM as summarized in Table 1. RSM is a mathematical tool used for planning experiments, developing polynomial models for predicting response, assessing the substantial effects of factors and optimizing the required function [16].

Box–Behnken design was chosen with three factors to design the experiment. The RSM was used to optimize the key variables influencing the extraction. The RSM of this extraction experiment was conducted to achieve the optimum degree extraction and optimum oil yield. In this statistical analysis, the relationship between the coded and the actual variables can be expressed by Eq. (3).

$$X_i = (T_i - P_0)\Delta T_i \quad (3)$$

where X_i is the independent variable, T_i is the actual independent value; P_o is the independent real value at the centre point and ΔT_i step change value. The oil yield is taken as the dependent variables. Therefore, the proposed model for the response is given by Eq. (4) [17, 18].

$$Y_i = \beta_0 + \beta_1 X_1 + \beta_2 X_2 + \beta_{11} X_1^2 + \beta_{22} X_2^2 + \beta_{12} X_1 X_2 \quad (4)$$

where Yi are independent variables, βo is the constant term, while β_1, β_2 are the linear coefficient, β_{11}, β_{22} are the squared terms and β_{12} is the interaction effect. STATISTICA 6.0 was used for analysis of variance (ANOVA), and multiple regression analyses of the data was obtained.

Results and Discussion

Experimental Data

The results of the experimental design (table not shown) shows the experimental design and predicted responses for oil yield from castor bean seed.

Regression Model

The results of experimental design and predicted responses (table not shown) were used to run ANOVA and Multiple Regression Analysis in STATISCA using polynomial model Eq. (2). This allows the optimum oil yield and other responses to be predicted. The coefficients of the model equation were used to predict the optimum oil yield by multiple regression analysis using STATISCA 6.0 and are shown as Eqs. (5–7).

$$Y_1 = 37.398 + 4.948X_1 - 2.341X_2 + 3.35X_1^2 + 1.667X_2^2 \quad (5)$$

$$Y_2 = 37.873 + 6.568X_1 - 2.618X_2 - 3.525X_1^2 + 1.992X_2^2 \quad (6)$$

$$Y_3 = 51.297 + 0.302X_1 - 0.805X_2 + 5.775X_1^2 - 1.142X_2^2 \quad (7)$$

with Y_1, Y_2 and Y_3 representing hexane, petroleum ether and ethanol solvent extractions and the responses for the tested variables in coded terms; time of extraction (X_1) and average particle size (X_2). The significance of each of the variable is calculated as F-value of the model from the mean square. The F-value for hexane, petroleum ether and ethanol are 14.16, 11.01 and 29.0 respectively, with a very low probability value of <0.000001, <0.000001 and <0.00002 respectively. This shows a very high significant effect on the yield. It has been reported that regression models with P values less than 0.05 indicates a statistical significant. Analysis of variance (ANOVA) of the model indicates a good model performance with an R^2 value of 0.9216, 0.9587 and 0.9005 for hexane, petroleum ether and ethanol respectively. An R^2 of at least 0.80 is required for a model to be fit [19]. More so, the probability values $P > F$ and less than 0.05 (table not shown) shows that the model terms are significant. Therefore, the percentage oil recovery indicates that the model is most significant, as the calculated F-values representing the three solvent of extraction are much greater than the table experimental value of $F_{(2,15,0.05)} = 3.68$.

Response Surfaces Analysis

The effect of both independent variables on the dependent variable can easily be spotted using response plots generated from the STATISTICA software. The combined effect of extraction time and average particle size are shown in Fig. 1a using hexane as solvent of extraction. The optimum oil yield % was obtained at extraction time of 3 h and particle size of 1 mm with the linear effect of extraction time being dominant. Figure 1b also shows the effect of extraction time against average particle size using petroleum ether as solvent of extraction. The illustration shows the same pattern as that of hexane, with an extraction time being linear and dominant. The time factor therefore, plays a significant role in determining the oil

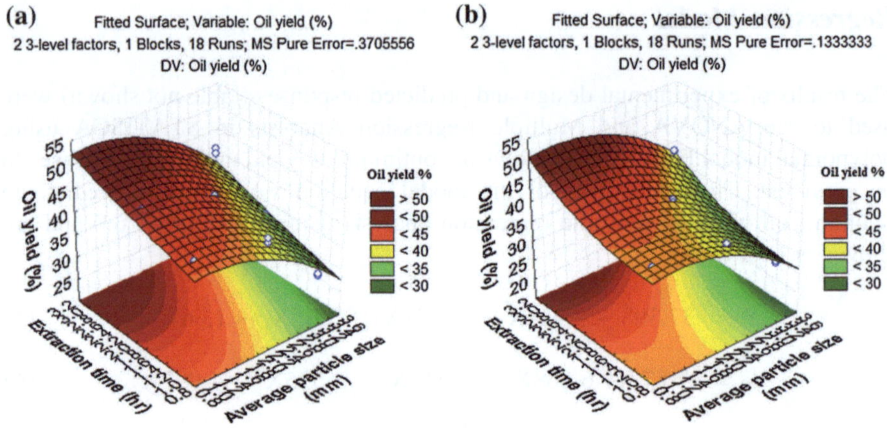

Fig. 1 3D response surface plots for oil yield % using **a** hexane and **b** petroleum ether as an extracting solvent versus extraction time and average particle size

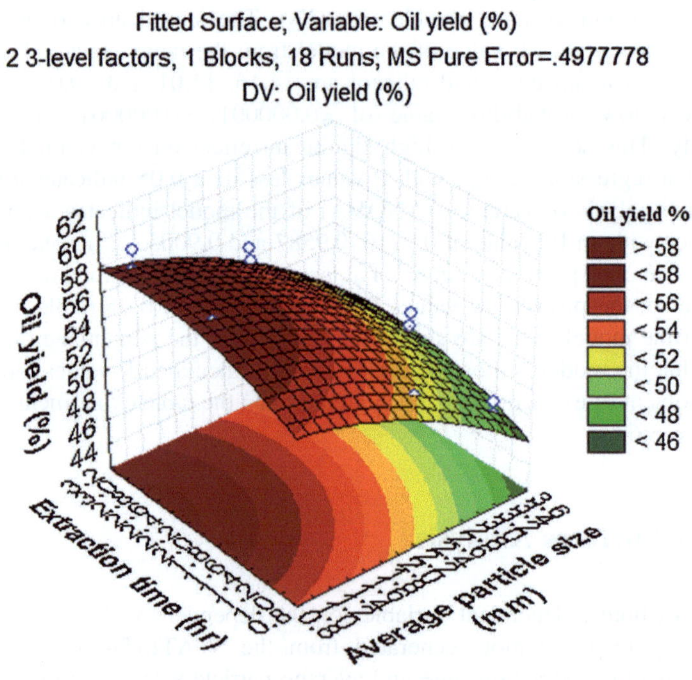

Fig. 2 3D response surface plots for oil yield % using ethanol as an extracting solvent versus extraction time and average particle size

yield. The effect of extraction time against average particle size is also shown in Fig. 2 using ethanol as an extracting solvent, indicates that the interaction effect of average particle size is linear and dominant over extraction time.

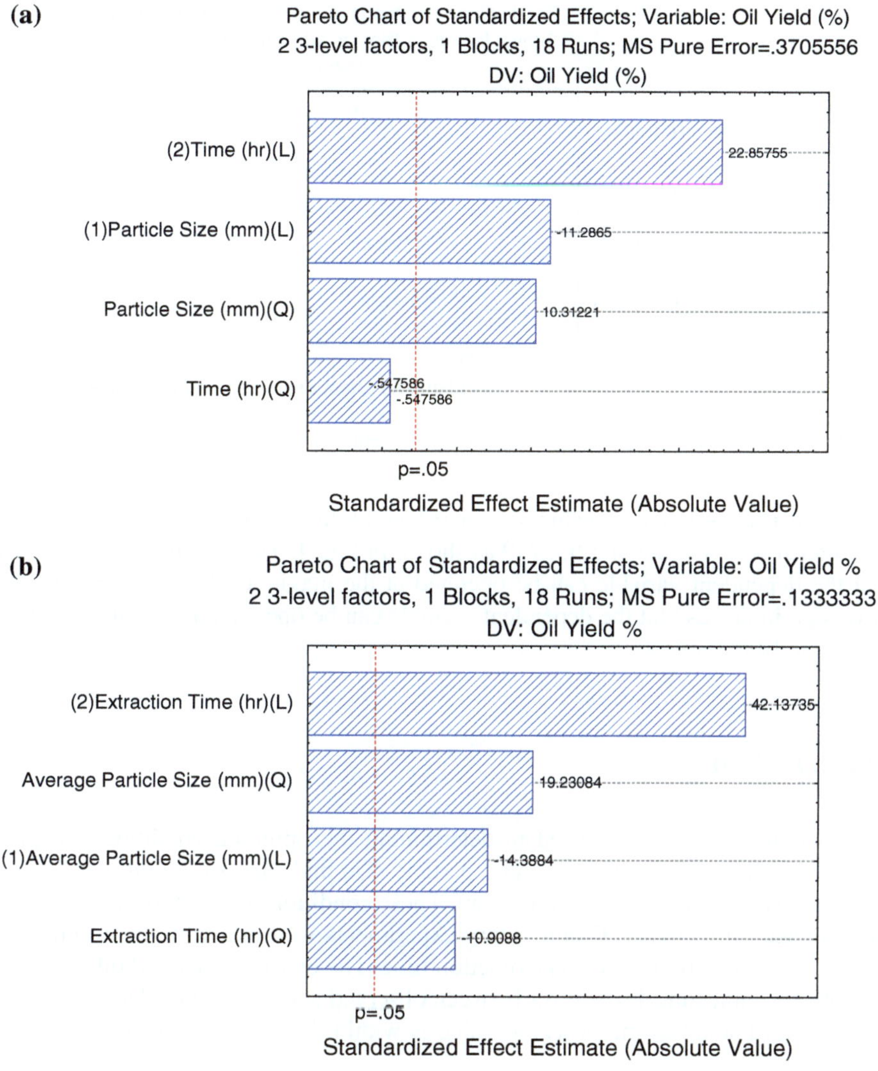

Fig. 3 Pareto Chart using **a** Hexane and **b** Petroleum ether as extracting solvent

Paretor Chart

The Pareto chart showed the most influence variable in the extraction process. Figure 3a, b, shows the most influence variable was time indicating linear term with values of 22.85755 and 42.137 using hexane and petroleum ether as the extracting solvent. Followed by the average particle size in linear and quadratic term with values of −11.2865 and −19.23084. However, ethanol shows a linear effect of

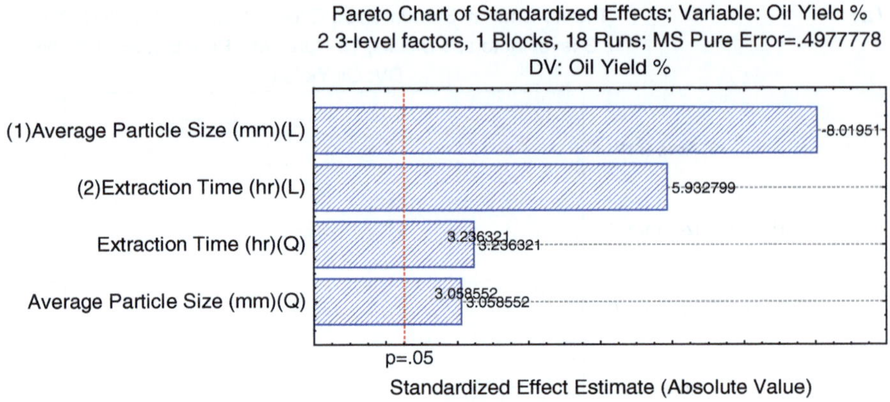

Fig. 4 Pareto chart using ethanol as an extracting solvent

average particle size with a value of −8.0195 and followed by the extraction time in linear term with a value of 5.932799 as shown in Fig. 4. The positive sign indicates that the dependent variable can be operated at the maximum level while the negative sign indicates that the dependent variable can be operated at a minimum level to get the best response.

Optimization

The models obtained were used in determining the optimum conditions for prediction of oil recovery at an average particle size and time of 1 mm and 2.53 h respectively for ethanol (table not show). These conditions corresponded to 57.8 % of predicted oil recovery. Confirmatory experiments were run with these optimum values, average particle size and extraction time. The observed value of oil recovery was 59.5 % indicating 1.7 % error between observed and predicted value. This error is considered small as the observed value is within the 5 % significance level.

Conclusion

Response surface methodology (RSM) was successfully applied for optimizing oil yield for solvent extraction. The linear term of extraction time significantly affected the oil recovery while using hexane and petroleum ether as an extracting solvent. In addition, the interactions between extraction time and average particle size had a significant effect on the oil recovery. The parametric analysis was adopted to find the optimum operating conditions of average particle size of 1 mm and an extraction time of 2.53 h. The experimental oil recovery of 50.9, 49 and 59.5 %

were achieved using the optimum conditions, which is compatible with the predicted value of 51.6, 49.2 and 57.8 % of hexane, petroleum ether and ethanol solvent respectively. Therefore, the model used in predicting the oil recovery gave a good fit with the experimental variables and hence the suitability of the RSM method for the optimal performance.

References

1. Othmer K (1979) Encyclopaedia of chemical technology. Wiley, New York
2. Oyinlola A, Ojo A, Adekoya LO (2004) Development of a laboratory model screw press for peanut oil expression. J Food Eng 64:221–227
3. Takashi Y (2006) Development trend of oil expression and degumming. Oleoscience 6:133–138
4. Faborode MO, Owolarafe OK, Lasisi AA, Kasali SA, Oguntuase KS (2003) Assessment of seed-oil extraction technology in some selected states of Nigeria. Technovation 23:545–553
5. Singh J, Bargale PC (2000) Development of a small capacity double stage compression screw press for oil expression. J Food Eng 43:75–82
6. Bamgboye A, Adejumo A (2007) Development of a sunflower oil expeller. CIGR E-J IX, Manuscript EE 06 015
7. Ibrahim A, Onwualu AP (2005) Technologies for extraction of oil from oil-bearing agricultural products: a review. J Agric Eng Technol 13:58–70
8. Abidakun OA (2010) Mechanical expression and lubricity of Irvingia gabonensis (Dika nut) oil. M.Sc. Thesis, Department of Mechanical Engineering, Obafemi Awolowo University, Ile Ife
9. Mwthiga G, Moriasi L (2007) A study of yield characteristics during mechanical oil extraction of preheated and ground soybeans. J Appl Sci Res 3:1146–1151
10. Anderson D (2005) A primer on oils processing technology. In: Shahidi F (ed) Bailey's industrial oil and fat products. Edible Oil and Fat Products: Processing Technologies, Wiley, Hoboken, pp. 1–56
11. Kemper TG (2005) Oil extraction. In: Shahidi F (ed) Bailey's industrial oil and fat products. Edible Oil and Fat Products: Processing Technologies, Wiley, Hoboken, pp. 57–98
12. Ali MF (2005) Edible oils, fats and waxes. In: Ali MF, El Ali BM, Speight JG (eds) Handbook of industrial chemistry: organic chemicals. McGraw-Hill, New York, pp. 86–121
13. Bas D, Boyaci IH (2007) Modeling and optimization I: usability of response surface methodology. J Food Eng 78:836–845
14. Giovanni M (1983) Response surface methodology and product optimization. Food Technol 37:41–45
15. Myers RH, Montgomery DC (1995) Response surface methodology: process and product optimization using designed experiments. Wiley, New York
16. Zhang Q, Tang L, Zhang J, Mao Z, Jiang L (2011) Optimization of thermal-dilute sulphuric acid pre-treatment for enhancement of methane production from cassava residues. Bioresour Technol 102:3958–3960
17. Shehu MS, Abdul Manan Z, Wan Alwi SR (2012) Optimization of thermo-alkaline disintegration of sewage sludge for enhanced biogas yield. Bioresour Technol 114:69–74
18. Cornell JA (1990) How to apply response surface methodology, revised edition. tammy griffin. Book Crafters, United State of America
19. Guan X, Yao H (2008) Optimization of viscozyme L assisted extraction of oat bran protein using respond surface methodology. Food Chem 106:345–350

Immobilization of Nano-sized TiO_2 on Glass Plate for the Removal of Methyl Orange and Methylene Blue

Lim Ying Chin, Lim Ying Pei, Razzana binti Rosli and Nur Hazira binti Mohd Atni

Abstract Titanium dioxide (TiO_2) is the most researched photocatalyst due to its potential for dye removals in wastewater treatment. However, the TiO_2 powder suffers from agglomeration which hindered its process of industrialization. Thus, this study was carried out to immobilize nano-sized TiO_2 on glass plate via doctor blade technique and the crystal structure and surface morphology of the samples were accomplished using XRD and FESEM. The effects of operational parameters such as effect of light and pH in the removal of methyl orange (MO) and methylene blue (MB) using the as-prepared TiO_2 films were investigated. MO achieved a removal of 95 %, while only 47 % removal efficiency for MB after 120 min under UV-light illumination. On the other hand, percentage removal for MO and MB are recorded as 65 and 28 % respectively under LED light irradiation. The photocatalytic removal of MO and MB was found to be mainly due to the adsorption-photocatalysis process, while the photolysis process is almost negligible. The kinetic study indicates that the photodegradation rate of MO and MB is fitted to the pseudo-first order rate model.

Keywords Adsorption · Photodegradation · Dyes · Titanium dioxide

L.Y. Chin (✉)
Faculty of Applied Sciences, Universiti Teknologi MARA,
40450 Shah Alam, Selangor Darul Ehsan, Malaysia
e-mail: yclim213@gmail.com

L.Y. Pei · R.b. Rosli · N.H.b. Mohd Atni
Faculty of Chemical Engineering, Universiti Teknologi MARA,
40450 Shah Alam, Selangor Darul Ehsan, Malaysia
e-mail: yingpei@salam.uitm.edu.my

R.b. Rosli
e-mail: razzanarosli@gmail.com

N.H.b. Mohd Atni
e-mail: haziraatni@gmail.com

M.A. Hashim (ed.), *ICGSCE 2014*, DOI 10.1007/978-981-287-505-1_13

Introduction

Large quantities of dyes are extensively used in the textile industries particularly those involved in the finishing processes. As such, the textile effluents contain large amount of organic compound that is not easily to be treated using conventional methods. Environmental pollution by organic dyes possesses severe ecological problem as the organic dyes are toxic to microorganisms and have long degradation time in the environment. Consequently, wastewater treatment methods such as adsorption, reverse osmosis, flocculation and biological degradation have been used to treat such effluents. Ultimately, advanced oxidation process (AOPs) for the destruction of synthetic organic species that are highly potent and strongly oxidizing radical allows the destruction of a wide range of organic chemical substrate with no selectivity [1]. Among the AOPs, heterogeneous photocatalysis employing TiO_2 has proven as efficient method for degrading both aquatic and atmospheric organic contaminants. TiO_2 has emerged as an excellent photocatalyst due to its non-toxicity, high corrosion stability, high photocatalytic efficiency and the photogenerated holes are highly oxidizing [2].

TiO_2 can be used either in suspension mode or immobilized it on a substrate as thin film. Practically, immobilization method is more preferable over the suspension mode as it saves unnecessary cost for catalyst separation and simplify the purification process. In addition, TiO_2 tends to agglomerate especially at high concentrations and this set a limitation towards continuous flow system [3]. Therefore, the objective of this work is to immobilize the TiO_2 suspension on the glass plate and to evaluate the photocatalytic activity of the immobilized TiO_2 samples in degradation of methyl orange (MO) and methylene blue (MB). Further more, the effects of different light source and initial dye solution pH were investigated.

Experimental

Materials

TiO_2 powder was obtained from Riedel-de Haën. Polyvinyl alcohol and formaldehyde were purchased from Sigma Aldrich. Hydrochloric acid (HCl) obtained from JT Baker and sodium hydroxide pellets (NaOH) purchased from Merck were used to adjust the solution pH. All the materials were used as received without further purification. Methyl orange purchased from Bendosen and methylene blue obtained from Unilab were selected as the model pollutant. The MO and MB solution was prepared through dissolving the MO and MB powder in ultra pure water (Direct-Q3 Water Purification System).

Preparation of Immobilized TiO_2

1.5 g of anatase TiO_2 powder was added to 15 mL of ultra pure water and the mixture was stirred for an hour to ensure homogeneity. In order to attach the prepared TiO_2 suspension to the substrate, the binder was prepared by mixing 5 g of polyvinyl alcohol to 80 mL of formaldehyde in water bath at 70 °C under continuous stirring until it became viscous and sticky. Next, glass plate with a dimension of 2.54 cm × 7.62 cm was weighed prior to coating. Then, the plates were applied with a thin layer of binder followed by TiO_2 suspension. The layered of binder and TiO_2 suspension was dried and the coated samples were weighed again to determine the amount of TiO_2 coated onto glass plate. Morphological characterization of the sample was carried out using a Field Emission Scanning Electron Microscope (FESEM, Carl Zeiss Supra 40 V). The structure of the sample was identified using an X-ray Diffractometer (XRD, Shimadzu D6000) using Cu Kα radiation (λ = 1.5406 Å).

Adhesion Strength Test

The strength test of the immobilized TiO_2 was carried out to evaluate the adhesion property of the binder as described in [4]. Briefly, the immobilized TiO_2 plate was immersed into a beaker containing pure water and was sonicated for 30 s using a 37 kHz ultrasonicator (Elmasonic S30/(H). The differences in weight of the glass plate before and after every interval of 5 s of ultrasonication were measured to obtain the percentage of TiO_2 particles adhere on the glass plate.

Dyes Removal

A custom-made glass photoreactor measured by 17 cm × 12 cm × 12 cm was used for all the photodegradation experiments. Five pieces of TiO_2 plate were used as photocatalyst and irradiated with an UV fluorescent lamp (Generic, 9 W) at a distance 10 cm from the top of the dye solution. Oxygen was continuously supplied into the solution using an aquarium pump (Regent calm, RC-002) with a flow rate of 37 cm^2/s. In typical experiment, 200 mL of MO and MB solution with initial concentration of 10 ppm was used as model pollutant. To investigate the effect of light source on the dyes removal, the UV light was replaced with LED lamp (Aletko, 9 W). The initial pH of the MO and MB solution was pH 6.2 and 4.7, respectively. It was adjusted by adding either 0.1 M HCl or 0.1 M NaOH to the desired values in order to study the influence of pH value (pH range 3–9) on the dye degradation.

5 mL of the samples were taken every 15 min for a period of 120 min. The concentration of MO and MB was determined by measuring the absorbance at λ = 465 nm and 650 nm for MO and MB, respectively using a DR 2800 HACH spectrophotometer. The photocatalytic decolorization efficiency was calculated using Eq. 1 and the kinetics was determined using the first-order kinetic equation as shown in Eq. 2:

$$\text{Dye removal}(\%) = (C_0 - C)/C_0 \times 100\,\% \quad (1)$$

$$\ln C/C_0 = k_1 t \quad (2)$$

where C is the concentration of substrate in solution at time t of illumination (ppm); C_0 is the concentration of the substrate in solution at time zero of illumination (ppm); and k_1 is the pseudo first-order rate constant (min^{-1}). The rate constant, k_1 was obtained from the slope of the straight-line plots of ln C/C_0 versus time, t.

Control Sets of Experiment

In order to verify the photocatalytic degradation of the samples, two sets of control experiments were carried out on MO and MB for 120 min. First set of experiment was photolysis (direct illumination of UV lamp without TiO_2). This is important to test the dye degradability under light exposure. The second set was adsorption (presence of TiO_2 without light illumination) with the aim to examine whether removal of dye occured via adsorption on TiO_2 surface. The MO and MB solution used for control experiments was acidic without adjusting the pH.

Results and Discussion

Characterization of TiO_2

Adhesion strength, Fig. 1 shows the percentage of TiO_2 remained at different sonication times. As can be seen from Fig. 1, the amount of TiO_2 decreased to about 67 % for the first 15 s and remained constant thereafter, suggesting good adherence of TiO_2 films on the glass plate. This result also implied that the mixture of polyvinyl alcohol and formaldehyde offers good adhesion and acted as good adhesive to strengthen the TiO_2 coating on the glass plate. Similar results using other inorganic binders such as sodium silicate and zinc phosphate were reported by Park et al. [5, 6].

XRD analysis, Fig. 2 shows the XRD pattern of TiO_2 sample. It is discernible that the sample exhibits crystalline anatase at 2θ = 25.2°, 36.9°, 37.8°, 38.5°, 48.0°,

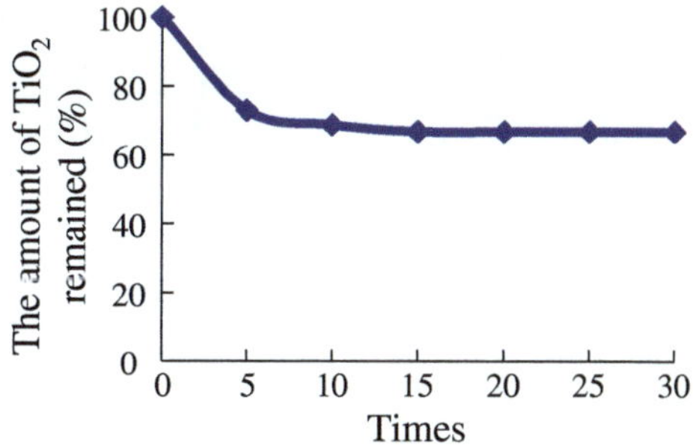

Fig. 1 Amount of TiO_2 (%) remained on the glass plate at different sonication time

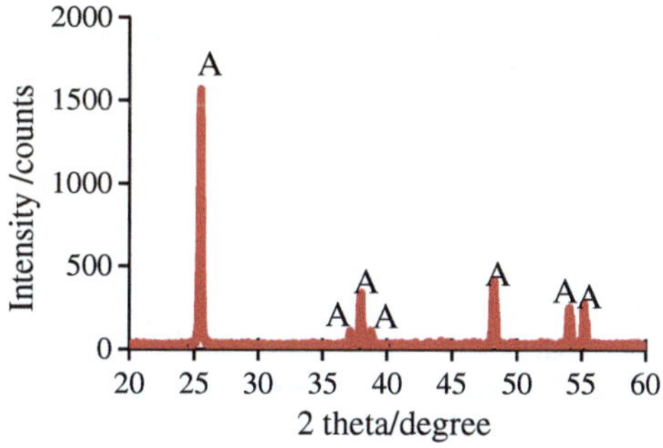

Fig. 2 XRD pattern of TiO_2 powder

53.9° and 55.0° corresponding to (101), (103), (004), (112), (200), (105) and (211) plane, respectively (JCPDS No. 21-1272). The broad diffraction peaks are mainly due to small tetragonal crystal size.

FESEM analysis, Fig. 3 shows the FESEM images for immobilized TiO_2 onto glass plate. At 5000× magnification, the glass plate was well covered with TiO_2 without any cracks. 50,000× magnification view of the sample revealed that the sample comprised of almost spherical-like structure with diameter approximately 95 ± 5 nm. TiO_2 nanoparticles are more desirable as this could provide higher surface area and more active site for photodegradation compare to that of bulk TiO_2.

Photodegradation of MO and MB, *Control tests*, The colour removal of MO and MB at different conditions is shown in Fig. 4. From Fig. 4, the photolysis rate for both MO and MB dye under photolysis is almost negligible (9 %), indicating that MO and MB is highly stable towards photolysis. However, the adsorption of MO was found to be 67.3 %, while for MB is 18.4 %. MO molecules are known to be anionic in nature and under acidic conditions, the positive charged TiO_2 surface being more available for the absorption of the MO molecules. For both MO and MB, the highest degradation rate was obtained through photocatalysis, indicating that the presence of UV light is necessary for the degradation of azo dye using TiO_2.

Effect of different light source, Fig. 5a depicts the degradation of MO and MB under UV light illumination. MO achieves a removal rate of 95 %, while MB only 47 % after 120 min of photodegradation. This result implied that with the presence

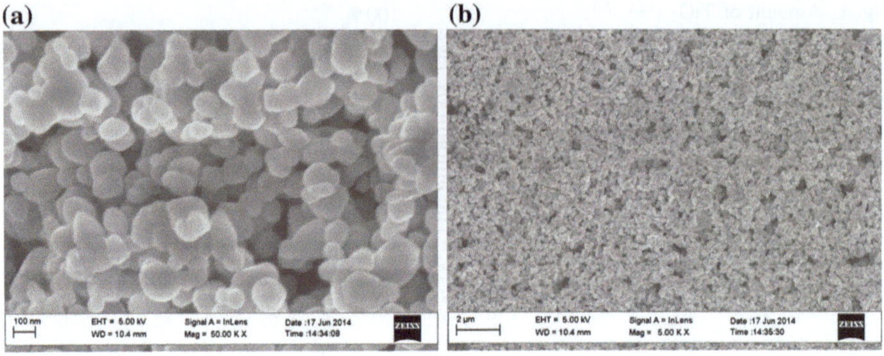

Fig. 3 FESEM images of immobilized TiO_2 onto glass plate with **a** 50000x and **b** 5000x magnification

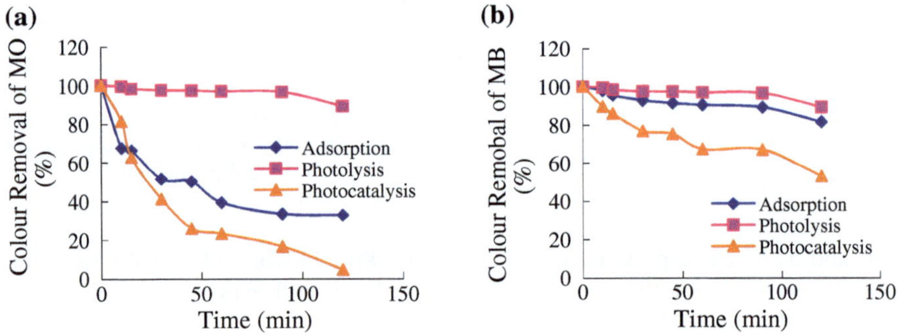

Fig. 4 Effect of different conditions on colour removal of **a** MO and **b** MB

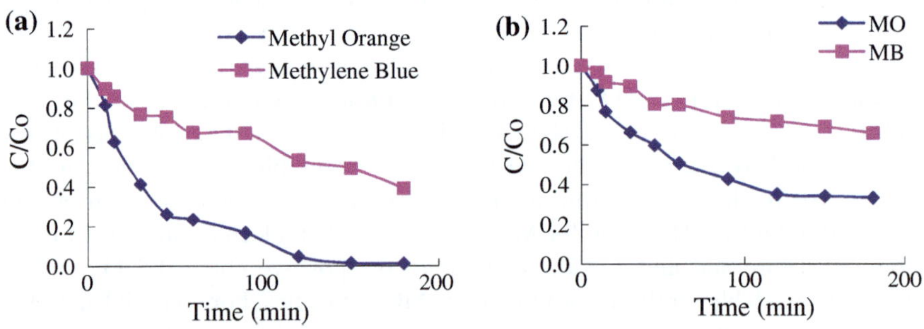

Fig. 5 Degradation of MO and MB as a function of irradiation time under illumination of **a** UV lamp and **b** LED lamp

of UV light, the photodegradation of both MO and MB has further enhanced by 30 % compared to the adsorption process. Under UV light illumination, electrons excite from a filled valence band to an empty conduction band, giving rise to electron–hole pairs. The photo-generated holes and electrons can migrate to the surface of TiO_2 particles and serve as redox sources that react with adsorbed reactants, leading to the formation of superoxide radical anions, hydrogen peroxide and hydroxyl radicals involved in the oxidation of dye. The mechanism of the photodegradation of dye by using TiO_2 is proposed as follows [7]:

$$TiO_2 + hv \rightarrow TiO_2(e^-, h^+)$$
$$h^+ + OH^- \rightarrow \cdot OH,$$
$$e^- + O_2 \rightarrow O_2^-$$

$$\text{dye} + O_2^- \text{ or } \cdot OH \rightarrow \text{peroxylated or hydroxylated intermediate} \rightarrow \text{degraded products}$$

Figure 5b depicts the degradation of MO and MB under LED light illumination. After 120 min, MO achieves a removal efficiency of 65 %, while MB yields 28 %. This showed that photodegradation of MO under LED illumination does not vary from the adsorption process as the energy from the LED light is not sufficient to generate the electron-hole pairs, However, LED does enhance the removal of MB with approximately 10 % of rises. Similarly to the one under UV light illumination, photodegradation of MB with LED illumination follows the first order kinetic. Table 1 shows the value of the rate constant and half time for both the MO and MB with different light source. As seen from the correction factors, R^2, the result shows that the photodegradation of MO and MB follows the first order kinetic.

Effect of different pH, The degradation of MO at different pH is shown in Fig. 6a. The ability of TiO_2 to degrade MO in alkaline medium was lower compared to that in acidic medium. It was observed that the decolorization efficiency increases with increase in pH, exhibiting maximum efficiency (95.4 %) at pH 6, beyond which the rate of degradation decreased at pH 9 (72 %). The initial pH of the dye solution did affect the degradation of MO and MB. Point of zero charge (pHpzc) of TiO_2 is 6.8. Therefore, the TiO_2 surface becomes positively charged with the species $TiOH_2^+$ at pH < pHpzc, negatively charged with the species TiO^- at pH > pHpzc and neutral with the species TiOH at pH = pHpzc [8]. Furthermore,

Table 1 Kinetics data for MO and MB under UV and LED light illumination

Light source	Dye	First order rate constant, k ($\times 10^{-3}$ min^{-1})	Half-life time, $t_{1/2}$ (min)	Correlation factor, R^2
UV	MO	26.3	26.3	0.95
	MB	5.2	133.3	0.98
LED	MO	7.7	90.0	0.82
	MB	2.7	256.7	0.87

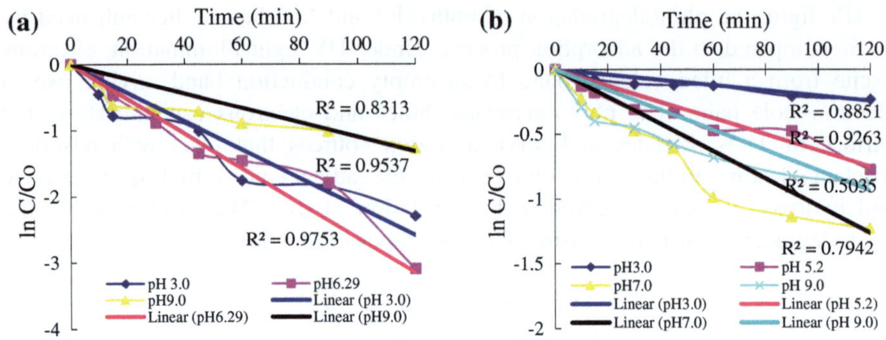

Fig. 6 Effect of pH on the degradation of **a** MO and **b** MB

protonation and deprotonation condition of the photocatalysts surface by taking TiO_2 as a model, can be shown as follows [9, 10]:

$$TiOH + H^+ \rightarrow TiOH_2^+ \text{ in acidic solution}$$

$$TiOH + OH^- \rightarrow TiO^- + H_2O \text{ in alkaline solution}$$

In acidic medium, the TiO_2 surface is positively charged. Since MO is an anionic dye, the electrostatic attraction between TiO_2 and MO can be achieved in acidic medium and thus results in higher percentage removal [11]. Oppositely, the surface of TiO_2 becomes negatively charges in alkaline medium. This causes a repulsive force between the surface of TiO_2 and MO, leading to the reduction of percentage removal.

Contrast trend was observed for the removal of MB. The ability of TiO_2 to degrade MB in alkaline medium was higher compared to that in acidic medium. The percentage of removal for MB was the highest at pH 7 (70.7 %) followed by pH 9 with a degradation efficiency of 58.4 %. Since MB is a cationic dye, the electrostatic attraction between TiO_2 and MB can be achieved in alkaline medium and thus results in higher percentage removal. As expected, the percentage removal for MB at pH 3 is only 20.5 % ascribed to repulsive force between the surface of TiO_2 and MB.

Conclusion

Nano-sized TiO_2 suspension has been successfully immobilized on the glass plates for the removal of azo dyes. The photodegradation of both MO and MB was found to be due to combination of photocatalysis and adsorption process, while the photolysis was almost negligible. Under illumination of different light source, UV-light irradiation has shown better degradation efficiency compared to LED for both MO and MB dye. The photo decolorization rate was the highest in pH 6 for MO and in pH 7 for MB with a maximum efficiency of 95.4 and 70.7 %, respectively. The kinetic data shows that the photodegradation of MO and MB follows the pseudo-first order reaction.

Acknowledgment The authors would like to acknowledge the financial support from the Universiti Teknologi MARA through Research Intensive Faculty Fund (RIF with No. 600/RMI/DANA 5/3/RIF 381/2012) and the Ministry of Education Malaysia via Fundamental Research Grant Scheme (FRGS/1/2014/SG01/UiTM/02/1).

References

1. Andreozzi R, Caprio V, Insola A, Marotta R (1999) Advanced oxidation processes (AOP) for water purification and recovery. Catal Today 53(1):51–59
2. Yu JC, Yu J, Zhao J (2002) Enhanced photocatalytic activity of mesoporous and ordinary TiO_2 thin films by sulfuric acid treatment. Appl Catal B Environ 36(1):31–43
3. Pal M, Serrano JG, Santiago P, Pal U (2007) Size-controlled synthesis of spherical TiO_2 nanoparticles: morphology, crystallization, and phase transition. J Phys Chem C 111 (1):96–102
4. Nawi MA, Zain SM (2012) Enhancing the surface properties of the immobilized Degussa P-25 TiO_2 for the efficient photocatalytic removal of methylene blue from aqueous solution. Appl Surf Sci 258(16):6148–6157
5. Park S, Choi GR, Lee JC, Kim YC, Lee JH (2012) Effects of zinc phosphate binder on the immobilization properties of photocatalytic ZnO nanopowders synthesized by a solution combustion method. J Ceram Process Res 13:S170–S173
6. Park S, Lee JH, Yoo K, Park HJ, Kim HS, Lee JC (2008) Adhesion properties of inorganic binders for the immobilization of photocatalytic ZnO and TiO_2 nanopowders. J Phys Chem Solids 69(5):1461–1463
7. Su Y, Yang Y, Zhang H, Xie Y, Wu Z, Jiang Y, Wang ZL (2013) Enhanced photodegradation of methyl orange with TiO_2 nanoparticles using a triboelectric nanogenerator. Nanotechnology 24(29):295401
8. Chiou CH, Wu CY, Juang RS (2008) Photocatalytic degradation of phenol and *m*-nitrophenol using irradiated TiO_2 in aqueous solutions. Sep Purif Technol 62:559–564
9. Habib MA, Ismail IMI, Mahmood AJ, Ullah MR (2012) Photocatalytic decolorization of brilliant golden yellow in TiO_2 and ZnO suspensions. J Saudi Chem Soc 16:423–429
10. Liao J, Lin S, Zhang L, Pan N, Cao X, Li J (2012) Photocatalytic degradation of methyl orange using a TiO_2/Ti mesh electrode with 3D nanotube arrays. ACS Appl Mater Interfaces 4 (1):171–177
11. Yang H, Zhang K, Shi R, Li X, Dong X, Yu Y (2006) Sol–gel synthesis of TiO_2 nanoparticles and photocatalytic degradation of methyl orange in aqueous TiO_2 suspensions. J Alloys Compd 413(1):302–306

Mercury Removal in Simulated Flue Gas by Oil Palm EFB Activated Carbon

Noor Hidayu Abdul Rani, Nor Fadilah Mohamad, Sharmeela Matali and Sharifah Aishah Syed A. Kadir

Abstract This study investigated the effects of simulated flue gas temperature and mercury inlet concentration on the mercury removal efficiency using activated carbon produced from oil palm empty fruit bunch (EFB) which is prepared using physical (steam) activation method. The elemental, proximate, porous structure and particle size were performed for the oil palm EFB activated carbon. At low temperature, performance of activated carbon showed a good potential for adsorption of mercury. However, as temperature increase from 90 to 180 °C, the efficiency of mercury was found to be decreased from 92 to 64 %. When the inlet mercury concentration was increased, the mercury removal efficiency is increased from 55 to 98 %. It was observed that mercury removal was favoured at lower temperature, such as 30 and 90 °C and higher mercury concentration, such as 0.192 ppm for virgin oil palm EFB activated carbon.

Keywords Mercury · Adsorption process · Activated carbon · Oil palm empty fruit bunch · Simulated flue gas

---

N.H. Abdul Rani (✉) · N.F. Mohamad · S. Matali · S.A. Syed A. Kadir
Faculty of Chemical Engineering, Universiti Teknologi MARA,
40450 Shah Alam, Malaysia
e-mail: hidayu_rani@yahoo.com

N.F. Mohamad
e-mail: fala5494@yahoo.com

S. Matali
e-mail: sharmeela@salam.edu.uitm.my

S.A. Syed A. Kadir
e-mail: drsharifah@salam.edu.uitm.my

M.A. Hashim (ed.), *ICGSCE 2014*, DOI 10.1007/978-981-287-505-1_14

Introduction

Mercury pollution is a global crisis encountered by society today, because it is a toxic and hazardous element that can severely threaten human health and the environment. The emissions of mercury in atmospheric come from flue gas, which from anthropogenic sources. Flue gas consists of three forms which are oxidized form (Hg^{2+}), particulate form (Hg^{p}) and elemental form (Hg^{0}). The elemental mercury (Hg^{0}) is the predominant form of atmospheric Hg, which has a long residence time in the atmosphere from 0.5 to 2 years and can be dispersed and deposited in remote places even 1000 km away from the sources [1, 2].

With the exception of mercury mining itself, the mercury emissions arise from mercury that is present as an 'impurity' in the fuel or raw material used. The main 'by-product' emissions are from sectors that involve combustion of coal or oil, production of pig iron and steel, production of non-ferrous metals, and cement production. So, due to the growth of these industries, concerns on the mercury emission should not be ignored as the level of mercury in the air will potentially increase. This concern has led to the development of technologies and researches for reduction of mercury in flue gas especially from the largest global single-known source of anthropogenic i.e. coal-fired power plant. However, existing pollution abatement technologies such as flue gas desulfurization and baghouses are ineffective for the capture of Hg^{0} because of its low melting point, high equilibrium vapor pressure and low solubility in water. Therefore, most Hg^{0} is released into the atmosphere. Recent researches have proved that application of activated carbon can be effective adsorbent for the removal of vapor phase mercury in a variety of conditions [3, 4].

However, using commercial activated carbon as adsorbent for the removal Hg^{0} will be increasing the cost of operating process because the adsorbent costs are relatively high. Therefore, many researchers have focused on alternative low cost adsorbents generated from agricultural wastes [5]. Extensive previous studies have shown that agricultural wastes such as corn cob, coconut shell, grain sorghum, olive stone, walnut shell, palm shell and date palm seed have been found to be suitable precursors owing to their high carbon and low ash contents [6]. Plus, they have unique properties, including large surface area, a high degree of surface reactivity, universal adsorption effect, and favorable pore size [7].

In this study, oil palm EFB is choosing to be utilized as activated carbon since it is abundantly available in Malaysia and has very low market value. Indeed, conversion of oil palm EFB to activated carbon would directly solve part of the environmental problem. So, this paper aims to investigate the performance of Hg^{0} adsorption on oil palm EFB activated carbon surface. In order to accomplish the targets, mercury adsorption tests have been conducted in a fixed bed stainless steel reactor under simulated flue gas and activated carbon was prepared under physical (steam) activation.

Experimental

Materials. Oil palm EFB was provided by Meru Palm Oil Mill Sdn Bhd in Klang, Selangor. The bio-char of oil palm EFB was produced using pyrolysis at temperature 500 °C.

Preparation of activated carbon. Thc activated carbon was prepared by physical (steam) activation process of the bio-char using nitrogen as the carrier in a tubular reactor. The superheated steam was injected at the flow rate 120 mL/h with the activation temperature at 764.5 °C and activation time of 77 min. This optimum operating condition was obtained for steam activation of bio-char using response surface methodology (RSM).

Characterization of activated carbon. Proximate analysis was performed using thermogravimetric analyzer, TGA (Mettler Toledo, United States) according to the ASTM D7582-10 and the results are shown in terms of moisture, volatile content, fixed carbon and ash contents. Ultimate analysis was carried out using CHNS-O Analyzer, (FlashEA, 1112 Series, United States) to determine carbon, hydrogen, nitrogen and oxygen (by difference) of the samples. BET (Brunauer, Emmett, and Teller) surface area and pore structure of the precursor and activated carbon was determined by automated gas adsorption Analyzer, AUTOSORB-1 (Quanta Chrome Instruments, USA).

Mercury adsorption process. The mercury removal efficiency of the virgin oil palm EFB activated carbon from gaseous phase was carried out in a fixed bed reactor. The schematic diagram of the experimental setup was shown in Fig. 1. The reactor was stainless steel reactor with 20 mm inner diameter and 20 cm long. A screen was fitted in the reactor to support the activated carbon. The concentration of mercury in the process gas was controlled by varying water bath temperature. The permeation device (VICI Metronics) was used to provide a vapor phase Hg^0 source

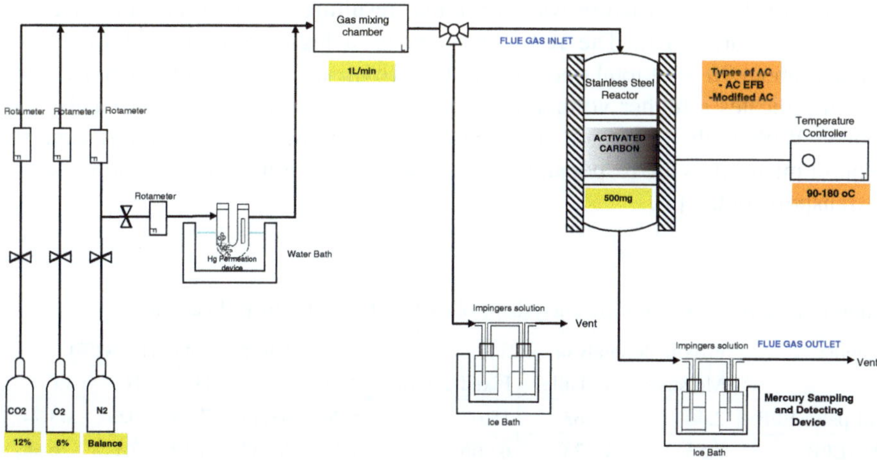

Fig. 1 Experimental setup for mercury adsorption process

with nitrogen (200 ml/min) as carrier gas. Premixed gas mixtures of carbon dioxide (12 %), oxygen (6 %) and nitrogen (balance) was passed through the tube and mixed together at gas mixing chamber for a certain time to ensure perfectly mixing is achieved. The total flow rate of flue gas is set to 1.0 L/min. After mixing, the simulated flue gas was flowing directly to stainless steel reactor and kept inside the reactor at selected temperatures, i.e. 30, 90, 140 and 180 °C for 2 h in order to determine the efficiency of mercury adsorption in flue gas. For virgin EFB activated carbon, 500 mg sample was used in the reactor. Impinger solution (4 W/V% $KMnO_4$, 10 V/V% H_2SO_4) was placed downstream of reactor for total mercury analysis. Atomic absorption spectrometry, AAS (Perkin Elmer, Lambda 25 UV Spectrophotometer) was used measure the concentration of mercury.

Result and Discussion

Characterization of activated carbon. The elemental and proximate analysis were performed in Table 1 for raw oil palm EFB and activated carbon. It showed that after activation, the carbon content of activated carbon increased from 48.48 to 68.32 wt%. This is because at high temperature, volatile content was released [8]. With high carbon content, there is high possibility the amount of active site is increased which is helped for mercury adsorption [9]. For proximate analysis, EFB has high fixed carbon and volatile contents. As a result, EFB is a good and viable options precursor for the production of activated carbon and with low ash content, surface area of activated carbon is increased and gave the higher adsorption efficiency [11].

The BET surface area, micropore area, total pore volume, and micropore volume of the activated carbon are shown in Table 2. It can be seen that the BET of activated carbon is higher compared to its raw material. This is due to at high temperature, fragile structure was generated, resulting in increased pore size and BET surface area [10]. The result also shows a high proportion of micropore volume (about 90 % of total pore volume) which means that oil palm EFB activated carbon indicates a higher volume of wide micropores and presence of small mesopores. With high proportion of micropores, high mercury adsorption capacity because micropores are responsible for mercury adsorption while mesopores serves as transport routes [3].

Table 1 Proximate and ultimate analysis of oil palm EFB and activated carbon

Sample	Proximate analysis (wt%)				Ultimate analysis (wt%)			
	Moisture	Volatile	Fixed carbon	Ash	C	H	N	O[a]
Oil palm EFB	2.44	73.63	18.67	5.26	48.48	7.14	0.64	43.74
AC-EFB	7.53	15.23	67.66	9.58	68.32	3.12	2.12	26.44

[a]Oxygen by difference

Table 2 Surface area and pore structure analysis of oil palm EFB

Sample	BET surface area (m^2/g)	Total pore volume (cm^3/g)	Micropore volume (cm^3/g)	Average pore diameter (Å)	Micropore percentage (%)
Oil Palm EFB	2.005	0.013	5.545×10^{-4}	265.400	4.27
AC-EFB	720.0	0.341	0.304	18.890	89.15

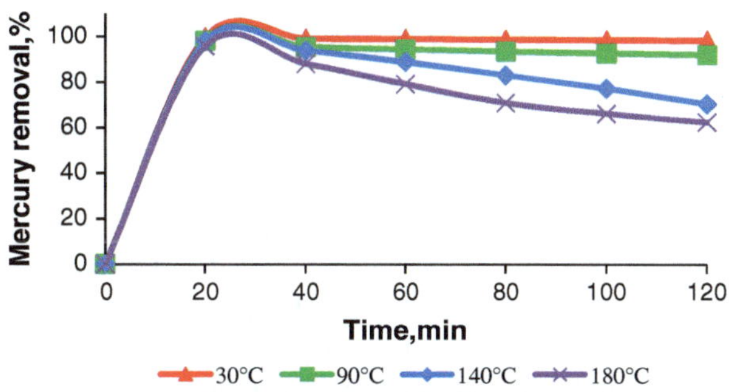

Fig. 2 Effect of flue gas temperature on mercury removal efficiency of AC-EFB (adsorbent amount = 500 mg; C_0 = 0.07–0.09 mg/L)

Effect of simulated flue gas temperature. Figure 2 shows the effect of on the Hg^0 removal efficiency of the oil palm EFB activated carbon. As the temperature is increased from 30, 90, 140 and 180 °C, mercury adsorption efficiency is decreased from 98, 92, 70 and 62 %. It is indicated a physisorption mechanism occurred due to van de Waals forces between the molecules Hg^0 and the oil palm EFB activated carbon which only favored at low temperature [4]. The result obtained is in similar to previous study made by De et al. [4], Shen et al. [11], and Zeng et al. [12] which proposed that physisorption was responsible for mercury removal by unimpregnated activated carbons [4, 11, 12]. Activated carbon is a porous matter which has a high BET surface area (720 m^2/g), so it has extremely excellent adsorption capacity. In this process, Hg^0 molecules easily penetrated through a bed of highly porous activated carbon and allowed a large mercury capacity to enter. Then, Hg^0 molecules diffused into pores of activated carbon and are adsorbed onto the surface of pores activated carbon by weak forces of intermolecular cohesion, resulting higher percentage mercury removal [13].

Effect of initial mercury concentration. Figure 3 shows the effect of inlet mercury concentration on the Hg^0 of the oil palm EFB activated carbon with varying water bath temperatures (30, 40 and 50 °C) in order to get the different value of inlet mercury concentration. According to previous research, the adsorption capacity of mercury increased as the inlet mercury concentration increased [4, 14, 15]. In this study, with an increase in inlet mercury concentration from 0.05 mg/L (30 °C) to 0.07 mg/L (40 °C), mercury removal efficiency at 90 °C increased from 55 to 92 % and continue rising to 98 % when the water bath temperature is increased to 50 °C.

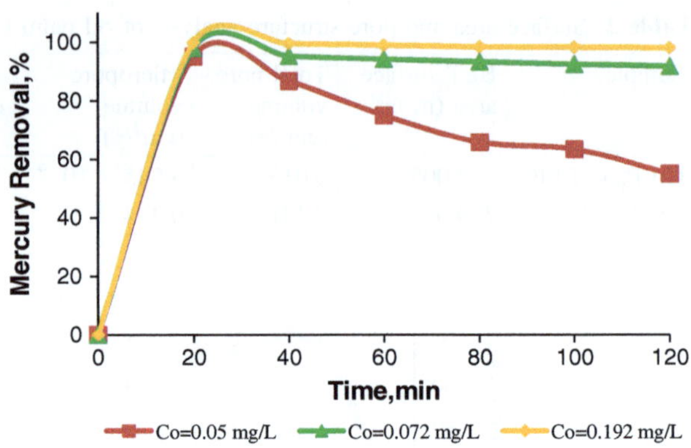

Fig. 3 Effects of inlet mercury concentration on mercury removal efficiency of AC-EFB (adsorbent amount = 500 mg; $T_{REACTOR}$ = 90 °C)

This may be caused by the fact that when concentration of inlet mercury in feed gas is higher, the atoms of Hg have more chances to interact and be attached to the carbon surface and finally increased the efficiency of adsorption [4]. According to Ping et al. [16], mercury molecules become more active when inlet concentration mercury increase because with high inlet concentration mercury, the gas phase driving force for mass transfer will be increased and make the reaction rate increases and finally leading to an attraction between the surface of activated carbon [16].

Conclusion

This work showed that oil palm empty fruit bunch (EFB) can be used as the good raw material with steam as the activating agent for the production of activated carbon. From the results, with a high temperature of flue gas indicated lower efficiency of mercury removal. This is due to the breaking of weak Van der Waals bond during physisorption process. When the inlet mercury concentration was increased, the mercury removal efficiency was increased. This may be caused by the fact that when concentration of inlet mercury in feed gas is higher, the mercury atoms have more chances to interact and be attached to the carbon surface. As a conclusion, for the unimpregnated activated carbon, the physical properties of activated carbon such as surface area, pore volume and pore size play an important role in mercury adsorption.

Acknowledgment This work was financially supported by the Research Management Institution (RMI) through Excellence Fund (600-RMI/DANA 5/3/RIF (311/2012)) and UiTM Shah Alam. The authors also thanks to Prof. Masaki Takaoka for his contribution in this research.

References

1. Li P, Feng XB, Qiu GL, Shang LH, Li ZG (2009) Mercury pollution in Asia: A review of the contaminated sites. J Hazard Mater 168:591–601
2. Hutchison A, Atwood D, Santilliann-Jiminez QE (2008) The removal of mercury from water by open chain ligands containing multiple sulfurs. J Hazard Mater 156:458–465
3. Yao Y, Velpari V, Economy J (2014) Design of sulfur treated activated carbon fibers for gas phase elemental mercury removal. Fuel 116:560–565
4. De M, Azargohar R, Dalai AK, Shewchuk SR (2013) Mercury removal by bio-char based modified activated carbons. Fuel 103:570–578
5. Tan ZQ, Xiang J, Su S, Zeng HC, Zhou CS, Sun LS, Hu S, Qiu JR (2012) Enhanced capture of elemental mercury by bamboo-based sorbents. J Hazard Mater 239:160–166
6. Tongpoothorn W, Sriuttha M, Homchan P, Chanthai S, Ruangviriyachai C (2011) Preparation of activated carbon derived from Jatropha curcas fruit shell by simple thermo-chemical activation and characterization of their physico-chemical properties. Chem Eng Res Des 89:335–340
7. Muyibi SA, Kamaldin N, Gombak J (2008) Production Of activated carbon from oil palm empty fruit bunches for removal of zinc. In: Presented at the twelfth international water technology conference, IWTC12. Alexandria, Egypt
8. Arami-Niya A, Wan Daud WMA, Mjalli FS, Abnisa F, Shafeeyan MS (2012) Production of microporous palm shell based activated carbon for methane adsorption: Modeling and optimization using response surface methodology. Chem Eng Res Des 90:776–784
9. Hidayu AR, Mohamad NF, Matali S, Sharifah ASAK (2013) Characterization of activated carbon prepared from oil palm empty fruit bunch using BET and FT-IR techniques. Proc Eng 68:379–384
10. Hidayu AR, Mohamad NF, Matali S, Sharifah ASAK (2014) Preparation and characterization of activated carbon made from oil palm empty fruit bunch. Key Eng Mater 594-595:44–48
11. Shen Z, Ma J, Mei Z, Zhang J (2010) Metal chlorides loaded on activated carbon to capture elemental mercury. J Environ Sci 22:1814–1819
12. Zeng H, Jin F, Guo J (2004) Removal of elemental mercury from coal combustion flue gas by chlorideimpregnated activated carbon. Fuel 83:143–146
13. US Environmental Protection Agency (2008) In: Adsorption, control of gaseous emissions (Chapter 4)
14. Takaoka M, Takeda N, Fujiwara T (2002) Control of mercury emissions from a municipal solid waste incinerator in Japan. J Air Waste Manag Assoc 52:931–940
15. Carey TR, Hargrove OW, Richardson CF, Chang R, Meserole FB (1998) Factors Affecting Mercury Control in Utility Flue Gas Using Activated Carbon. J Air Waste Manag Assoc 48:1166-1174
16. Ping F, Chao-ping C, Zi jun T (2012) Experimental study on the oxidative absorption of Hg0 by KMnO4 solution. Chem Eng J 198–199:95-102

Production of Glucose, Galactose and Mannose from the Skins of Durian and Mangosteen

Zamila Mohd Senu, Maryam Husin, Abd Rashid Li, Rusnah Samsuddin, Mohd Radzi Ahmad, Nik Roslan Nik Abd Rashid and Nur Zalikha Mohd Taza

Abstract In this study, production of glucose, galactose and mannose from the skins of tropical fruits, durian and mangosteen was carried out. Those sugars were obtained by acid hydrolysis of micro-crystalline cellulose and polysaccharide rich-fractions of the samples. The micro-crystalline cellulose (MCC) rich-fractions were purified using nitric acid-acetic extraction while polysaccharide rich-fractions were obtained using aqueous-acetone precipitation method. The hydrolysates consist of glucose, galactose, mannose and other sugar residues were analyzed its total sugar/carbohydrate contents and monosaccharide composition using phenol-sulfuric acid and pre-column derivatization HPLC techniques, respectively. The results indicated that yields of glucose that obtained from the MCC of durian skins is 14.23 % of the dried skin of durian compared to 8.52 % of glucose from the MCC of mangosteen skins. The yields of glucose, galactose and mannose from the polysaccharides of the durian and mangosteen skins are very low, which are 0.98, 0.70 and 0.16 % of the

Z.M. Senu (✉) · M. Husin · R. Samsuddin · N.R. Nik Abd Rashid
Faculty of Applied Sciences, Universiti Teknologi MARA, 40450 Shah Alam, Selangor, Malaysia
e-mail: zamila.msenu@gmail.com

M. Husin
e-mail: marya911@salam.uitm.edu.my

R. Samsuddin
e-mail: rusna386@salam.uitm.edu.my

N.R. Nik Abd Rashid
e-mail: nikro954@salam.uitm.edu.my

A.R. Li · M.R. Ahmad · N.Z. Mohd Taza
Phytochemistry Program, Natural Products Division, Forest Research Institute Malaysia (FRIM), 52109 Kepong, Selangor, Malaysia
e-mail: abdrashid@frim.gov.my

M.R. Ahmad
e-mail: radzi@frim.gov.my

N.Z. Mohd Taza
e-mail: nurzalikha@frim.gov.my

M.A. Hashim (ed.), *ICGSCE 2014*, DOI 10.1007/978-981-287-505-1_15

dried skin of durian, respectively. While 0.23 % glucose, 0.18 % galactose and 0.17 % mannose were determined from the polysaccharide fraction of mangosteen skins. Results proposed that durian and mangosteen skins are feasible for the production of bioethanol from plant wastes.

Keywords Fruit skins · Acid hydrolysis · Hydrolysates · Polysaccharides · Micro-crystalline cellulose

Introduction

Waste generation has increased steadily around the world for a number of reasons. Tonnes of waste are being thrown away and buried in the landfill every day. Agriculture produces several million tonnes of solid waste annually which are disposed through various methods. In view of this, investigations have been focused on the reuse of agri-food waste that contains numerous organic matters into food ingredients, biofuel and other value-added applications [1]. These organic wastes are converted from low value matters into valuable products.

Durian and mangosteen are the most popular and important seasonal tropical fruits throughout Southeast Asia. The popularity of these fruits is continuously increasing due to its delicious taste, good smell, nutrition and medicinal values. Nowadays, both durian and mangosteen have been accepted by people around the world and were commercialized into wide ranges of food products such as jam, ice-cream, durian crepe, syrup and candy. In 2011, Malaysia produced 363,421 and 21,555 metric tonnes of durian and mangosteen with the plantation areas of 76,399 and 4,268 ha respectively [2]. During its season, waste from both fruits such as skins and low quality fruits becomes abundant. The waste if not disposed properly will be harmful to the environment. Disposal or burning the waste products usually was undertaken by farmers and consumers which subsequently caused air, water and soil pollutions.

Cellulosic materials are the main components of plant material composed of lignin, hemicellulose and cellulose. Both hemicellulose and cellulose are comprised of long chains of sugar molecules and are enclosed rigidly by lignin which contains no sugar and can be extracted by acid hydrolysis [3–6]. Recently, researchers have been focused on cellulose hydrolysis from the agricultural waste such as rice husk [7], sisal pulp [8], orange mesocarp [9], oil palm biomass [3], rice straw, wheat straw and corn stalks [10]. Polysaccharides cellulose obtained from the cellulosic biomass can produce cellulosic ethanol (or bioethanol) [8]. The cellulosic ethanol is produced from the natural yeast fermentation on soluble sugars obtained from polysaccharide cellulose via acid or enzymatic hydrolysis.

This cellulosic ethanol could be used as one of the alternatives to petroleum-based fuel. Ethanol that was blended with gasoline (85 % ethanol–15 % gasoline) has caused the increment in oxygen content, causing better hydrocarbon oxidation and diminishing greenhouse gasses [11]. Nowadays, with the price of petroleum worldwide almost doubled during the last ten years and the use of petroleum also contributed towards the accumulation of greenhouse gases in the atmosphere [12], the introduction of cellulosic ethanol as petroleum-based fuel is an added advantage.

The purpose of this study is to extract micro-crystalline cellulose (MCC) and polysaccharide rich-fractions from dried skins of durian and mangosteen. Then both fractions were treated with diluted acid hydrolysis to obtain glucose, galactose and mannose. The total contents of those sugars were determined using pre-column derivatization HPLC method combined with phenol-sulfuric acid method. This study has focused on those sugars because natural yeast can only ferment glucose, galactose and mannose into bioethanol [13].

Experimental

Mangosteen and durian skins were obtained from the fruit market in Shah Alam, Selangor. The fresh skins obtained were washed and cut into smaller pieces then dried at 105 °C for 24 h. The dried skins were finely ground before storing at room temperature for further analysis. Methanol, acetic acid, nitric acid, hydrochloric acid and sulfuric acid, used for hydrolysis were purchased from Merck, Malaysia. For HPLC analysis, glucose, galactose, arabinose, mannose and other monosaccharides (Sigma-Aldrich) were used as standards.

Isolation of carbohydrates. The fine powdered skins were subjected to repeated extractions with methanol followed by aqueous extraction using Soxhlet apparatus. The water extract was collected and then treated with acetone before freeze dried to obtain polysaccharide rich-fraction. 80 % of acetic acid (20 ml/g biomass) and 65 % of nitric acid (2 ml/g biomass) were added to the dried skin residue and then heated for 1 h at 90–100 °C. The reaction mixture was filtered and neutralized with distilled water. Dried cellulose was hydrolyzed with 2.5 N of hydrochloric acid in the ratio 1:20 (cellulose:liquor) at 90–100 °C for 1 h. The mixture was allowed to cool at room temperature, then filtered and washed with distilled water before being dried at room temperature. The mass of MCC rich-fraction was recorded upon drying. Concentrated sulfuric acid was used to hydrolyze the carbohydrates prior to heating at 120 °C for 120 min. The hydrolyzed product was neutralized by using 0.5 M sodium hydroxide until the pH was 7.0. The salt was removed and the sugar was extracted by using rotary evaporator.

Sample analysis. The analysis of monomeric sugars was carried out on a Series LC-10A HPLC system (Shimadzu, Kyoto, Japan) equipped with an aminopropyl-silyl column and refractive index detector (RID-10A). The separation was done using acetonitrile/water, 7:3 (v/v) as a mobile phase at 80 °C of column temperature. The sugar extract was hydrolyzed with phenol solution and sulfuric acid prior to heating and homogenized at 30 °C for 30 min in an incubating shaker. The mixture was then read under UV absorbance 493 nm using UV-Vis spectrophotometer in triplicate for each sample. External calibration standards were performed for the quantification of sugars.

Result and Discussion

The MCC and polysaccharide rich-fractions composition of dried durian and mangosteen skins were shown in Table 1. Durian skins contained approximately 13.93 % of polysaccharide and 25.69 % of MCC. A similar result was reported on orange mesocarp that was treated with 2.5 N of hydrochloric acid where the yield of MCC was 25.30 % [9]. Meanwhile, mangosteen skins contained 15.09 % of polysaccharide and the MCC content in mangosteen skins was 11.59 % less than those in durian skins.

Typical HPLC chromatograms for monomeric sugar in MCC hydrolysate of durian skins and mangosteen skins were shown in Fig. 1. Glucose and rhamnose were two primarily sugars that contained in MCC hyrolysate of the durian and mangosteen skins. Results for actual percentage of glucose obtained from pre-column derivatization HPLC and phenol-sulfuric acid method for MCC of both fruit skins were shown in Table 2. From 25.69 % of the MCC rich-fraction composition of durian skins, 98.60 % of glucose and 1.40 % of rhamnose were obtained

Table 1 MCC and polysaccharide rich-fractions composition of dried skins of durian and mangosteen

Sample (%)	Durian skins	Mangosteen skins
MCC	25.69	14.10
Polysaccharide	13.93	15.09

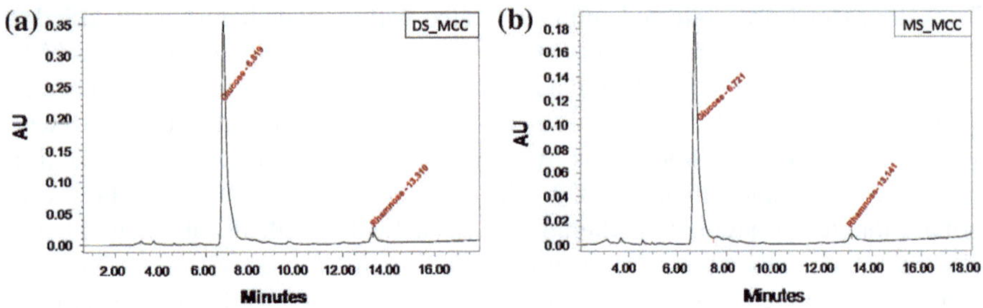

Fig. 1 HPLC chromatograms for monomeric sugar in MCC rich-fraction hydrolysate of **a** durian skins; **b** mangosteen skins

Table 2 Actual percentage of sugar compound in MCC of durian skins and mangosteen skins

Sample (%)	MCC of durian skins	MCC of mangosteen skins
Glucose percentage[a]	98.60	97.72
Rhamnose percentage[a]	1.40	2.28
Total sugar content[b]	61.30	56.69
Actual percentage of glucose	60.44	55.40

[a]Result obtained from the pre-column derivatization HPLC method
[b]Result obtained from phenol-sulfuric acid method

and the actual percentage of glucose was 60.44 % of the total sugar content (61.30 %). It was found that the glucose yield in this study was higher than those reported by Unhasirikul (50.71 g/L) on sulfuric acid hydrolysis of durian skins with the total sugar was 53.18 g/L [14]. Mangosteen skins contained 14.10 % of MCC where 97.72 % of it was glucose and the other 2.28 % was rhamnose. The actual percentage of glucose in the MCC of mangosteen skins was 55.40 % of the total sugar content. The results showed that glucose was the main composition in the MCC of both durian and mangosteen skins. As glucose was mainly produced by acid hydrolysis of cellulose [14, 15], galactose, xylose and mannose were not detected in both hydrolysates.

For hydrolysis of polysaccharides of durian skins and mangosteen skins, glucose, galactose, rhamnose and mannose were detected by pre-column derivatization HPLC techniques as shown in Fig. 2. The other major compound detected was galacturonic acid. From the HPLC analysis, glucose, galactose, rhamnose and mannose percentage in polysaccharide of durian skins obtained were 23.60, 16.00, 4.20 and 3.90 % respectively where the actual percentage of glucose, galactose and mannose were 7.28, 4.94 and 1.20 % of the total sugar content respectively. However, the actual percentage of glucose, galactose and mannose in polysaccharide of mangosteen skins obtained were slightly less than those in durian skins which are 1.51, 1.20 and 1.17 % respectively of total sugar content as shown in Table 3.

Yield of glucose, galactose and mannose in the MCC and polysaccharide hydrolysates of durian skins and mangosteen skins were tabulated in Table 4. The highest glucose yield was reported from the MCC of durian skins which was 14.23 % of weight of dried sample. Similar to results for polysaccharide of both fruits skins, the yield of fermentable sugars were high in durian skins compared to those in mangosteen skins. The yield of glucose, galactose and mannose in the polysaccharide hydrolysate of durian skins were 7.28, 4.94 and 1.20 % respectively. The monomeric sugars yield from both MCC and polysaccharide of mangosteen skins were less than the yield from durian skins where the glucose from MCC was 8.52 % while yield from the polysaccharide for glucose, galactose and mannose were 1.51, 1.20 and 1.17 % of weight of dried sample respectively.

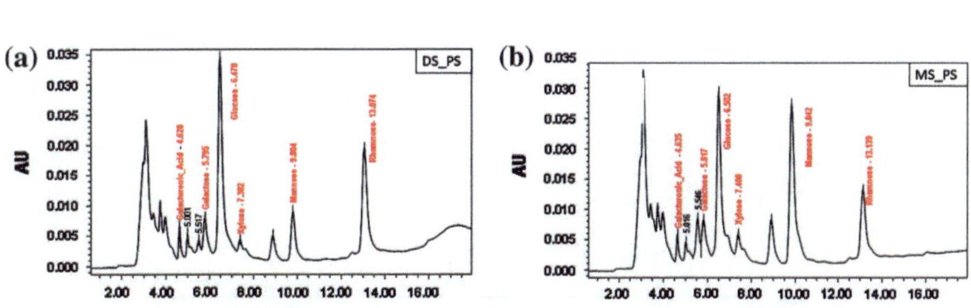

Fig. 2 HPLC chromatograms for monomeric sugar in polysaccharide rich-fraction hydrolysate of **a** durian skins; **b** mangosteen skins

Table 3 Actual percentage of sugar compound in polysaccharide of durian skins and mangosteen skins

Sample (%)	Polysaccharide of durian skins	Polysaccharide of mangosteen skins
Galacturonic acid percentage[a]	52.30	44.70
Glucose percentage[a]	23.60	21.00
Galactose percentage[a]	16.00	16.70
Rhamnose percentage[a]	4.20	1.30
Mannose percentage[a]	3.90	16.20
Total sugar content[b]	30.86	7.21
Actual percentage of glucose	7.28	1.51
Actual percentage of galactose	4.94	1.20
Actual percentage of mannose	1.20	1.17

[a]Result obtained from the pre-column derivatization HPLC method
[b]Result obtained from phenol-sulfuric acid method

Table 4 Yield of glucose, galactose and mannose in the MCC and polysaccharide hydrolysates of durian skins and mangosteen skins

Sample (%)	Durian skins		Mangosteen skins	
	MCC	Polysaccharide	MCC	Polysaccharide
Glucose	14.23	7.28	8.52	1.51
Galactose	–	4.94	–	1.20
Mannose	–	1.20	–	1.17

The fermentable monosaccharide sugars were obtained from both MCC and polysaccharide rich-fractions via diluted acid hydrolysis treatment. When the yield of monomeric sugars from these two rich-fractions are summed, the yield of fermentable glucose, galactose and mannose from the durian skins were 21.51, 4.94 and 1.20 % respectively while the yield from the mangosteen skins were 10.03, 1.20 and 1.17 % respectively as shown in Fig. 3. From another study [16], the

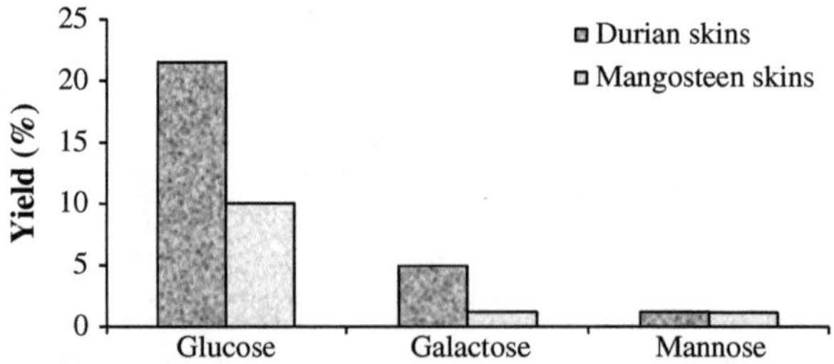

Fig. 3 Total yield of glucose, galactose and mannose in the MCC and polysaccharide hydrolysate of durian skins and mangosteen skins

maximum glucose yield from two-stage diluted sulphuric acid-catalyzed hydrolysis of bark-rich sawmill was 13.60 % by combining the yield from first and second stage hydrolysis while the yield of galactose and mannose was 10.50 % of the original dry feedstock. The glucose yield was higher in this study might be attributed to effective extraction of rich-fraction of MCC by nitric acid-acetic extraction and polysaccharide rich-fraction by aqueous-acetone precipitation method. Glucose was the main product obtained from hydrolysis of the durian skins [14] as well as the mangosteen skins. Those sugars were very important in the production of bio-ethanol from plant wastes. Besides glucose which was widely known, galactose and mannose could also be converted into bioethanol with the presence of microorganism.

Conclusion

Durian skins and mangosteen skins were successfully fractionated into MCC and polysaccharide rich-fractions by nitric acid-acetic extraction and aqueous-acetone precipitation method, respectively. Both of MCC and polysaccharide rich-fractions were effectively hydrolyzed to produce fermentable sugars. It was noted that those fractions contained fermentable sugars. As an approximation, sugars could be considered as originating not only from MCC but also from the polysaccharide of the fruit skins of durian and mangosteen. Acid hydrolysis of MCC and polysaccharides rich-fraction of durian skins and mangosteen skins produced mainly glucose followed by galactose and mannose. Based on the phenol-sulphuric acid and pre-column derivatization HPLC techniques, the maximum yield of glucose, galactose and mannose from the durian skins were 21.51, 4.94 and 1.20 % respectively while the yield from the mangosteen skins were 10.03, 1.20 and 1.17 % respectively. Durian skins and mangosteen skins showed a high potential to be converted to monomeric sugars and those sugars could be further utilized for the production of cellulosic ethanol via natural yeast fermentation.

Acknowledgment This study was financially supported by the Research Excellence Fund—Research Intensive Faculty (600-RMI/DANA 5/3/RIF (567/2012)) of Universiti Teknologi MARA. A special thanks to support staff of Natural Product Division of Forest Research Institute Malaysia for laboratory assistance.

References

1. Makris DP, Boskou G, Andrikopoulos NK (2007) Polyphenolic content and in vitro antioxidant characteristics of wine industry and other agri-food solid waste extracts. J Food Compos Anal 20(2):125–132
2. DOA (2011) Fruit crop statistics, 1st edn. Peninsular Malaysia

3. Mohamad MH, Eichhorn SJ, Hassan A, Jawaid M (2013) Isolation and characterization of microcrystalline cellulose from oil palm biomass residue. Carbohydr Polym 93(2):628–634
4. Das K, Ray D, Bandyopadhyay NR, Sengupta S (2010) Study of the properties of microcrystalline cellulose particles from different renewable resources by XRD, FTIR, Nanoindentation, TGA and SEM. J Polym Environ 18(3):355–363
5. Terinte N, Ibbett R, Schuster KC (2011) Overview on native cellulose and microcrystalline cellulose I structure studied by X-Ray diffraction (WAXD): comparison between measurement techniques. Lenzinger Berichte 89:118–131
6. Oliveira RL, Silva Barud H, Assunção RMN, Silva Meireles C, Carvalho GO, Filho GR, Ribeiro SJL (2011) Synthesis and characterization of microcrystalline cellulose produced from bacterial cellulose. J Therm Anal Calorim 106(3):703–709
7. Johar N, Ahmad I, Dufresne A (2012) Extraction, preparation and characterization of cellulose fibres and nanocrystals from rice husk. Ind Crops Prod 37(1):93–99
8. Paula MP, Lacerda TM, Zambon MD, Frollini E (2012) Adding value to the Brazilian sisal: acid hydrolysis of its pulp seeking production of sugars and materials. Cellulose 19(3):975–992
9. Ejikeme PM (2007) Investigation of the physicochemical properties of microcrystalline cellulose from agricultural wastes I: orange mesocarp. Cellulose 15(1):141–147
10. Nuruddin M, Chowdhury A, Haque SA, Rahman M, Farhad SF, Sarwar Jahan M et al (2011) Extraction and characterization of cellulose microfibrils from agricultural wastes in an integrated biorefinery initiative. Cellul Chem Technol 45:347–354
11. Hill J, Nelson E, Tilman D, Polasky S, Tiffany D (2006) Environmental, economic, and energetic costs and benefits of biodiesel and ethanol biofuels. Proc Natl Acad Sci USA 103 (30):11206–11210
12. Hossain ABMS, Fazliny AR (2010) Creation of alternative energy by bio-ethanol production from pineapple waste and the usage of its properties for engine. Afr J Microbiol Res 4 (9):813–819
13. Wallheimer B (2010) New yeast can ferment more sugar, make more cellulosic ethanol. http://www.purdue.edu/newsroom/research/2010/100607MosierYeast.html. Accessed 17 Dec 13
14. Unhasirikul M, Narkrugsa W, Naranong N (2013) Sugar production from durian (Durio zibethinus Murray) peel by acid hydrolysis. Afr J Biotechnol 12(33):5244–5251
15. Jahan MS, Saeed A, He Z, Ni Y (2010) Jute as raw material for the preparation of microcrystalline cellulose. Cellulose 18(2):451–459
16. Kim KH, Tucker M, Nguyen Q (2005) Conversion of bark-rich biomass mixture into fermentable sugar by two-stage dilute acid-catalyzed hydrolysis. Bioresour Technol 96 (11):1249–1255

Reaction Kinetics Study for Microwave Energy Pretreated *Jatropha Curcas L* In-Situ Transesterification

Sintayehu Mekuria Hailegiorgis, Shuhaimi Mahadzir and Duvvuri Subbarao

Abstract In the present work, in-situ methanolysis of microwave irradiation heat pre-treated *Jatropha curcas l* oil in the presence of alkaline benzyltrimethylammonium hydroxide (BTMAOH) as a phase transfer catalysis (PTC) was investigated. Combined use of microwave heat pretreatment of *Jatropha curcas l* seed particles and alkaline BTMAOH as a PTC drastically reduced the reaction time from 240 to 30 min. The yield of fatty acid methyl ester (FAME) was also increased from 49.7 to 93.5 %. Reaction kinetics was investigated to study the extent of reaction rate. Reaction kinetic study demonstrated that the order of the reaction is a first order reaction at all reaction temperature under investigation. Activation energy of the reaction was found to be 21,641 J/mole. Combined use of microwave heat treatment of *Jatropha curcas l* seed particles and PTC is promising to enhance the reaction rate of in-situ methanolysis.

Keywords *Jatropha curcas l* · Phase transfer catalysis · Microwave irradiation · Benzyltrimethylammonium hydroxide · Fatty acid methyl esters

Introduction

The rising concerns of energy demand and the dependence on fossil fuels along with its environmental impact have aggravated the search for renewable alternative fuels. Biodiesel is one of the most promising such options to substitute petroleum based diesel [1]. Biodiesel production is based on transesterification of vegetable oils and fats through the addition of lower alcohols such as methanol and a catalyst, giving glycerol as a co-product. The main feedstocks for the production of biodiesel were edible oil sources such as rapeseeds, sunflower seeds, soy seeds and palm oil

S.M. Hailegiorgis (✉) · S. Mahadzir · D. Subbarao
Chemical Engineering Department, Universiti Teknologi PETRONAS, 31750 Bandar Seri Iskandar, Tronoh, Perak, Malaysia
e-mail: sintayehu.h@petronas.com.my

M.A. Hashim (ed.), *ICGSCE 2014*, DOI 10.1007/978-981-287-505-1_16

situ methanolysis assisted by alkaline BTMAOH as a PTC 93.5 % w/w maximum FAME yield was achieved in 30 min as compared to 89.8 % FAME yield observed in 90 min for microwave untreated seeds of the same reaction condition. From this result, it can be deduced that microwave treatment of seed particles has significant effect to increase the rate of reaction and reduce the reaction time from 90 min to about 30 min while increasing the yield of FAME.

Empirical Reaction Kinetics

Jatropha curcas l oil is mainly a mixture of triglycerides of C_{16} to C_{18} fatty acids with an average molecular weight of 878 g/mol. Fatty acid triglycerides can be further transesterified with lighter alcohols in presence of a catalyst (alkaline/acidic/enzymatic materials) to produce biodiesel and glycerol. The reaction can be slow as the alcohols and oils are not very soluble. Soap may also be formed by the undesirable fatty oils reaction with alkalis. In spite of the heterogeneity due to limited solubility among the different phases, estimation of an effective empirical rate constant assuming the system to be a pseudo homogeneous phase can provide useful information for the reactor design calculations. The expected overall reactions are as shown in Eq. (1)

$$\text{Triglycerides} + 3\ \text{Alcohol} \underset{}{\overset{\text{NaOH, BTMAOH}}{\rightleftarrows}} 3\ \text{Biodiesel} + \text{Glycerol} \qquad (1)$$

One mole of triglycerides requires 3 mol of alcohol to produce 3 mol of biodiesel and one mole of glycerol. The undesirable saponification reaction which can produce soap needs to be suppressed to improve the economics. The alkaline transesterification reaction is mildly exothermic and reversible in nature though the reverse reaction is very slow. Use of excess alcohol can drive the reaction to completion. Dependence of the reaction kinetics on concentration of reactants and temperature through activation energy needs to be experimentally measured and correlated by empirical of the reaction rate Eq. (2).

$$-\frac{\mathrm{d[TG]}}{\mathrm{d}t} = \bar{k}\,[TG]^n \qquad (2)$$

where [TG] is triglycerides concentration, $\bar{k}$ apparent reaction rate constant and t is reaction time, n order of the reaction.

The apparent rate constant $\bar{k}$ can depend on concentration of catalyst, ratio of alcohol to oil, temperature and level of mixing [9]. Molar concentration of triglycerides, diglycerides, monoglycerides and FAME estimated from the chromatograms using the internal reference standards with reaction time are shown in Fig. 2.

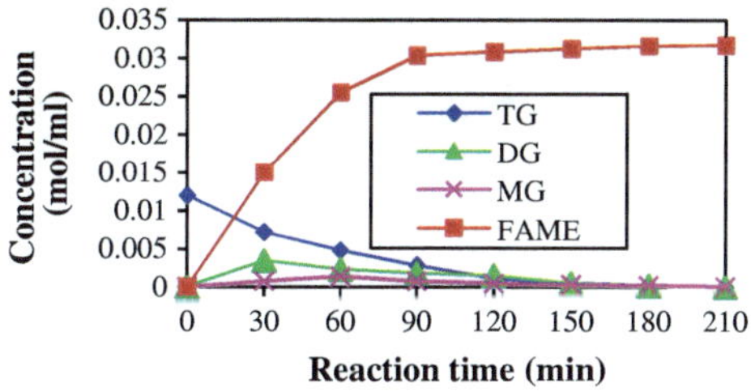

Fig. 2 Reaction profiles of TG, DG, MG and FAME and FAEE as a function of reaction time

Table 1 The reaction rate equation for triglycerides conversion at different reaction temperature

Temp, K	Rate equation, $-r_{[TG]}$, (mol/ml.min)	1/T	R. order, n	Rate constant, k
303	$-r_{TG} = 0.04328[TG]^{0.8587}$	0.0033	1	0.04328
313	$-r_{TG} = 0.04856[TG]^{0.9704}$	0.0032	1	0.04856
323	$-r_{TG} = 0.05498[TG]^{1.1768}$	0.0031	1	0.05498
333	$-r_{TG} = 0.06479[TG]^{1.264}$	0.0030	1	0.06479

It can be seen that diglycerides and monoglycerides are intermediates which are formed and get converted to biodiesel simultaneously. The overall reaction rate of triglycerides conversion was estimated by differential analysis [10] to obtain the order of reaction and rate constant as per Eq. (2) for triglycerides conversion during in-situ methanolysis at various temperatures and the corresponding reaction rate equations are shown in Table 1.

The result shows that the order of the reaction is a first order for the range of operating conditions investigated. With the first order reaction, rate constants were re-evaluated at each temperature for further kinetics analysis and presented in Table 1. Reaction rate constant is a temperature dependent term that can be represented by Arrhenius' law as shown in Eq. (3);

$$k = k_A\, e^{\frac{-Ea}{RT}} \tag{3}$$

where E_a is activation energy of the reaction, k_A is frequency factor and R is universal gas constant.

Arrhenius plots are presented in Fig. 3. The activation energy of the reaction was found to be 21,641 J/mole. A similar investigation reported by Marjanović et al. [11] demonstrated that the activation energy of base-catalyzed sunflower oil transesterification reaction was in the range of 8300–35,100 J/mole. However, Doell et al. [12] reported in their work the activation energy of trans-methylation of soybean oil as 63,000 J/mol. Present values are relatively less and this could be due to diffusion effects inherent in in-situ transesterification reactions. The empirical

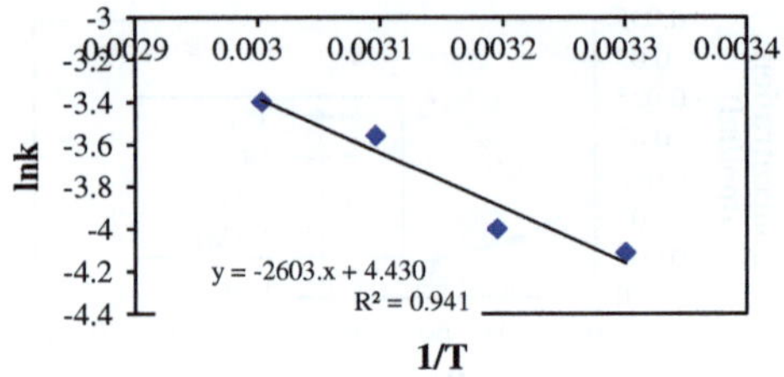

Fig. 3 Arrhenius' plots of lnk versus 1/T

rate equation obtained using experimental results can be rewritten in terms of reaction temperature and conversion as a function of reaction time as;

$$-r_{TG} = 84[TG]_o\, e^{\left(-\frac{21641}{8.314T}\right)}(1 - X_{TG}) \quad (4)$$

Validation of the Rate Equations

The reactor used in the present work is a batch reactor; the performance report for a batch reaction can be evaluated based on resident time of the reaction using Eq. (5).

$$t = TG_0 \int_0^{X_{TG}} \frac{dX_{TG}}{-r_{TG}} \quad (5)$$

For a first order equation

$$X_{TG} = 1 - e^{-kt} \quad (6)$$

Then, the rate constant, k for the present work is

$$k = 84\, e^{\left(-\frac{21641}{8.314T}\right)} \quad (7)$$

The results of triglycerides conversion, X_{TG} obtained using Eq. (6) and substituting the values of reaction rate constant k of Eq. (7) for in-situ methanolysis as a function of reaction time, t at different reaction temperatures compare well with experimental observations as shown in Fig. 4.

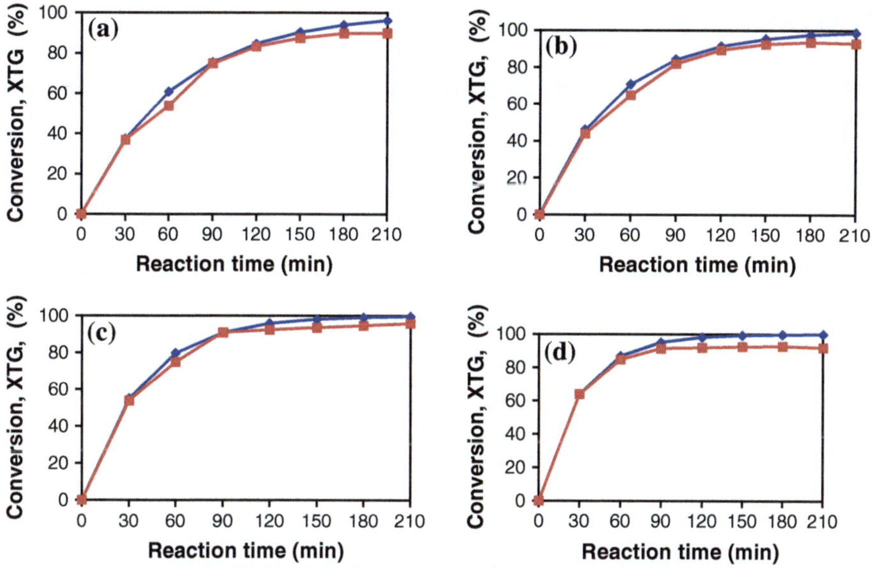

Fig. 4 Comparisons of experimentally achieved conversion of triglycerides (*filled square*) with batch reactor performance Eq. (6) (*filled diamond*) at different reaction temperatures. **a** at 30 °C, **b** at 40 °C, **c** at 50 °C, **d** at 60 °C

Conclusion

In-situ methanolysis of *Jatropha curcas l* seed was investigated. To increase the extractability of oil and enhance the reaction rate, the seed was pretreated with microwave irradiation. Benzyltrimethylammonium hydroxide (BTMAOH) was used as a phase transfer catalyst to enhance the rate of reaction. Fast conversion of triglyceride to fatty acid methyl esters was observed when the reaction was assisted by BTMAOH as a PTC for a reaction conducted using seeds pretreated with microwave energy. The reaction kinctics was investigated to study the extent of reaction rate. The reaction kinetic study indicated activation energy of the reaction is 21,641 J/mole. Microwave heat treatment of *Jatropha curcas l* seed and use of BTMAOH as a PTC is promising to enhance the reaction rate of in-situ methanolysis.

Acknowledgment The authors would like to thank Universiti Teknologi PETRONAS, Malaysia for financial support, laboratory facilities and encouragement to carry out this work.

References

1. Leung Dennis YC, Xuan Wu, Leung MKH (2010) A review on biodiesel production using catalyzed transesterification. Appl Energy 87:1083–1095
2. Lin L, Cunshan Z, Vittayapadung S, Xiangqian S, Mingdong D (2011) Opportunities and challenges for biodiesel fuel. Appl Energy 88:1020–1031
3. Hailegiorgis SM, Mahadzir M, Subbarao D (2011) Enhanced in-situ etanolysis of *Jatropha curcas l* in the presence of phase transfer catalyst. Renew Energy 36:2502–2507
4. Harrington KJ, D'Arcy-Evans C (1985) A comparison of conventional and in-situ methods of transesterification of seed oil from a series of sunflower cultivars. J Am Oil Chem Soc 62:1009–1013
5. Haas MJ, Scott KM, Marmer WN, Foglia TA (2004) In-situ alkaline transesterification: an effective method for the production of fatty acid esters from vegetable oils. J Am Oil Chem Soc 81:83–89
6. Jianli Zeng, Xiaodong W, Zhao Bing, Sun Jingcan, Wang Yuchun (2009) Rapid in-situ transesterification of sunflower oil. Ind Eng Chem Res 48:850–856
7. Hailegiorgis SM, Mahadzir M, Subbarao D (2013) Parametric study and optimization of in-situ transesterification of *Jatropha curcas l* assisted by benzyltrimethylammonium hydroxide as a phase transfer catalyst via response surface methodology. Biomass Bioenergy 49:63–73
8. American Standard Test Method (ASTM) (2002) Test methods for determination of free and total glycerin in B-100 biodiesel methyl esters. ASTM Designation D6584-08
9. Hailegiorgis SM (2013) Enhanced in-situ transesterification of *Jatropha curcas l* in the presence of alkaline phase transfer catalyst. PhD thesis, Universiti Teknologi PETRONAS, Malaysia
10. Levenspiel O (1999) Chemical reaction engineering, 3rd edn. Wiley, New York
11. Marjanović AV, Stamenković OS, Todorović ZB, Lazić ML, Veljković VB (2010) Kinetics of the base-catalyzed sunflower oil ethanolysis. Fuel 89:665–671
12. Doell R, Konar SK, Boocock DGB (2008) Kinetic parameters of a homogeneous transmethylation of soybean oil. J Am Oil Chem Soc 85:271–276

Induction Time of L-Isoleucine Crystallization with the Presence of Electric Field

Nik Salwani Md Azmi, Nornizar Anuar, Noor Fitrah Abu Bakar and Mohammad Akmalhakim Zakaria

Abstract Nucleation signifies the start of phase transition where nuclei of crystalline forms in the solution prior to crystal growth. There is an increasing demand in investigating reliable method to measure the nucleation rate in order to control the process an it has been reported that electric field has been successfully used to control the nucleation consequently the crystallization process. Hence, it becomes the objective of this paper to examine the effect of electric field on the induction time crystallization of L-isoleucine. The experiment was conducted in a 200 mL jacketed reactor where constant stirring was provided by retreat curve impeller and the temperature was regulated using two refrigerated baths. The induction time in this work was found to be lower than system without the application of electric field due to the hydrophobicity of L-isoleucine molecule and competition to form hydrogen bond with water when electrolysis occurs. The interfacial tension calculated shows no significant change with concentration while nucleation rate increases as the supersaturation increases. The critical radius calculated also proves to be smaller and decreases when supersaturation increases. The result also suggests the presence of new crystalline phase for the crystals produced at 48 g/L and supersaturation ratio of 1.08 and 1.14.

Keywords Interfacial tension · Nucleation rate · Critical radius · Polymorphism

N.S. Md Azmi · N. Anuar (✉) · N.F. Abu Bakar · M.A. Zakaria
Universiti Teknologi MARA Malaysia, 40450 Shah Alam, Selangor, Malaysia
e-mail: nornizar@salam.uitm.edu.my

N.S. Md Azmi
e-mail: niksalwanimdazmi@gmail.com

N.F. Abu Bakar
e-mail: drnoorfitrah@gmail.com

M.A. Zakaria
e-mail: akmalhakim_z@ymail.com

M.A. Hashim (ed.), *ICGSCE 2014*, DOI 10.1007/978-981-287-505-1_17

Introduction

L-Isoleucine is an amino acid with many applications in various sectors, such as pharmaceutical, food and fine chemical industries. Along with other types of amino acid, it is a constituent of infusions and special dietary products [1]. L-isoleucine is classified as an essential amino acid, in which the body cannot produce and has to be acquired through dietary supplementation. L-isoleucine is known as the most hydrophobic amino acid and exists as zwitterions in solution [2], making it an interesting material to be studied. L-isoleucine helps in maintaining optimal level of blood sugar and energy [3], and improves the nutrition of hepatic disease patient [4]. It exists in crystalline form meaning that it can be produced through crystallization process.

Crystallization refers to the formation of solid crystal from a homogenous solution [5, 6]. It is vital in the pharmaceutical industry as it functions as a separation process and frequent acts as final step in manufacture of active pharmaceutical ingredients (APIs) [7]. Supersaturation is the driving force of crystallization process. A supersaturated solution is metastable, meaning that crystal does not necessarily formed in the solution as a certain period of time is needed before the appearance of the first nucleus can takes place. The metastability of a solution is inversely proportional to the supersaturation, which means that the higher the supersaturation, the faster the appearance of a crystal from a clear solution [5], hence the faster the nucleation process occurs. Crystal nucleation is an important step in crystallization process as it denotes the start of phase transition where nuclei of crystalline forms in the solution before continually growing out to form larger crystal [8]. The demand for developing a reliable method to measure the nucleation rates is growing as it provides validation to nucleation theories, and control over the crystal product quality in industrial crystallization process can be attained [9]. It is a well known fact that a stable crystal grows in the metastable zone region, which is kinetically inactive [10]. For the nucleation to occur in the metastable region, external forces needs to be introduced and electric field has been successfully used to induce the nucleation process as reported by previous researchers [11–13]. Hence, it becomes the main focus of this paper to investigate the effect of electric field on the crystallization of L-isoleucine through induction time experiment.

Theory

Referring to the basis theory of the classic homogenous nucleation for the supersaturated solution, the rate of nucleation, J can be expressed in the form of Arrhenius reaction velocity [14]:

$$J = A\exp\left(\frac{-\Delta G}{kT}\right) \tag{1}$$

Since the Gibbs free energy, ΔG is the summation between the volume free energy, ΔG_V (=$4/3\pi r\Delta G_v$) and the surface free energy, ΔG_S (= $4\pi r\gamma$), and taken into account the Gibbs-Thomson relation, Eq. (1) becomes Eq. (2):

$$J = A\exp\left(-\frac{16\pi\gamma^3 v^2}{3k^3T^3(\ln S)^2}\right) \tag{2}$$

It is assumed that the induction time, t_{ind} to be inversely proportional to nucleation rate, J, (t_{ind} α $1/J$) and by applying logarithm to both side of the Eq. (2), it can be rewritten as Eq. (3):

$$\ln\frac{1}{t_{ind}} = \ln A - \frac{16\pi\gamma^3 v^2}{3k^3T^3(\ln S)^2} \tag{3}$$

where t_{ind} is the induction time, γ is the interfacial tension, S is the supersaturation, k is the Boltzmann constant, T is the temperature, A is the pre-exponential factor, and $v = M/(\rho N_A)$ is the molecular volume in which M is the molecular weight, ρ is the density and N_A is the Avogadro's number with value of 6.023×10^{23} no./mol. Meanwhile, the number of molecules in a critical radius, N^* can be determined using Eq. (4) which is derived and rearrange from [15, 16]:

$$N^* = 4\pi r^{*3}/3v \tag{4}$$

where r^* is the critical radius size and can be calculated using Eq. (5) [14]:

$$r^* = 2\gamma v/kT\ln S \tag{5}$$

Experimental

Material, L-isoleucine with molecular formula of $C_6H_{13}O_2N$ and purity of higher than 99 % was chosen as the study material. It was purchased from Merck and distilled water was used to make up the solution. Carbon electrodes were used to transfer the current to the solution. Pre-treatment of electrodes was conducted by immersing the electrodes in acetone to remove any organic impurities and rinsing them with distilled water, prior to induction time experiment.

Induction time crystallization, L-isoleucine was dissolved in a jacketed reactor containing 200 mL distilled water where constant stirring was produced by retreat curve impeller. The solution was heated to temperature 15 °C higher than the saturation temperature for 60 min. 5 V electric current was supplied to the solution via electrodes and the solution was cooled to the saturation temperature and maintained at that temperature for 60 min. Then the solution was drastically cooled to the desired temperature and was kept constant until the first crystal formed and the induction time was recorded.

Characterization of product crystal, The product crystal was characterized using X-ray powder diffraction (XRPD) with Cu Kα radiation at 40 mA, 40 kV, rate of 0.01°/s and 2θ between 3° and 40°.

Results and Discussion

Induction time is defined as the elapse time between the creation of supersaturation and the formation of solid phase [8, 17, 18]. Ghader et al. [17] stated that induction time is dependent on temperature, interfacial tension, external factors (i.e impurities and seed), and supersaturation, although the later is the most influential factors in determining the induction time [14]. Figure 1 shows the relationship between the induction times with supersaturation. The result clearly shows that the relationship for each of the concentrations are made up of two distinguishable straight line with different slopes based on the system nucleating at low and high supersaturations as reported in Anuar et al. [19]. From experiment, it was observed that shorter induction time was achieved when the supersaturation increases. This occurrence is probably due to higher supersaturation leads to higher coagulation of clusters and smaller critical size of nuclei hence successfully reduces the induction time [14, 17].

It is also clear that the induction time reduces when the solution concentration increases. At higher solution concentration, more supply of fresh solution containing particles is available in the solution, leading to the prolonged lifetime of

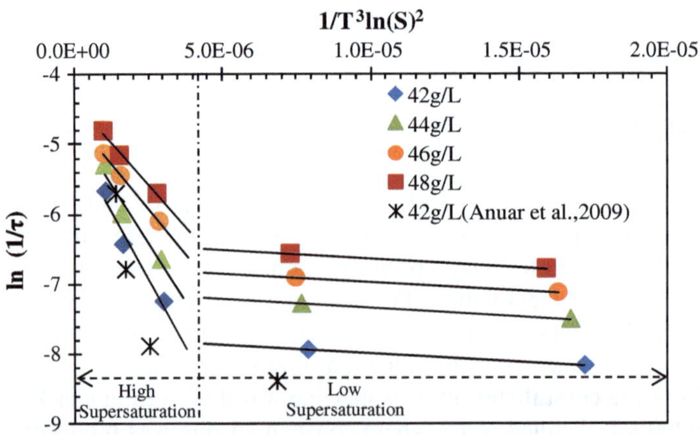

Fig. 1 The relation between the induction time and supersaturation of L-isoleucine at 5 V electric field, showing nucleation occurs at high supersaturation and low supersaturation system

nuclei, which then enhances the crystal growth. The data was compared with the data obtained from Anuar et al. [19] for isothermal crystallization of L-isoleucine without the application of electric field which was also plotted in Fig. 1. It shows that the induction time for the solution system with the presence of electric field is lower compared to the induction time without electric field. This indicates that the electric field has the capability to increase the onset of crystallization. This is in agreement with a study by Koizumi et al. [12], where the nucleation rate for crystal formation indeed increases with the application of electric field. The application of electric field causes the electrolysis process to occur. Water exists as H_3O^+ and OH^- ions in solution due to self-ionisation. Since H_3O^+ and OH^- ions have higher affinity to be discharged compared to the L-isoleucine molecule, H_3O^+ and OH^- ions are easily reduced and oxidised at the cathode and anode, forming H_2 and O_2 gas, respectively. L-isoleucine on the other hand, is a hydrophobic molecule due to the presence of the hydrophobic alkyl group [2] and it tends to aggregates close to each other in to reduce the contact with water. When H_3O^+ and OH^- ions are discharged, this creates a competition for L-isoleucine to form hydrogen bond with water near the vicinity of the electrodes, hence it is easier for L-isoleucine aggregation to nucleate. Once the nucleation process starts, it will grow rapidly to form crystal.

The nucleation phenomena and induction time data is very useful in determining the interfacial tension of solid-liquid. The interfacial tension, γ was estimated from the slope of Eq. (3) ($16\pi\gamma^3v^2/3k^3$) and as depicted in Fig. 1. The slope was determined from the graph in Fig. 1 and by applying the appropriate value into the slope equation, the interfacial tension, γ was determined and shown in Fig. 2a. It can be seen in Fig. 2a that there is no significant change in the interfacial tension of L-isoleucine across the concentration, however the value of the high supersaturation system indeed is higher compared to the low supersaturation system. The average interfacial tension values calculated in this work for low and high supersaturation is 0.61 and

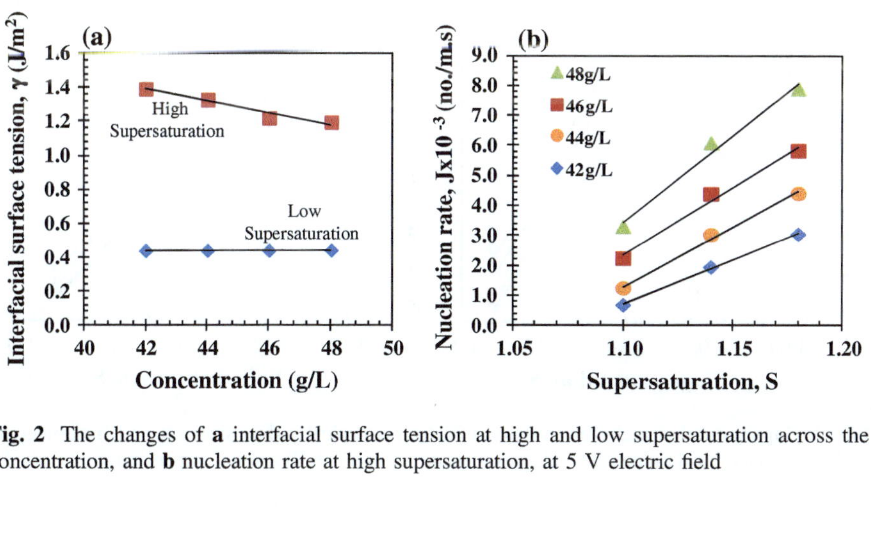

Fig. 2 The changes of **a** interfacial surface tension at high and low supersaturation across the concentration, and **b** nucleation rate at high supersaturation, at 5 V electric field

1.44 J/m^2 respectively. It is stated in [20] that large value of interfacial tension indicates that a large time is needed for the nucleation to occur. This seems to be the opposite of what has been happening in this work since at higher supersaturation, the induction time proves to be decreasing as depicted in Fig. 1. Meanwhile, the nucleation rate calculated (Fig. 2b) shows an increasing trend as the supersaturation increases at high supersaturation system and increases as the concentration increases. High nucleation rates means that the formation of nucleation in the solution is faster and these nuclei generally have relatively lower probability to grow into large crystal compared to the lower number of nuclei formed under similar conditions [15, 21]. At higher level of supersaturation, aggregation of molecules to build up new cluster reaching critical size is faster than to lose it as the molecules in the solution has become energetically favorable once the radius reach or exceed the critical radius [22].

Figure 3a, b shows the dependency of the critical radius size of nuclei and number of molecules in a critical radius size on supersaturation. From Fig. 3a, b, the graph clearly shows that at low supersaturation the critical radius increases when the concentration increases from 42 to 48 g/L, but at high supersaturation the critical radius size decreases as the concentration increases. The values for the critical radius size at 48 g/L for high and low supersaturation obtained in this work ranges between 10.13 – 13.74Å and 8.23 – 10.80Å respectively while for 42 g/L, it ranges between 11.38 – 15.43Å and 6.59 – 8.64Å for high and low supersaturation. The values of critical radius size were compared with [19] and it is proven that the critical radius is smaller due to the application of electric field. In classical nucleation theory for crystals formation, aggregates of molecules needs to form a nucleus with size exceeding the critical radius, r^* which in turn reduces the total free energy and hence the nucleus increases in size to form a crystal [14]. Theoretically, the application of electric field further reduces the free energy of formation as described

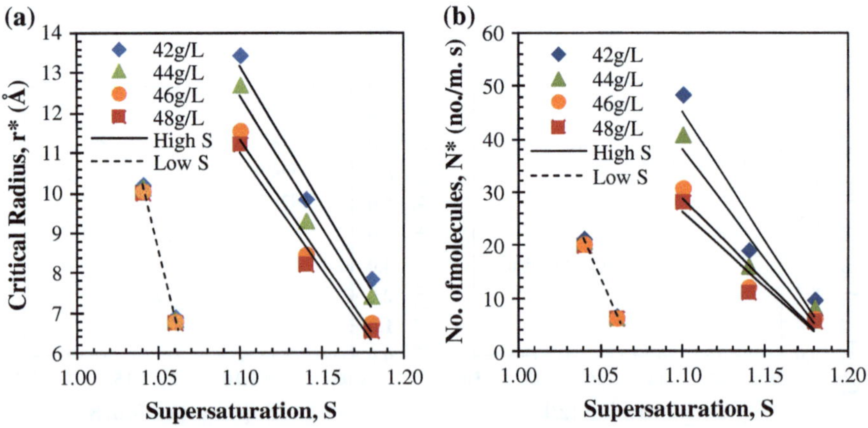

Fig. 3 Supersaturation and concentration effect on **a** critical radius size, r^*, and **b** number of molecules in a critical radius, N^*, at high supersaturation (High S) and low supersaturation (Low S)

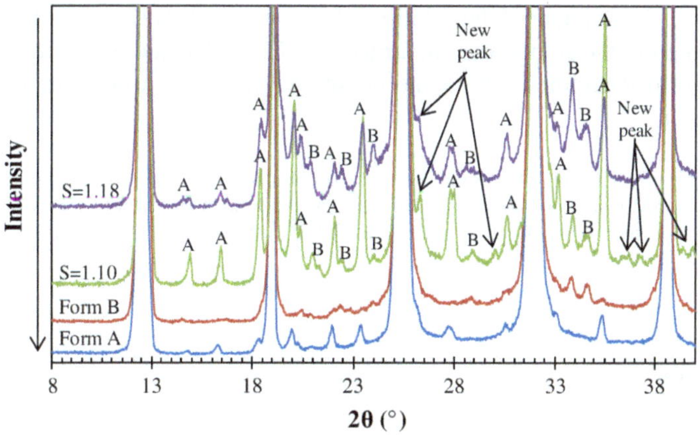

Fig. 4 Presence of peaks of Form A and Form B as well as the additional new peaks for crystal produced in 48 g/L solution concentration at 1.08 and 1.14 supersaturation

in Saban et al. [23] making it easier for the nuclei to nucleate. Figure 3a also shows that at high and low supersaturation, the critical radius size decreases when the supersaturation increases where the same observation is reported in [15, 21]. Meanwhile, the number of molecules in a critical radius depends on the critical radius value (Fig. 3b).

Polymorphism is a phenomenon where material is able to form more than one form of crystal structure [2]. L-isoleucine exists in two form which is Form A and Form B as reported in [19]. In this work, the crystals grown in high supersaturation at 1.08 and 1.14 were characterized and compared with the existing data obtained from [19]. Not all crystal can be collected especially at low supersaturation due to the difficulties in growing the crystal. Figure 4 shows the XRPD profile for crystal produced in a 48 g/L solution concentration at different supersaturation. The necessary data is plotted and the peaks are shown in the figure. The data in this work shows the crystals collected consists of mixtures of Form A, Form B and possibly new crystalline phase. The results shows new peaks are present at approximately 2θ of 26.18° at supersaturation of 1.08, and at 26.24°, 30°, 36.64°, 37.12°, 39.36° and 39.88° at supersaturation of 1.14. The presence of the new peak could mean that new crystalline phase is formed in the solution due to the presence of electric field.

Conclusion

Electric field has been proven to have significant impact in the crystallization of L-isoleucine. In this study, it can be concluded that electric field is capable to increase the onset of crystallization, hence the crystallization process can be achieved faster even at higher solution temperature. Electric field also is capable to induce the nucleation rate of the solution. However, the XRPD result shows that the crystal formed in the solution are in a mixture of L-isoleucine polymorphis form of Form A, Form B and new crystalline phase, and this pose a problem to industrial crystallization.

Acknowledgment The authors would like to express their gratitude to Universiti Teknologi MARA and the Ministry of Education, Malaysia for funding of this work. This work has been financially supported under the grant no. 600-RMI/FRGS 5/3(92/2013) and 600-RMI/ERGS 5/3 (29/2012).

References

1. Eggeling L, Morbach S, Sahm H (1997) The fruits of molecular physiology: engineering the l-isoleucine biosynthesis pathway in *Corynebacterium glutamicum*. J Biotechnol 56:167–182
2. Florence AT, Attwood D (2006) Physiochemical principles of pharmacy, 4th edn. Pharmaceutical Press, London
3. Kohlmeier M (2003) Isoleucine. In: Kohlmeier M (ed) Nutrient metabolism. Academic Press, London, pp 377–383
4. Xie XL, Wei M, Yunoki T, Kakehashi A, Yamano S, Kato M, Wanibuchi H (2012) Long-term treatment with L-isoleucine or L-leucine in AIN-93G diet has promoting effects on rat bladder carcinogenesis. Food Chem Toxicol 50:3934–3940
5. Pessôa Filho PA, Medeiros Hirata GA, Watanabe ÉO, Miranda ÉA (2011) 2.46—precipitation and crystallization. In: Moo-Young M (ed) Comprehensive biotechnology, 2nd edn. Academic Press, Burlington, pp 651–663
6. O'Ciardha CT, Frawley PJ, Mitchell NA (2011) Estimation of the nucleation kinetics for the anti-solvent crystallisation of paracetamol in methanol/water solutions. J Cryst Growth 328 (2011):50–57
7. Chen J, Sarma B, Evans JMB, Myerson AS (2011) Pharmaceutical crystallization. Cryst Growth Des 11:887–895
8. Kulkarni SA, Kadam SS, Meekes H, Stankiewicz AI, ter Horst JH (2013) Crystal nucleation kinetics from induction times and metastable zone widths. Cryst Growth Des 13:2435–2440
9. Jiang S, ter Horst JH (2010) Crystal nucleation rates from probability distributions of induction times. Cryst Growth Des 11:256–261
10. Hammadi Z, Veesler S (2009) New approaches on crystallization under electric fields. Prog Biophys Mol Biol 101:38–44
11. Koizumi H, Fujiwara K, Uda S (2009) Control of nucleation rate for tetragonal hen-egg white lysozyme crystals by application of an electric field with variable frequencies. Cryst Growth Des 2420–2424
12. Koizumi H, Tomita Y, Uda S, Fujiwara K, Nozawa J (2012) Nucleation rate enhancement of porcine insulin by application of an external AC electric field. J Cryst Growth 352:155–157
13. Azmi NSM, Anuar N, Abu Bakar NF, Shukri UR, Mokhlis MH (2014) Solubility study of electrocrystallized L-isoleucine in aqueous phase. J Appl Sci 14:1403–1408
14. Mullin JW (2001) Crystallization, 4th edn. Elsevier Butterworth-Heinemann, Oxford
15. Abdel-Aal EA, El-Shazly AN, El-Shahat MF (2013) Crystal nucleation of nano crystallite strontium malonate without and with additives. Cryst Growth Des 13:4395–4401
16. Boomadevi S, Dhanasekaran R, Ramasamy P (2002) Investigations on nucleation and growth kinetics of urea crystals from methanol. Cryst Res Technol 37:159–168
17. Ghader S, Manteghian M, Kokabi M, Mamoory RS (2007) Induction time of reaction crystallization of silver nanoparticles. Chem Eng Technol 30:1129–1133

18. Flaten EM, Seiersten M, Andreassen J-P (2010) Induction time studies of calcium carbonate in ethylene glycol and water. Chem Eng Res Des 88:1659–1668
19. Anuar N, Daud WRW, Roberts KJ, Kamarudin SK, Tasirin SM (2009) An examination of the solution chemistry, nucleation kinetics, crystal morphology, and polymorphic behavior of aqueous phase batch crystallized L-isoleucine at the 250 mL scale size. Cryst Growth Des 9:2853–2862
20. Wantha L, Flood AE (2012) Nucleation kinetics of the γ-polymorph of dl-methionine. Chem Eng Technol 35:1024–1030
21. El-Shall H, Jeon J-H, Abdel-Aal EA, Khan S, Gower L, Rabinovich Y (2004) A study of primary nucleation of calcium oxalate monohydrate: I-effect of supersaturation. Cryst Res Technol 39:214–221
22. McPherson A, Kuznetsov YG (2014) Mechanisms, kinetics, impurities and defects: consequences in macromolecular crystallization. Acta Crystallographica Section F 70:384–403
23. Saban KV, Thomas J, Varughese PA, Varghese G (2002) Thermodynamics of crystal nucleation in an external electric field. Cryst Res Technol 37:1189–1199

Gas Permeation Performance of Poly (lactic acid) Asymmetric Membrane for O_2/N_2 Separation

F. Mohamed, H. Hasbullah, W.N.R. Jamian, A.R.A. Rani, M.F.K. Saman, W.N.H. Salleh and R.R. Ali

Abstract Membrane gas separation has been one of the most favorable techniques used in industry due to its efficiency and low operating cost. One of the main problems in using membrane in gas separation is the non-biodegradable nature of the membrane material. In this study, PLA membrane was fabricated for oxygen/nitrogen separation system. The dope solution was prepared by using dichloromethane as a solvent with different concentration of 15, 17.5, 20 and 25 wt%. The membrane was cast at a fixed casting speed and solvent evaporation time of 0.0413 ms^{-1} and 60 s, respectively by using pneumatically-controlled casting system via dry/wet phase inversion technique. The prepared membrane morphology, mechanical strength and gas permeation performance was characterized using scanning electron microscope (SEM), tensile testing machine and soap bubble meter, respectively. Based on SEM result, the increase in dope solution concentration was proven to decrease the membrane porosity and increase its compactness that leads to the improvement of membrane gas permeation performance and mechanical strength. However, the further increase of concentration led to membrane brittleness with low elasticity causing the membrane become hard to handle for further testing.

Keywords Asymmetric membrane · Biodegradable polymer · Gas separation · Dope solution concentration

F. Mohamed · H. Hasbullah (✉) · W.N.R. Jamian · W.N.H. Salleh
Advanced Membrane Technology Research Centre (AMTEC), FPREE,
Universiti Teknologi Malaysia, 81310 Johor Bahru, Johor, Malaysia
e-mail: hasrinah@petroleum.utm.my

F. Mohamed · H. Hasbullah · W.N.R. Jamian · A.R.A. Rani · M.F.K. Saman · W.N.H. Salleh
Faculty of Petroleum & Renewable Energy Engineering (FPREE),
Universiti Teknologi Malaysia, 81310 Johor Bahru, Johor, Malaysia

R.R. Ali
Faculty Chemical Engineering, Universiti Teknologi Malaysia, 81310
Johor Bahru, Johor, Malaysia

M.A. Hashim (ed.), *ICGSCE 2014*, DOI 10.1007/978-981-287-505-1_18

Introduction

Gas separation involves number of techniques to separate gases, either to give multiple products or to purify a single product. Amine scrubbing technology had been established over 60 years ago for removal of hydrogen sulphide and CO_2 from gas streams in the oil and chemical industries. Commercially, it is the most well establish techniques available for CO_2 capture. However, this technique is mainly being practiced in gas streams which are chemically reducing, the opposite of the oxidizing environment of a flue gas stream [1]. Another method is the cryogenic gas separation, where separation occurs by cooling and condensation processes and commercially used for streams that already have high concentrations, typically more 90 % but not for more dilute streams [1].

Membrane gas separation is another method in industry that becoming more popular these days. A gas separation membrane operates by allowing one component in a gas stream to pass through faster than the other component. In order to achieve a high product purity and yield (by selectivity) and a high throughput (by permeability) in specific application with substantial market potential, permeability and selectivity balance of membrane with respect to specific compositions of the gases, vapors, or liquids to be separated require an extremely high degree of optimization [2]. Several membranes with different characteristics may be required to separate high-purity gas, while solvent assisted membranes are being developed to combine the best features of membranes and solvent scrubbing. Much development is required before membranes could be used on a large scale for capture in power stations.

Among the current concerns regarding of membrane separation are its biodegradability where the convenient polymer membrane gives a major drawback, which is difficult to degrade or dispose when it reached the limit of usage. A polymer based on a C–C backbone tends to resist degradation, whereas heteroatom-containing polymer backbones might possess biodegradability tendency. Therefore, a biodegradable membrane is introduced in this study. This research is conducted by using a commercially available and stable biodegradable polymer that is poly(lactic acid) (PLA). Biodegradable polymers such as PLA are used in medical application such as controlled drug releasing agent, plain membranes for guided tissue regeneration and surgical fixation to avoid a second operation to remove them [3].

In the preparation of biodegradable thin layer asymmetric membrane made from poly(lactic acid) biopolymer, the selection of solvent, its temperature and concentration, casting support and shear rate are among the important factors that influence the membrane structure, tensile strength, stability, and gas separation performance. Thus, for this study, the main objective is to investigate the effect of concentration of PLA on the morphology, gas permeation, thermal and mechanical properties of the biodegradable membrane.

Experimental

Materials, Poly(lactic acid) (PLA) was used as the membrane material provided by Shanghai Guanghe Biotech Co. Ltd. Dichloromethane (DCM) was used as solvent for poly(lactic acid) polymer and supplied by MERCK Schuchardt and tap water was used as coagulant medium.

Dope preparation, Prior to the dope solution preparation, the polymer was dried overnight in a vacuum oven at about 60 °C. Firstly, DCM was poured into a bottle and stirred. Then, PLA polymer was slowly added to avoid agglomeration and the mixing was continued for about 24 h to ensure a homogenous solution was prepared. The dope solutions were prepared with different concentration of 15, 17.5, 20, 25 wt% PLA. Then, the solution was degassed using an ultrasonic bath to remove trapped microbubbles prior to casting process.

Preparations of asymmetric flat sheet membrane, Asymmetric flat sheet membranes were prepared according to the dry/wet phase inversion process. The polymer solution was cast on a clean glass plate using a unique pneumatically controlled membrane casting system. The casting knife consists of a steel blade, which rested onto two runners, arranged to form a precise gap between the blade and glass plate. The solvent evaporation time and casting speed for the pneumatically controlled casting system was fixed at 60 s and 0.0413 ms^{-1} respectively for producing 13 in. length membranes. After the additional freestanding evaporation step, the membranes were immersed in the water bath at room temperature (~28 °C) (wet phase inversion) and remained there for about 24 h. Finally, the produced membranes were air dried under the atmospheric conditions for another 24 h before storing them in sealed plastic bags prior to testing.

Scanning Electron Microscopy (SEM), Scanning electron microscope was used to observe the morphology and structure of prepared membrane. Prior to positioning on a metal holder for gold coating using sputter coating operated under vacuum, these samples were cryogenically prepared. The SEM micrographs of both surface and cross-section of the membranes were taken.

Tensile test, The mechanical property of membranes can be obtained by measuring its tensile strength. Universal Testing Machine (UTM, Lloyd Instruments, UK) was employed to test the tensile strength of the membrane sample using ASTM standard for sample preparation. Samples were measured at a rate of 2 mm/min strain.

Gas permeation test, Gas permeation of pure oxygen (O_2) and nitrogen (N_2) were measured using soap bubble meter at 28 °C and pressures of 6 7, 8, 9 and 10 bar.

Result and Discussion

Membrane morphology, The morphology and structure of four different weight percentage of PLA 15, 17.5, 20, and 25 wt% were observed using scanning electron microscopy. The PLA asymmetric membrane surface and cross sectional structure were shown in Figs. 1 and 2 respectively. Form Fig. 1a–c, as the concentration increased from 15 to 20 wt%, the porosity of the membrane surface decreased. The decrease in membrane surface porosity might be due to the presence of more PLA molecules that cause the membrane intermolecular forces to increase resulting the membrane surface to be packed and well aligned. Besides, Hachisuka (1996) suggested that asymmetric membranes prepared using a low polymer concentration tends to have finger voids or very porous structures [4]. However, as the concentration was further increased from 20 to 25 wt%, the membrane surface showed the appearance of pinholes that may due to the increase in dope solution viscosity that cause by the presence of too much PLA molecules. The increase in dope solution viscosity may lead the difficulties for the solvent to leave from the membrane surface and non-solvent to penetrate from the bottom of the membrane. At first, the penetration of non-solvent leaves no traces on the membrane surface, however, when the membrane start to harden, the non-solvent penetration through the membrane may leaves traces on membrane surface in the form of pinhole.

The membrane cross sectional structure that shown in Fig. 2a–d revealed that membrane porosity decrease as the increase in PLA concentration. The increase in PLA concentration may cause the bonding between PLA molecules to increase. The increase in membrane intermolecular forces may cause the molecule to align themselves to a greater molecular orientation during membrane casting [5]. The significant increase of the dope solution viscosity cause by the increase in PLA concentration may creates the difficulty of non-solvent penetration into the membrane during the wet phase inversion process [6], thus reducing the membrane

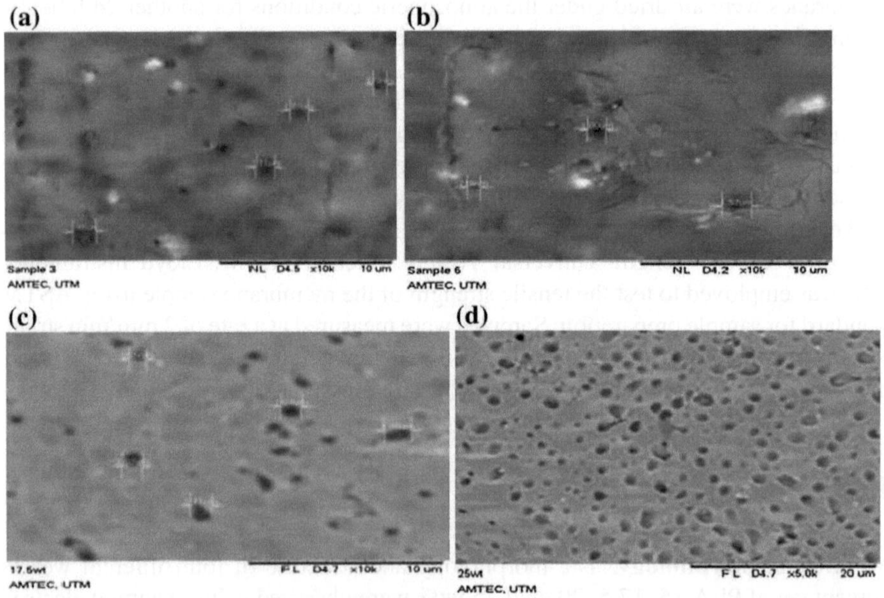

Fig. 1 Micrographs of the PLA membrane surface with difference concentration of **a** 15 %; **b** 17.5 %; **c** 20 %; and **d** 25 %

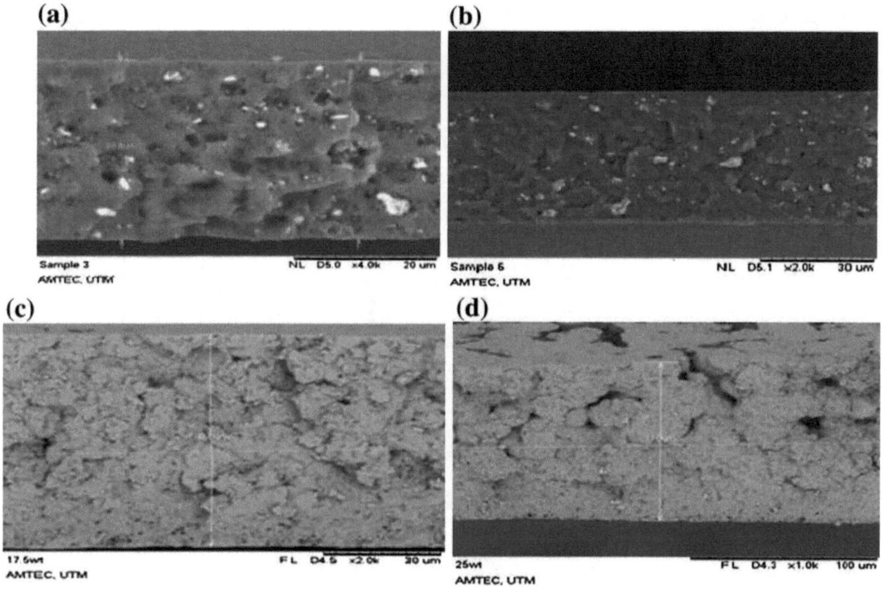

Fig. 2 Micrographs of PLA membrane cross sectional structure with difference concentration of **a** 15 %; **b** 17.5 %; **c** 20 %; and **d** 25 %

porosity. Another possibility that creates the difficulty of non-solvent penetration is the agglomeration of the PLA polymer during membrane casting.

Membrane tensile strength, The polymer concentration was varied at 15, 17.5. 20 and 25 wt% to determine the effect of concentration on membrane tensile strength. Figure 3 shows the relationships between PLA concentration on membrane tensile strength and young's modulus. As shown in the Fig. 3, the increase in PLA concentration from 15 to 17.5 wt% had increase the membrane tensile strength. These increases may due to the membrane compactness cause by the greater membrane molecular alignment [7]. As the concentration increase, it is expected to increase the membrane tensile strength. However, as the PLA concentration further increase to 20 and 25 wt%, the membrane tensile strength decrease. The decrease in membrane tensile strength may due to the improper membrane formation resulted by the polymer agglomeration causing the membrane to be brittle.

Membrane gas permeation performance, Increasing the concentration during dope preparation is expected to increase the skin layer thickness with tighter support structure, thus, decreasing the pressure-normalized flux but improving the overall gas separation performance in term of gas pair selectivity [8]. Figure 4 shows the gas separation performance of 15 wt% PLA asymmetric membrane. PLA membrane with concentration of 17.5 wt% is expected to perform better in gas permeation test than 15 wt%. Although the 17.5 wt% PLA membrane has tensile strength higher compared to membrane with other concentration, the elasticity of the membrane is lower thus the membrane is highly brittle and make it currently not

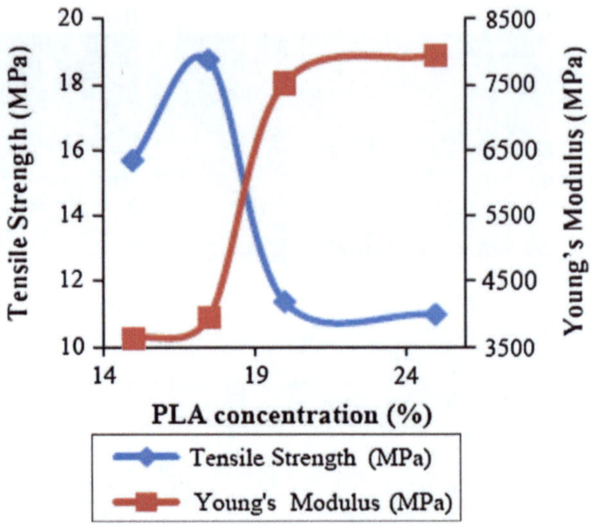

Fig. 3 Effect of concentration on mechanical properties of PLA membrane

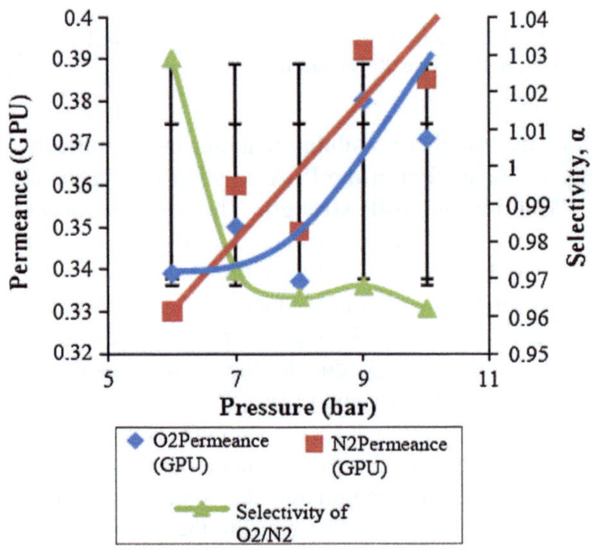

Fig. 4 Gas separation performance for PLA at concentration 15 %

readily to be further tested in gas separation process. As mentioned above, the membrane prepared with 20 and 25 wt% PLA was not properly formed, and it is proven to possess low tensile strength as well as severely low elasticity. Hence, these membranes were also difficult to handle and could not withstand the pressure used in the gas permeation test. Thus, the only sample was able to be tested was the membrane with concentration of 15 wt%. The permeance for all gas tested are fluctuated with varied pressure in the range of 6–10 bar. The gas permeance for all gases increase as the increase in pressure might be due to the increasing in the driving force for all gasses to pass through the membrane. At pressure 6 bar, the O_2 gas permeance in higher than N_2 gas may be owed to the O_2 gas characteristic of smaller kinetic diameter than N_2 gas as accordance to Graham's Law that make O_2 travel pass through the membrane 3–4 times faster than N_2. As the pressure is further increased, the N_2 gas permeance is higher than O_2 gas. This happens may be due to the prominent membrane surface defects that cause the gases to permeate faster through the membrane. As the pressure increase, the permeance of all gases increase, however, the O_2/N_2 gasses selectivity decrease. The decrease in selectivity is as accordance to the trade of between permeability and selectivity as discussed by Robeson et al. where the increase in gases permeability will decrease the selectivity of the membrane [9]. The separation factors were not commercially desirable but the potential of PLA as a membrane for gas separation cannot be denied.

Conclusion

The development of innovative biopolymers has been underway for a number of years and continue to be an area of interest for the modern scientists. Biodegradable material used for membrane gas separation is the new research interest concerning the issue of disposing the damaged non-biodegradable membranes into the environment. Thus, finding an environmental friendly material suitable for membrane production is crucial such as poly(lactic acid) that is very popular in the food packaging and medical application industry. The effect of PLA concentration on morphology, mechanical properties, and gas separation performance has been investigated in this study. The results shows that the prepared PLA membrane consists of skin layer on top and porous support layer on the bottom. The increase in PLA concentration might increase the bonding between membrane molecules thus reducing the membrane porosity, however the further increase on PLA concentration causing the dope solution viscosity to increase thus increasing the membrane surface defect. Based on the tensile test, the optimum PLA concentration for the sample to act as a gas separation membrane is at 15 wt% along with the optimum tensile strength. The tensile strength of the membrane is expected to increase with the increase in PLA concentration, however higher PLA concentration have cause the membrane to be more brittle with a reduction of membrane elasticity and currently not readily to be tested for gas permeation performance. The separation factor of PLA membrane with concentration of 15 wt% produced was not commercially desirable; however it shows a huge potential as a membrane material for gas separation with added benefit of being biodegradable. The membrane is recommended for further testing by coating the membrane using silicon rubber coated material (PDMS).

Acknowledgment We gratefully acknowledge the financial support from the Universiti Teknologi Malaysia and Ministry of Education (Grant No: 04H70 and 4L116).

References

1. Gupta M, Coyle I, Thambimuthu K (2003) CO_2 capture technologies and opportunities in Canada: strawman document for CO_2 capture and storage (CC & S) technology roadmap. CANMET Energy Technology Centre, Natural Resources Canada
2. Ulbricht M (2006) Advanced functional polymer membranes. Polymer (Guildf) 47:2217–2262
3. Wei XW, Guo G, Gong CY, Gou ML, Qian ZY (2011) Biodegradable polymers: research and applications. In: A handbook of applied biopolymer technology: synthesis, degradation and applications, no. 12, Chengdu, China, pp. 365–387
4. Hachisuka H (1996) New type asymmetric membranes having almost defect free hyper-thin skin layer and sponge-like porous matrix. J Memb Sci 116:265–272
5. Hasbullah H, Be N, Ibrahim N (2013) Mindel S-1000 based asymmetric membranes for O_2/N_2 separation: effect of polymer concentration. Chem Eng Trans 32:2035–2040
6. Chung TS, Shieh JJ, Lau WWY, Srinivasan MP, Paul DR (1999) Fabrication of multi-layer composite hollow fiber membranes for gas separation. J Memb Sci 152:211–225
7. Ghasem N, Al-Marzouqi M, Duidar A (2012) Effect of PVDF concentration on the morphology and performance of hollow fiber membrane employed as gas–liquid membrane contactor for CO_2 absorption. Sep Purif Technol 98:174–185
8. Pesek SC, Koros WJ (1993) Aqueous quenched asymmetric polysulfone membranes prepared by dry/wet phase separation. J Memb Sci 81:71–88
9. Robeson LML (1991) Correlation of separation factor versus permeability for polymeric membranes. J Memb Sci 62:165–185

Particles Mixing in a Fluidized Bed by Using Digital Image Processing and Thief Probe

Syaidatul Akma Mohd Zuki, Norazah Abd Rahman, Ihsan Mohd Yassin, Nurul Asyikin Md Zaki and Mohamad Shafiq Izzudin Suhaimi

Abstract Particle size plays a major role in segregation phenomena in mixing process. In pharmaceutical industry, it is essential to have a proper analysis to make sure that the good product is well mixed and to prevent segregation. In this study, Digital Image processing has been used as an alternative method to thief probe, to study the effect of particle size on mixing and segregation process. The experiment started with execution of image acquisition process of mixing process. The images taken were further analyzed using Image Processing Tool in MATLAB software. The experiments were done with two different arrangement of particle in the bed using air velocity of 0.94 ms^{-1}. The mixture was considered to be well mixed when the intensity of red of colour histogram is approximately constant. For the layered position, the good mixing time for set 1, 2, 3 and 4 were achieved at 62, 84, 86 and 74 s respectively. Then, for random position, homogeneous mixture were achieved at 62, 76, 82 and 72 s for set 1, 2, 3 and 4 respectively. This finding has been quantified by using Lacey Mixing Index where the value obtained was nearly to 1.0 which showed uniform mixing.

S.A. Mohd Zuki · N. Abd Rahman (✉) · N.A. Md Zaki · M.S.I. Suhaimi
Faculty of Chemical Engineering, Universiti Teknologi MARA, 40450 Shah Alam, Selangor Darul Ehsan, Malaysia
e-mail: noraz695@salam.uitm.edu.my

N. Abd Rahman
CoRe of Frontier Materials and Industry Applications, Universiti Teknologi MARA, 40450 Shah Alam, Selangor Darul Ehsan, Malaysia

N.A. Md Zaki
CoRe of Green Technology and Sustainable Development, Universiti Teknologi MARA, 40450 Shah Alam, Selangor Darul Ehsan, Malaysia

I. Mohd Yassin
Faculty of Electrical Engineering, Universiti Teknologi MARA, 40450 Shah Alam, Selangor Darul Ehsan, Malaysia

I. Mohd Yassin
CoRe of Advanced Computing and Communication, Universiti Teknologi MARA, 40450 Shah Alam, Selangor Darul Ehsan, Malaysia

M.A. Hashim (ed.), *ICGSCE 2014*, DOI 10.1007/978-981-287-505-1_19

Keywords Digital image processing · Spherical polystyrene · Fluidization · Lacey mixing index · Segregation · Intensity

Introduction

Particle mixing is a process of combining two or more particles to form a mixture. Typically, the particles involved in the process have different characteristic or properties; namely density, moisture ratio, size, porosity and shape. The difference in their properties lead to a condition called segregation, where the mixture potentially not to be mixed properly or unmixed [1]. Other than that, if a well-mixed mixture continuously undergoes the mixing process, they have a high tendency to segregate unless the process is stopped when the quality of mixing has been achieved.

Some applications of mixing require simple combination of particles, while others demand a high degree of homogeneity. The segregation process causes problem in some industries such as in pharmaceutical industry, where mixing is considered as a critical process. This is because mixing process is used in its formulation where different chemical substances such as active drug and inactive additives are mixed to produce a final product. For an orally taken drug, segregation of the mixture will produce poor distribution of the drug which has been incorporating into a tablet or a capsule [2]. Therefore, the mixing process has to be properly monitored to ensure a high degree of homogeneous mixture and prevent segregation process, thus achieving high product quality. The mixing process has to be stopped as soon as the mixing efficiency is achieved to optimize cost and time, as well as to prevent segregation process.

Over the years, the study of mixing is gradually evolving due to its importance in many industries. Two methods are generally available to identify mixing efficiency; namely invasive and non-invasive. Invasive method is a traditional manual procedure that involves continuous monitoring of the mixing process [3]. Weaknesses of the invasive method are disruption of the mixing process due to regular sampling intervals, and long mixing times. Because of these weaknesses, non-invasive methods such as Discrete Element Model (DEM) simulation [4, 5], Magnetic Resonance Imaging (MRI) [6–8], Positron Emission Particle Tracking (PEPT) [9, 10] and Digital Image Processing (DIP) [11–13] have been used. Recently, DIP has shown great potential to be studied further because of its advantage of economy, simplicity and non-invasiveness [14].

In this study, the two methods have been used where thief probe and a web camera have been used to represent an invasive and non-invasive method respectively. The main objective is to show that simple image processing technique like Multivariate Analysis RGB images produced from low cost web camera can be used to assess mixing performance for spherical polystyrene with similar sizes but different colours.

Materials and Methods

The material that has been used in the study was two different sizes of free flowing spherical polystyrene with same shape as well as particle density. Spherical polystyrene particles were sieved using mechanical sieve to obtain the particle sizes of 4.05 and 5.50 mm where both were coloured by red artificial dye. Fluidized bed with column diameter of 20 cm and height of 21 cm made from perspex was used in this study. Air was flown with velocity of 0.94 ms^{-1} from the bottom of the column. The gas distributor with pore size of 63 μm was installed in the bottom of the column to distribute the air during the fluidization process. 20 grams of white and red polystyrene sphere were used with weight ratio of 50:50. It was observed that the most suitable height for the placement of the web camera was at 25 cm from the bottom.

The polystyrene spheres had been flown upward by two different position. The red sphere polystyrenes were laid out at the bottom of the bed, followed by white polystyrene spheres spread over the top. For the layered position, the red polystyrene spheres were laid out evenly before the white polystyrene spheres were placed on top. Then, for the random position, the white polystyrene spheres were placed at the top of red polystyrene spheres without laying them out evenly. Figure 1 shows the illustration of the two positions for the placement of the polystyrene spheres. Four sets of experiment were conducted for each position and Table 1 shows the details of the arrangement used in the experiment.

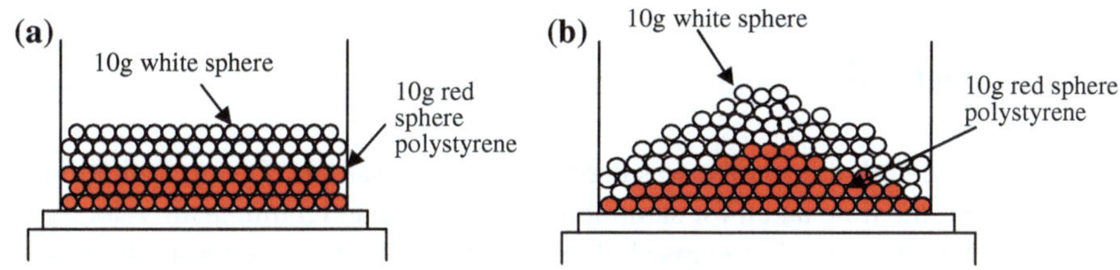

Fig. 1 Illustration of **a** layered position **b** random position

Table 1 Sets of experiment

Set of experiment	Above (white) (mm)	Below (red) (mm)
1	4.05	4.05
2	4.05	5.50
3	5.50	4.05
4	5.50	5.50

Non-invasive Method

For image acquisition system, a c920 Logitech web camera was used to capture the mixing process. The mixing image was captured every 2 s and saved by executing a MATLAB script to capture the images. The resolution of the image was 1920 x 1080 pixels with 24-bit color depth, and red, green and blue (RGB) representation. The acquired images undergo several image processing steps in MATLAB, namely cropping and histogram analysis. Maximum histogram points (describing the highest number of intensity occurrence) for red colors were recorded across time. The color histogram represents the color changes of the mixing image. The step of analysis is shown in Fig. 2.

This study focuses on the red, green and blue color histograms. The color histogram represented the color value at x-axis which has value from 0 to 255 and the frequency of the color value. The color value indicates the degree of intensity of a particular color. A high pixel value (near 255) indicates that the color intensity is high, and vice versa. For this study, the three primary colors are red, green and blue. But, as the polystyrene sphere was dyed with red, the study focused on the red color histogram for further analysis. From the red color histogram, the maximum point is recorded which represented the color value and the maximum intensity of that color value. As mentioned by Mohd Zuki et al. [15] the mixture is homogeneous when the distribution of the histogram is constant. It can be applied on maximum point of the histogram. In order to ease the mixing analysis by using image processing, maximum point of the histogram has been used to obtain the mixing time.

Invasive Method

For invasive method, thief probe was used to obtain the sample. Due to limitation of the equipment, the time taken for sampling are from 1, 2 and 3 min at air velocity of 0.94 ms^{-1}. After the time has elapsed, thief probe has was to obtain a small amount of sample at several locations in the fluidized bed. This method is carried out to support the finding from non-invasive method. In this study, Lacey mixing index [16] has been used to quantify the degree of mixing or segregation.

$$M = \frac{S_0^2 - S^2}{S_0^2 - S_R^2} \tag{1}$$

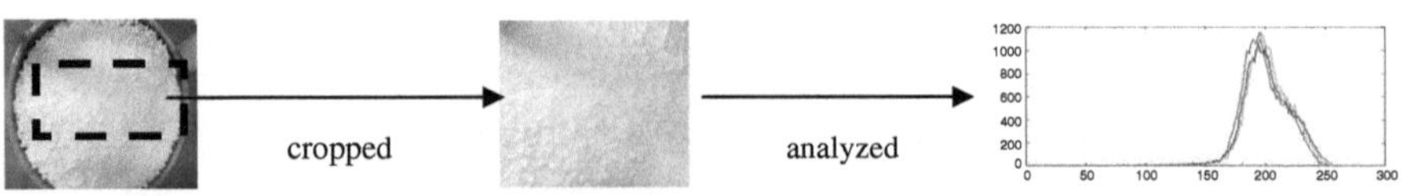

Fig. 2 Step of image analysis

where S^2 is the actual variance, S_0^2 and S_R^2, respectively, represent the variances of the completely segregated and well mixed states, given by

$$S_0^2 = p \cdot q \tag{2}$$

$$S_R^2 = \frac{p \cdot q}{N} \tag{3}$$

where p and q are the volume fractions of white and red particles in a mixture and N is the equivalent number of particles in the sample.

Result and Discussion

Figures 3 and 4 shows the graphs of maximum intensity of red value versus time of mixing process for 4 sets of experiment in layered and random positions respectively. Those four graphs started with high maximum frequency of red value (colour value of 165–195) in range of 704–807 Hz at mixing time of 0 s. This is because at this time, particles were not fluidized and the fluidized bed acted as a fixed bed. The mixture was completely unmixed or completely segregated. The image at this point showed the white polystyrene spheres at the top of the fluidized bed. The maximum points of the red, green and blue histogram are approximately the same. In theory, white color represented the same value of those three primary

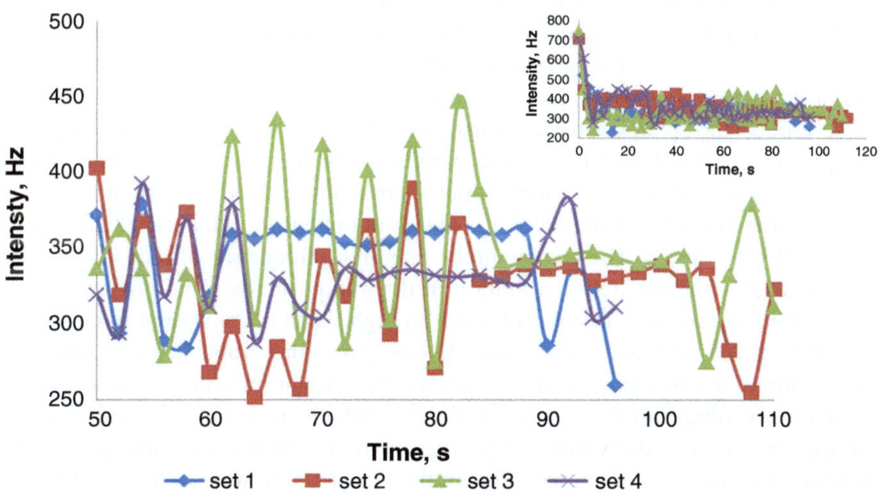

Fig. 3 Intensity versus time for layered position

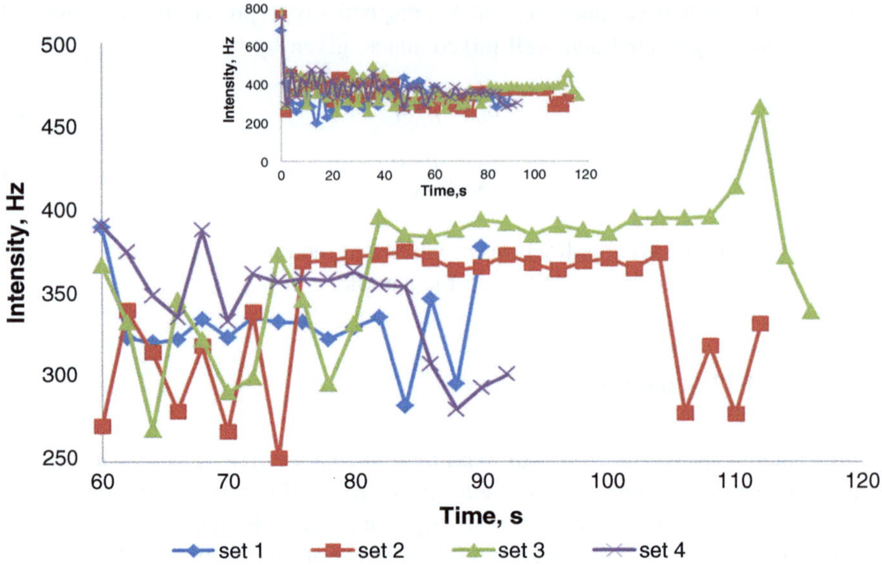

Fig. 4 Intensity versus time for random position

colors. If the images have high value of any one of the three colors, it will project that color in the image.

As the mixing time was increased, red polystyrene particles started to move to the top part of the bed and the maximum frequency of red value decreased. The movement of red particles from the bottom layer is influenced by the wake of air bubbles which induces mixing process and at the same time, white particles which were at the top moved down due to drag force which induces segregation. The air flowed around the spherical particles and made them moved and this movement is anticipated to be the factor to mix the particles. The images that have been taken during this process have two colors, which were white and red. Based on the range of color value of red for white color, the maximum intensity in that range have been plotted in Figs. 3 and 4. The change in the maximum frequency of red is caused by the change of white and red distribution in the mixing image. The image that has more white polystyrene spheres will result in high value of maximum intensity for red value of 165–195 and vice versa. At one point, the graph begins to show a constant trend. It indicated that the image have the similar intensity of the red value. Based on the observation, it was concluded that the mixture had started to be homogenous because the white and red polystyrene spheres were spread equally. Therefore, the intensity of the red value in the image taken, were approximately the same. The constant value had been taken as the time when the particle mixes well [15]. At this state, a dynamic equilibrium state is achieved where the mixing process balances the segregation process after several rearrangements of the particles [16]. The results of the experiments have been summarized in Fig. 5.

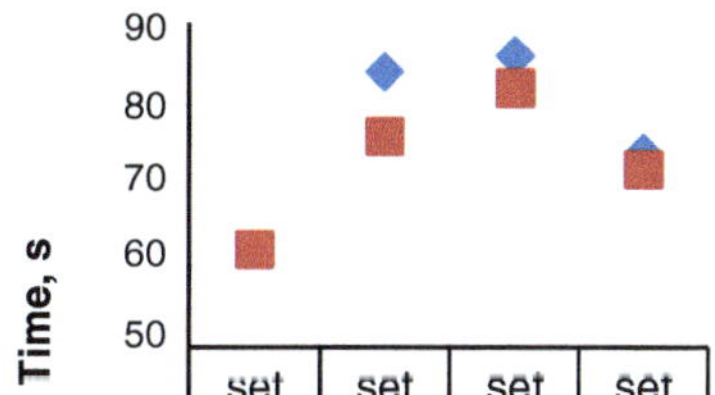

	set 1	set 2	set 3	set 4
◆ layerred position	62	84	86	74
■ random position	62	76	82	72

Fig. 5 Mixing time

The mixing times for two different position of material placement for 4 sets of experiment were shown in Fig. 5. The mixing times for this experiment were in the range of 62–86 s. As shown in Fig. 5, mixing time for set 2 and 3 are slightly higher than set 1 and 2. The different in particle size for white and red polystyrene in set 2 and 3 give more time for the mixture to be well mixed. As mentioned by Hogg [17], the difference in particle size give a main problem for mixing process. This factor will lead to segregation phenomena as the process continues. The small particles have the capability to fill the void age between the bigger particles in the expanded fluidized bed. This filling process was assisted by gravitational force as the small particle move downward during the process. The process is known as percolation.

Layered positioning gave higher mixing times compared to random positioning for each set of experiment. The mixing and segregation process also can be affected by the interaction between particle and the drag force. When the sphere polystyrenes were filled in the fluidized bed as the random position, the particles are concentrated in the middle of the bed like heap formation. There are no forces applied during the filling process. The particles flowed downward freely and formed the heap. At the middle of the heap which is a dense region, the drag force of the particles are less than the gravitational force because there is no space for air to flow upward [16]. The air bubbles take some time to flow upward after rearranging the particles. However, the interaction between particles in random position is less as compared to layered position since no force is applied to the particles. So, longer time is needed to fluidize the particles for layered position as the force take its effect during the filling process. The particle is flattened before the mixing process start. Furthermore, for layered position, the dense regions are not only in the center of the bed but spread out all over the area. So, the time taken for air bubble to move upward is longer.

Figure 6 shows the effect of time on the mixing index for the two arrangements when the superficial gas velocity is 0.94 ms^{-1}. Due to limitation of the equipment, the times taken to do the sampling were above 1 min. According to results obtained from image processing, uniform mixture has already been achieved at the selected time. Hence, mixing index showed value close to 1.

Fig. 6 Lacey mixing index

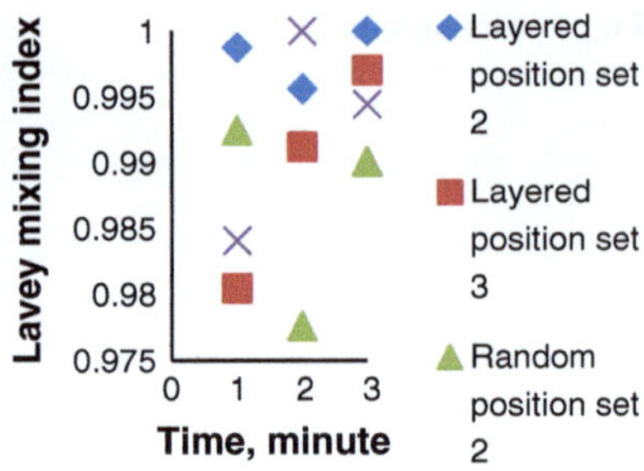

Conclusion

The graphs of maximum intensity of red value versus mixing time were plotted for four sets of experiment for two different position of the placement of polystyrene spheres; which were layered and random position. It was observed that the graph starts to become constant at one point, where it indicates the time for well mixed mixture. For the layered position, the well mixing time for set 1, 2, 3 and 4 are 62, 84, 86 and 74 s respectively. Then, for random position, the well mixing time is 62, 76, 82 and 72 s for set 1, 2, 3 and 4 respectively. This finding has been quantified by using Lacey Mixing Index where the value obtained was near to 1.0 which showed uniform mixing at selected time of operation. The experiment also can be further studied to solve the segregation problem in mixing process. The well mixed mixture is expected to segregate if the mixing process is continued further for a certain length of time.

Acknowledgement The authors would like to thank the Research Management Institute of Universiti Teknologi MARA for Excellence Fund (600-RMI/DANA5/3/RIF(90/2012)) for the financial support, Faculty of Chemical Engineering and Universiti Teknologi MARA.

References

1. Susana L, Canu P, Santomaso AC (20100) Development and characterization of a new thief sampling device for cohesive powders. Int J Pharm 416:260–267
2. Allen T (1981) Powder technology theory. Chapman & Hall, London
3. Kaye BH (1997) Powder mixing. Chapman & Hall, London
4. Chaikittisilp W, Taenumtrakul T, Boonsuwan P, Tanthapanichakoon W, Charinpanitkul T (2006) Analysis of solid particle mixing in inclined fluidized beds using DEM simulation. Chem Eng J 122:21–29
5. Di Renzo A, Di Maio FP, Girimonte R, Formisani B (2008) DEM simulation of the mixing equilibrium in fluidized beds of two solids differing in density. Powder Technol 184:214–223
6. Richardson JC, Bowtell RW, Mäder K, Melia CD (2005) Pharmaceutical applications of magnetic resonance imaging (MRI). Adv Drug Deliv Rev 57:1191–1209
7. Hardy EH, Hoferer J, Kasper G (2007) The mixing state of fine powders measured by magnetic resonance imaging. Powder Technol 177:12–22

8. Müller CR, Holland DJ, Sederman AJ, Mantle MD, Gladden LF, Davidson JF (2008) Magnetic resonance imaging of fluidized beds. Powder Technol 183:53–62
9. Rafiee M, Bakalisa S, Fryer PJ, Ingram A (2011) Study of laminar mixing in kenics static mixer by using positron emission particle tracking (PEPT). Procedia Food Sci 1:678–684
10. Stellema CS, Vlek J, Mudde RF, de Goeij JJM, van den Bleek CM (1998) Development of an improved positron emission particle tracking system. Nucl Instrum Methods Phys Res Sect A 404:334–348
11. Berthiaux H, Mosorov V, Tomczak L, Gatumel C, Demeyre JF (2006) Principal component analysis for characterising homogeneity in powder mixing using image processing techniques. Chem Eng Process 45:397–403
12. Le Coënt AL, Rivoire A, Briançon S, Lieto J (2005) An original image processing technique for obtaining the mixing time: the box-counting with erosions method. Powder Technol 152:62–71
13. Muerza S, Berthiaux H, Massol-Chaudeur S, Thomas G (2002) A dynamic study of static mixing using on-line image analysis. Powder Technol 128:195–204
14. Abdul Rahman N, Mohd Zuki SA, Mohd Yassin I (2012) A review of image processing technique in particle mixing analysis. In: Proceeding of 2012 IEEE 8th international colloquium on signal processing and its applications
15. Mohd Zuki SA, Abdul Rahman N, Mohd Yassin I (2013) Particle mixing analysis using digital image processing technique. J Appl Sci 14:1392–1396
16. Feng YQ, Xu BH, Zhang SJ, Yu AB (2004) Discrete particle simulation of gas fluidization of particle mixures. AIChE J 50:1713–1728
17. Hogg R (2009) Mixing and segregation in powders: evaluation. Mech Processes Powder Part J 27:3–17

mechanism. In this mechanism, gas molecule is separated by their size and shape. As a result, composite membrane with porous filler usually have high permeability and selectivity which is beyond the Robeson upper bound because porous filler has concise apertures [13].

Thus, the presence of highly selectively porous filler in polymer matrix will result in a composite membrane with a better permselectivity than that of the pure polymer membrane since the porous filler in the composite membrane selectively allow the desire component to pass through its pores. However, the result of this study is not consistent with the aforementioned theory since only the permeability of the composite membrane with 3 wt% agarwood is higher than that of Psf membrane instead of both permeability and selectivity. The possible explanation for this finding is the formation of non-selective voids that result from lack of adhesion at the agarwood-polymer interface. The presence of these voids increase the permeability and cause the apparent selectivity of the composite membrane to reduce. Therefore the agarwood become ineffective since gas molecule pass through the less resistance interfacial voids instead of passing through the pore of the agarwood [13].

Thermal stability test, In order to identify the effect of bentonite and agarwood waste powder on thermal stability of Psf composite membrane, TGA was carried out. The thermal stabilities as a function of weight (%) were evaluated via TGA with heating rate of 10 °C/min and the result are shown in Fig. 2. One decomposition step can be observed for Psf membrane and 3 wt% bentonite composite membrane. Meanwhile, the decomposition of 3 wt% agarwood composite membrane occurred in 3 steps. The first step is associated with the dehydration of moister that was trapped inside the agarwood. The second step illustrates the decomposition of hemicellulose and lignin that exist in the agarwood waste powders that are not in intimate contact with the polymer matrix [12]. This is because, agarwood waste powder is a lignocellulosic material with a thermal stability lower than that of pure polymers [14]. The third step on the other hand represents decomposition of the composite [12].

The present of bentonite in the polymer matrix result in a membrane with better performance in term of thermal stability since the total mass loss of the composite membrane at 1000 °C was less than the Psf membrane. The Psf membrane decomposed to a of mass of 30.8 % from 100 % at 1000 °C. In other word, Psf

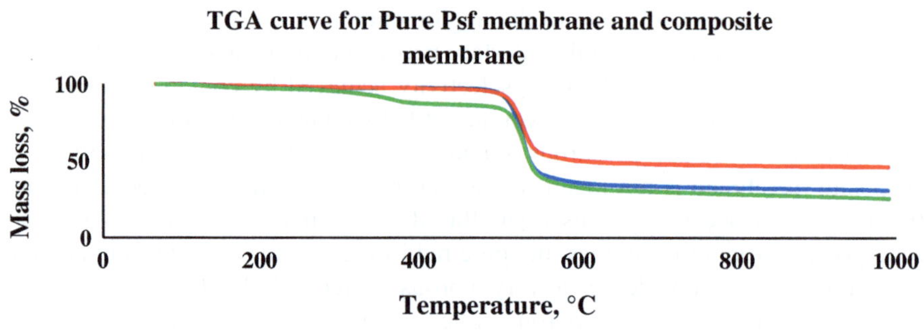

Fig. 2 TGA curve of Psf membrane and composite membrane

membrane loss 69.2 % of its total mass at 1000 °C. Meanwhile when 3 wt% of bentonite was added to Psf matrix, the resulted membrane loss only 54 % of its total mass at 1000 °C. This result was observed probably because the presents of bentonite in the polymer matrix act as barrier that obstruct the diffusion of oxygen molecule into the composite membrane. Other than that, bentonite dispersed in the Psf matrix result in a membrane with tortuous in structure which cause volatilization to hold up [15].

On the other hand, a decrease in thermal stability is observed for composite membrane with 3 wt% of agarwood waste powder since its mass loss at 1000 °C was 7.1 % higher compare to the Psf membrane. This may result from the decomposition of hemicellulose and lignin that exist in the agarwood waste powders that are not in intimate contact with the polymer matrix that occur before the rest of the composite decompose [12]. When this occurs, some free radicals are created which in turn accelerate the failure of the polymer at higher temperature [14]. Agarwood waste powders are not in intimate contact with the polymer matrix due to poor compatibility of the agarwood with the polymer. This is because polymer are hydrophobic, while agarwood is hydrophilic. As a result, the polymer and agarwood will not adhere to each other effectively [16].

Mechanical properties, The tensile strength and elongation at break for Psf membrane and composite membrane was compared to determine the effect of adding filler additive to Psf membrane on the mechanical strength of the Psf membrane. The measurements done to determine the mechanical strength parameters show that the Psf membrane had a tensile strength of 7 MPa, and elongation at break of 10.1 %. Generally, the addition of bentonite to pure Psf membrane is expected to result in membrane with a lower elongation at break since the membrane will become more brittle [17]. Ironically, the addition of bentonite had result a better membrane in term of elongations at break which corresponded to an increase of 78 % in a membrane with 3 wt% of bentonite as shown in Fig. 3.

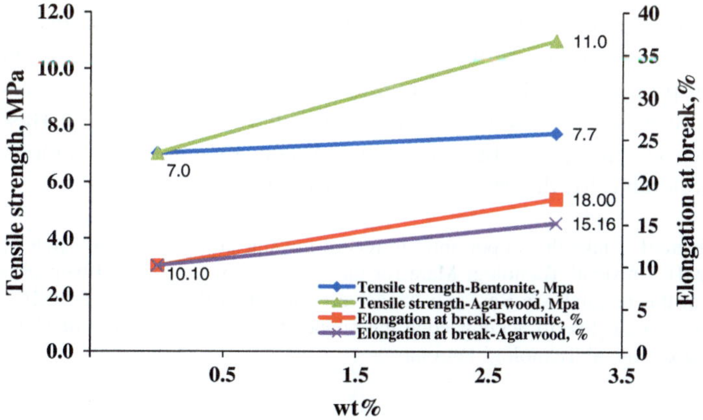

Fig. 3 Effect of bentonite on tensile strength and elongation at break for Psf membrane and composite membrane

Such result was obtained because there is a possibility that the clay platelets rearrange in the direction of the deformation enable more deformations [18]. Besides that, the membrane with 3 wt% of bentonite can take a higher amount of tensile stress before it break compared to Psf membrane, correspond to a value of 7.7 MPa in a membrane with 3 wt% of bentonite. The increase in tensile strength is associated with the increase in the surface area of interaction between clay and polymer matrix due to the insertion of the polymer chains between the silicate layers [18]. The addition of agarwood waste powder also enhances the mechanical properties of Psf membrane as shows in Fig. 3. The tensile strength of the Psf membrane increased to 4 MPa with the presence of 3 wt% agarwood waste powder in the polymer matrix. The agarwood waste powder may be regarded as acting efficiently if it carries a relatively high proportion of the externally applied load. This can result in higher strength. On the other hand, elongation at break of membrane with 3 wt% of agarwood waste powder is 15 %. This shows that the presences of agarwood waste powder increase the ability of the membrane to deform by enhancing the mobility of the polymer chains. As a result, it is easier for the segment of the membrane to slip past each other [12].

Conclusion

This research reveals that the addition of 3 wt% bentonite to polymer solution will enhance the performance of membrane for gas separation in terms of gas permeability, mechanical properties and thermal stability. However the 3 wt% bentonite composite membrane achieved a lower CO_2/CH_4 Selectivity compared to Psf membrane. Besides that, this study shows that a composite membrane with 3 wt% agarwood waste powder has a lower thermal stability compared to Psf membrane. Other than that, a higher gas permeability compared to pure Psf membrane was observed for 3 wt% agarwood composite membrane although CO_2/CH_4 Selectivity of 3 wt% agarwood composite membrane is lower than that of Psf membrane. Nevertheless, the addition of 3 wt% agarwood waste powder improved the mechanical properties of Psf membrane since a greater value of tensile stress and elongation at break was recorded compared to pure Psf membrane. Based on the performance and the properties of the membrane, Psf-bentonite composite membrane is preferable for the separation of CH_4/CO_2.

Acknowledgment I take this opportunity to record my sincere thanks to Research Management Institute (RMI), Universti Teknologi Mara for the cooperation, support and help throughout the completion of this research via Research Intensive Faculty 600-RMI/DANA 5/3/RIF (310/2012). Besides that, I would like to acknowledge the cooperation and sponsorship of Universiti Sains Malaysia and Universiti Teknologi Petronas.

References

1. Nguyen QT, Sublet J, Langevin D, Chappey C, Marais S, Valleton JM, Poncin-Epaillard F (2010) CO_2 Permeation with pebax®-based membranes for global warming reduction. Membr Gas Sep 9:255–277
2. Shimekit B, Mukhtar H (2012) Natural gas purification technologies—major advances for CO_2 separation and future directions. InTech Open Access Publisher
3. Aresta M, Tommasi I (1997) Carbon dioxide utilisation in the chemical industry. Energy Convers Manage 38:S373–S378 (Supplement(0))
4. Moulijn JA, Makkee M, van Diepen AE (2013) Chemical process technology. Wiley, Hoboken
5. Liu Ye, Wang Rong, Chung Tai-Shung (2001) Chemical cross-linking modification of polyimide membranes for gas separation. J Membr Sci 189(2):231–239
6. Mulder M (1996) Basic principles of membrane technology, 2nd edn. Kluwer Academic Pub, Dordrecht
7. Cong Hailin, Radosz Maciej, Towler Brian Francis, Shen Youqing (2007) Polymer—inorganic nanocomposite membranes for gas separation. Sep Purif Technol 55(3):281–291
8. Momeni SM, Pakizeh M (2013) Preparation, characterization and gas permeation study of PSf/MgO nanocomposite membrane. Braz J Chem Eng 30(3):589–597
9. Deng L, Kim TJ, Sandru M, Hägg MB (2009) PVA/PVAm blend FSC membrane for natural gas sweetening. In: Paper presented at the proceedings of the 1st annual gas processing symposium. Elsevier, Amsterdam
10. Zulhairun AK, Ismail AF, Matsuura T, Abdullah MS, Mustafa A (2013) Asymmetric mixed matrix membrane incorporating organically modified clay particle for gas separation. Chem Eng J 241:495–503
11. Anadão P, Priscila Sato, Fumie L, Wiebeck Hélio, Valenzuela-Díaz F Rolando (2010) Montmorillonite as a component of polysulfone nanocomposite membranes. Appl Clay Sci 48 (1–2):127–132
12. Salemane MG, Luyt AS (2006) Thermal and mechanical properties of polypropylene—wood powder composites. J Appl Polym Sci 100(5):4173–4180
13. Aroon MA, Ismail AF, Matsuura T, Montazer-Rahmati MM (2010) Performance studies of mixed matrix membranes for gas separation: a review. Sep Purif Technol 75(3):229
14. Nemati M, Khademieslam H, Talaiepour M, Bazyar B, Khakifirooz A (2013) Studying the effect of nanoclay on thermal properties of wood flour/recycled polystyrene composite. J Basic Appl Sci Res 3(5):789–794
15. Lu H, Hu Y, Li M, Chen Z, Fan W (2006) Structure characteristics and thermal properties of silane-grafted-polyethylene/clay nanocomposite prepared by reactive extrusion. Compos Sci Technol 66(15):3035–3039
16. Özmen N, Çetin NS, Mengeloğlu F, Birinci E, Karakuş K (2012) Effect of wood acetylation with vinyl acetate and acetic anhydride on the properties of wood-plastic composites. BioResources 8(1):753–767
17. Nigam V, Setua DK, Mathur GN, Kar KK (2004) Epoxy-montmorillonite clay nanocomposites: synthesis and characterization. J Appl Polym Sci 93(5):2201–2210
18. Anadao P, Montes RR, Larocca NM, Pessan LA (2013) Influence of the clay content and the polysulfone molar mass on nanocomposite membrane properties. Appl Surf Sci 275:110–120

Comparative Study Between Isolated *Xanthomonas Campestris* from Rotten Cabbage and *Xanthamonas Campestris* Culture by YDC Media as Substrate

Mohibah Musa, Ku Halim Ku Hamid and Miradatul Najwa Muhd Rodhi

Abstract *Xanthomonas campestris* is a gram-negative bacterium which causes black rot disease in varieties of brassicas which include cabbage, cauliflower, broccoli, turnip and radish. The cabbage samples were collected locally from Cameron Highlands, Pahang Darul Makmur. The bacteria were locally isolated and cultured to produce pure species of the bacteria which later several analysis and a series of fermentation was done in order to obtain Xanthan gum. Fermentation procedure conducted through shake flask fermentation. Five different culture and media combination were fermented to compare the final results of each substrate. On specific time intervals e.g.: every 2 h sampling, several analysis were conducted to determine bacteria cell mass and glucose concentration of the fermentation samples. After the completion of the fermentation, Xanthan gums were recovered through precipitation by isopropanol.

Keywords Fermentation · Grinded rotten cabbage · Xanthomonas campestris · Xanthan gum · YDC

Introduction

In East Africa, the black rot disease attacking cabbages is still a nightmare to the farmers. The plant pathogen has increasingly become harder to manage. On main reason is due to the cultivation of barely prone to disease seeds to the plant pathogen [1]. Despite of the damage that the bacteria are capable of, it also is an answer

M. Musa (✉) · K.H. Ku Hamid · M.N. Muhd Rodhi
Faculty of Chemical Engineering, Universiti Teknologi MARA, Selangor, Malaysia
e-mail: mohibah@salam.uitm.edu.my

K.H. Ku Hamid
e-mail: kuhalim@salam.uitm.edu.my

M.N. Muhd Rodhi
e-mail: miradatul@salam.uitm.edu.my

M.A. Hashim (ed.), *ICGSCE 2014*, DOI 10.1007/978-981-287-505-1_21

to the useful xanthan gum that is widely used. Usually, the bacteria are cultured in a batch fermenter under respective conditions. The xanthan gum is recovered by precipitation and it is produced by the fermentation of *Xanthomonas Campestris*, the Gram negative plant-pathogenic bacterium. [2].

The fermentation of microbes can be done either in shake flask or in bioreactor. In addition, these complex exopolysaccharides are containing D-glucose, D-mannosyl, and D-glucuronyl acid residues [3]. This is the answer for it that remarkably found to be suitable for many temperature, pH and ionic strength [4]. These properties are causing the gum to be more stabilized. Moreover, xanthan gum's ordered confirmation is very good for the stability. This stability also can be enhanced with presence of salt to optimize the xanthan gum functions [5]. In addition, they are so versatile till that they are being produced for many applications. Often, the xanthan gum are added for the food application, personal care application, industrial applications, animal feed, and as well as in pharmaceuticals. Xanthan gum is everywhere to be used daily.

Xanthan gum is not only acts as a stabilizer in most salads dressings and sauces, but also in dairy products colloidal and solid materials where they are found to be suitable to be stabilized by xanthan gum. Also, xanthan gum also found to be reasonable to be used in frozen foods and beverages. In fact, it is possible to undergo syneresis for the frozen foods. Syneresis is the process of extraction liquid from respective gel with xanthan gum. In beverages, xanthan gum is good at suspending the fruit pulps in order to give better taste and flavours. Apart of that, xanthan gum gives desired texture in many ice-creams by enhances the binding of free water for the product stability. Meanwhile, xanthan gum can improve the starch cohesions in bakery that benefits for extending the products shelf-life where the moisture preservation. It also prevents gas entrainment, where tiny bubbles always form in cake making [6]. On the hands, it is very common for xanthan gum in cosmetics and pharmaceuticals application. As for instance, xanthan gum plays an important role as the thickener and stabilizer in the making of personal care product like creams. As a result the xanthan gum gives gentle and soft feel as due to shear thinning flow behaviour of xanthan gum. The shear thinning flow behaviour is also significant for toothpaste which acts as the binder that improves the uniform dispersion on teeth and better rinsing during brushing of teeth. Next, xanthan gum is also good in drilling fluids as thickener which are used in large quantities. The fluids capable to bring out the solid cuts by drilling bit [7].

Nevertheless, xanthan gum production has its own condition to be achieved where different sources and a few other limitations will affect the production itself. It is wide opportunity for local people as if it is found that the sources are readily available in order to produce local product as it is profitable to the nation economy. Hence the development of locally produced xanthan gum is not impossible. The research upon xanthan gum is initiate by the lab scale is necessary especially for the preliminary research where the many basic properties of xanthan gum production can be obtained [8].

Experimental

Materials, Growth media, malt extract, yeast extract, peptone, glucose and all apparatus used was provided by Universiti Teknologi MARA (UiTM) Shah Alam, Malaysia. The research has been conducted in Bioprocess Laboratory, UiTM Shah Alam and the equipment such as incubator shaker and spectrophotometer were provided by UiTM Shah Alam. Rotten cabbage samples for the research were collected locally from Cameron Highland, Pahang Darul Makmur.

Preparation of Growth Media, A media need to be prepared in order to grow the bacteria. Yeast-Malt (YM) Agar in the form of solid and Yeast-Malt (YM) Broth in the form of liquid was used for the growth. The composition of both media as follows in Table 1.

Isolation of the *Xanthomonas Campestris*, For the first series, the black rot spot were aseptically cut in small pieces and put on top of prepared YM agar. The sample were labelled and incubated for 48 h at 28 °C. After the incubation completed for 48 h, the colonies formed on the sample were observed which later inoculated into YM agar and YM broth. Both samples inoculated in YM agar and YM broth were then incubated once again for 48 h at 28 °C. For the second series, small pieces of rotten cabbage were dipped in YM Broth which then later incubated for 48 h at 28 °C. The samples were not inoculated instantly instead, 100 ml of 24 h YM Broth cultures were pippetted onto YM Agar before spread the culture thoroughly. The samples then taken for incubation for 48 h at 28 °C. After 48 h, the culture inoculated into YM Agar and YM Broth.

Preparation of Fermentation Media, YDC media prepared consist of 10 g of yeast extract, 5 g glucose and 5 g calcium carbonate. 20 g rotten cabbage was grinded and mixed with 100 L of distilled water for fermentation substrate.

Innoculum Development, Samples from 48 h YM agar plate were inoculated into 3 mL YM broth Media. The broth then will be incubated at 28 °C for 48 h. 100 μL of the previous sample cultures were pippetted into universal bottle containing 10 mL of YM broth media. The culture incubated in the incubator shaker for 48 h at 28 °C. For further bioreactor study, 40 mL of prepared culture in the universal bottle will be transferred into 400 mL flask of production media which then will be incubated in the incubator shaker at 28 °C, agitation speed of 200 rpm for 48 h.

Table 1 Composition of growth media

	YM broth	YM agar
Yeast	3 g/L	3 g/L
Malt	3 g/L	3 g/L
Peptone	5 g/L	5 g/L
Glucose	10 g/L	10 g/L
Distilled water	1 L	1 L
Agar		20 g/L

Table 2 Substrate for fermentation process

Sample	Substrate
1	YDC Media + Collection Culture
2	Grinded Rotten Cabbage
3	Grinded Rotten Cabbage + Collection Culture
4	Isolated Culture + YDC Media
5	Isolated Culture + Grinded Rotten Cabbage

Fermentation, Prior the fermentation process, fives type substrates for fermentation of xanthan gum were prepared for further fermentation. The combination of media and culture to form certain substrate can be refer on Table 2.

Result and Discussion

The effect of bacteria cell dry mass over time of fermentation, In comparison with theoretical growth of bacteria versus time, growth trends of all 5 samples (Fig. 1) satisfied with theoretical one although there are slight differences in term of the exponential phase of the bacterial growth. It takes longer time for samples to reach its maximum growth. The bacteria activities are at its peak between 32th and 36th h. After 36th h the bacteria activity start to decline due to limiting factor such as nutrients provided in the fermentation broth. Samples 3 show the best results among others.

The composition of the samples which consist of grinded rotten cabbage and collection of xanthamonas campestris culture may provide rough hypothesis why it is the best among others. Based on the trends, the usage of grinded rotten cabbage in sample 2, 3 and 5 shows that grinded rotten cabbage may be the main contributor to *Xanthamonas Campestris* cell dry mass as all three samples leave both sample 1 and 4 behind. This may not suggest that YDC media is not suitable for the growth

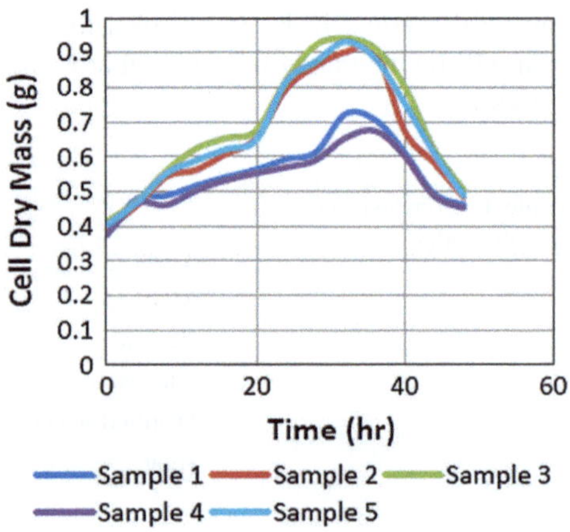

Fig. 1 Profile of cell dry mass at different substrate formula

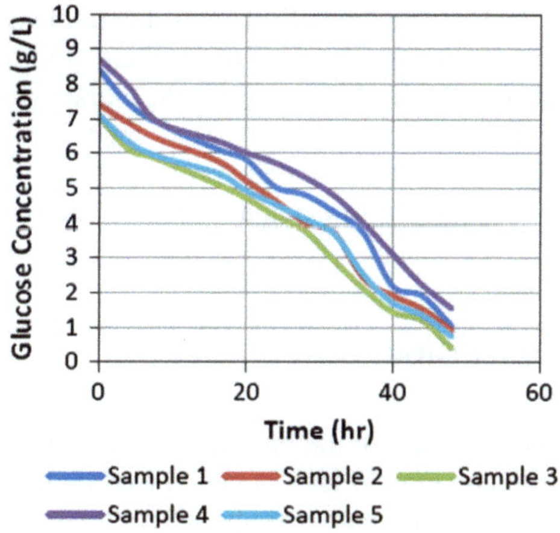

Fig. 2 Profile of xanthan gum produced at different substrate formula

of the bacteria, this is because the grinded rotten cabbage may consist of vast amount of the bacteria during the initial process of fermentation compare to substrate bases with YDC media.

The effect of glucose concentration level over time of fermentation, Glucose analyses are important to determine the activities of bacteria in fermentation broth. Theoretically, as the activity of the bacteria increase, the glucose concentration of substrate decreases. Figure 2 shows the trend of glucose content of the substrates. The trends meet the theory stated as the time passed the glucose content concentration declines. The declination of glucose content satisfied with the increasing of cell dry mass of the bacteria.

The Effect of Xanthan Gum Produced over Time of Fermentation, Based on the Fig. 3, all 5 samples shows great increasing amount of xanthan gum by the time passed up until it reaches its maximum amount of xanthan gum produced which is in between 32 and 36 h. As the time passed on 36th h, amount of xanthan gum start

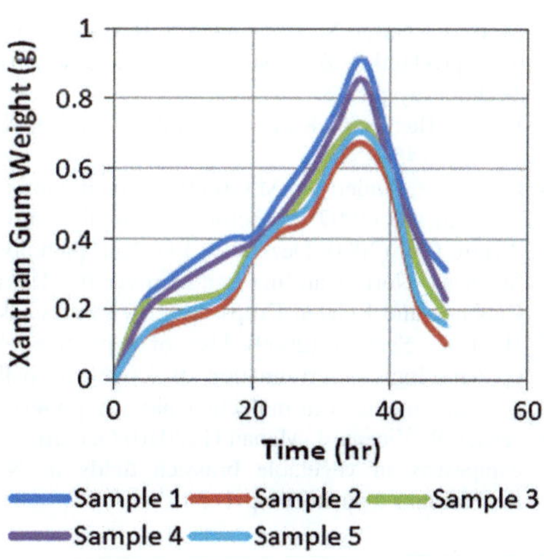

Fig. 3 Profile of glucose concentration at different substrate formula

to decline dramatically. This is due to the decreasing amount of cultured bacteria to precede the production of xanthan gum. Sample 1 show the best productivity among others while sample 2 the worst among others. The trend shows both sample 1 and 4 which bases on YDC media provide better condition for xanthan gum production. YDC media provide better condition for the fermentation process despite having low amount of bacteria as mentioned before in Fig. 2.

Conclusion

Cell dry mass of the bacteria will increase as the fermentation proceeds up until its peak condition which is called exponential phase of bacteria growth. The bacteria reach its maximum activity at 32nd to 36th h and then will decline dramatically due tu nutrient deficiency. Glucose content of the substrate trends over fermentation period provide perfect example of nutrient used up by the bacteria. Xanthan gum production increasing as the time passed up until 32nd to 36th h and the production will decreases due to insufficient amount of bacteria for further fermentation. YDC media bases substrate proved to be better condition for the xanthan gum production.

Acknowledgment The authors would like to thank the Ministry of Education Malaysia for financial support via Fundamental Research Grant Scheme (FRGS). The authors also grateful acknowledge use of services and facilities of Chemical Engineering Faculty and Universiti Teknologi MARA.

References

1. Massomoa S, Mabagalaa R, Swaib I (2003) Evaluation of varietal resistance in cabbage against the black rot pathogen, Xanthomonas campestris pv. campestris in Tanzania. Plant Pathogens 315–324
2. Moosavi- Nasab M, Pashangeh S, Rafsanjani M (2010) Effect of fermentation time on xanthan gum production from sugar beet molasses. World Academy of Science, Engineering and Technology pp 1244–1247
3. Ielpi L, Becker A, Katzen F, Puhler A (1998) Xanthan gum biosynthesis and application. Life Science 145–152
4. Ciullo PA, Andersson M (2003) Xanthan gum, a clearly better stabilizer. SÖFW J Sonderdruck
5. Katzbauer B (1997) Properties and applications of xanthan gum. Sci Biol 59:82–96
6. Averre CW (2000, December 14) Plant pathology extension, College of Agricultural and Life Sciences, North Carolina State University. Retrieved December 20, 2012, from Black Rot of Cabbage and Related Crops: http://www.ces.ncsu.edu/depts/pp/notes/oldnotes/vg16.htm
7. Hiett K, Seal B (2009) Use of repetitive element palindromic PCR (rep-PCR) for the epidemiologic discrimination of foodborne pathogens. Molecular epidemiology of microorganisms, methods in molecular biology pp 49–58
8. Jensen B, Vicente J, Manan H (2010) Occurrence and diversity of *Xanthomonas Campestris* pv. campestris in vegetable brassica fields in Nepal. Plant Disease, University of Warwick institutional repository pp 1–34

Part III
Energy and Renewable Energy

1-Pyrenebutyric Acid Functionalized Reduced Graphene Oxide (1-Pb-Rgo) Energy Storage

Ellie Yi Lih Teo, Mashitah Mohd Yusoff and Kwok Feng Chong

Abstract Supercapacitors are a class of energy storage device which has high energy density and high power density. As a material with unique 2D structure as well as outstanding physical properties such as high electrical conductivity and large surface area, graphene demonstrates great potential to be the electrode material for supercapacitors. Despite graphene showing theoretical surface area as high as 2630 m^2/g, results acquired showed that not all the surface area were utilized. This could be due to the tendency of the graphene layers to restack. In this work, 1-pyrenebutyric acid (1-PB) was anchored to graphene with the pyrenyl group via π-π stacking to prevent the restacking of graphene layers. The successful functionalization of 1-PB on the hydrophobic surface of rGO was characterized with UV-Vis Spectroscopy and Fourier Transformed Infrared Spectroscopy (FTIR). The electrochemical performance of 1-PB-rGO was studied through cyclic voltammetry (CV), galvanostatic charge-discharge (CD) and electrochemical impedance spectroscopy (EIS). Using 6 M KOH as the electrolyte, we obtained an enhanced specific capacitance for 1-PB-rGO. These findings indicates that the non-covalent functionalization of 1-PB on rGO enhances the capacitive storage ability and it show potential as an electrode material in the energy storage application.

Keywords Reduced graphene oxide · Non-covalent functionalization · Energy storage

E.Y.L. Teo (✉) · K.F. Chong
Faculty of Industrial Sciences and Technology, Universiti Malaysia Pahang, Lebuhraya Tun Razak, 26300 Gambang, Kuantan, Pahang, Malaysia
e-mail: ellie_teo@hotmail.com

K.F. Chong
e-mail: ckfeng@ump.edu.my

M.M. Yusoff
Research and Innovation Department, Universiti Malaysia Pahang, Lebuhraya Tun Razak, 26300 Gambang, Kuantan, Pahang, Malaysia
e-mail: mashitah@ump.edu.my

M.A. Hashim (ed.), *ICGSCE 2014*, DOI 10.1007/978-981-287-505-1_22

Introduction

In the modern era of technology where portable electronics are a current trend together with the depletion of fossil fuel have led to the search for a dependable and durable energy storage solution. Among the various solutions which are being explored, supercapacitor have gained the attention of the research community. The supercapacitor is a class of energy storage device which has a high energy density as compared to conventional capacitors and a higher power density compared to batteries.

Since its discovery by Andre Geim and Konstantin Novoselov, graphene has become a sensational material due to its exceptional properties. Graphene is a single layer of sp^2 bonded carbon atoms arranged in honeycomb crystal lattice. It possess exceptional properties such as very high specific surface area ($\sim$2630 m^2/g) and high electrical conductivity which fits the criteria as the electrode material of the supercapacitor [1]. This results in graphene to be highly explored in the application of energy storage.

However graphene tends to agglomerate as a result of the Van der Waals interaction. In order to prevent this, functionalization of graphene is performed by the covalent or non-covalent modification technique in order to improve their solubility, self-assembly properties and in applications [2]. The non-covalent functionalization if preferred as it does not disrupt the conjugation of the graphene sheet and improves its stability [3]. A few works have been reported on the non-covalent functionalization of graphene with pyrene derivatives. Xu et al. [2] functionalized graphene with 1-pyrenebutyric acid through a simultaneous functionalization and reduction method. Supercritical carbon dioxide was used in the research of Li et al [4] to assist in the functionalization process. While there are a few literatures which reported on the different methods of functionalization, only Ghosh et al. (2012) and An et al. (2010) reported on the application of pyrene derivatives functionalized graphene as energy storage device. They studied on the effect of 1-pyrenecarboxylic acid functionalization of graphene on its capacitive storage ability and reported specific capacitance value of 200 and 120 F/g respectively [1, 5].

In this work, 1-pyrenebutyric acid, a pyrene derivative which is reported to interact with graphite with the pyrene group via π stacking is used to functionalize reduced graphene oxide (rGO). 1-PB was anchored to the hydrophobic surface of reduced graphene oxide (rGO) through a simple sonication process at room temperature. Reduced graphene oxide was successfully functionalized with 1-pyrenebutyric acid (1-PB) with this method. The application of 1-pyrenebutyric functionalized reduced graphene oxide (1-PB-rGO) is studied electrochemically with cyclic voltammetry (CV), galvanostatic charge-discharge (CD) and electrochemical impedance spectroscopy (EIS).

Experimental

Chemicals

All chemicals used in this experiment were purchased from Sigma-aldrich unless otherwise stated and used as received.

Graphene Oxide Preparation

Briefly, 4.0 g of graphite powder was added into a mixture of concentrated sulfuric acid, H_2SO_4 (30 mL), potassium persulfate, $K_2S_2O_8$ (6.0 g) and phosphorus pentoxide, P_2O_5 (6.0 g). The mixture was heated at 80 °C and kept stirred for 6 h using oil bath. After 6 h, the solution was cooled and diluted with 2 L of deionized water. The solution was filtered using Whatman filter paper and the residue was dried naturally. The pre-treated graphite was added into concentrated H_2SO_4 (300 mL), followed by the addition of potassium permanganate, $KMnO_4$ (35 g) gradually with stirring. The mixture was stirred at 35 °C for 4 h and diluted with 2 L of water by keeping the temperature under 50 °C. Hydrogen peroxide, H_2O_2 (100 mL) was then added to the mixture drop by drop. The solution was then filtered and washed with 1:10 hydrochloric acid, HCl aqueous solution (2 L) followed by deionized water (2 L). The resulting solid was filtered and dried at room temperature.

Reduction of Graphene Oxide

60 mg graphite oxide was dispersed and sonicated in water (50 mL) for 4 h to exfoliate it. The pH of the solution was then adjusted to pH 10. Hydrazine monohydrate (1.2 mL) was added to the solution and it was refluxed for 24 h at 80 °C.

Functionalization of 1-Pyrenebutyric Acid (1-PB) on Reduced Graphene Oxide (RGO)

10 mg of 1-PB and 10 mg of rGO were dispersed in 5 mL of ethanol, sonicated for 1 h stirred mechanically overnight. The resulting solution was filtered, washed with deionized water and dried in oven at 80 °C for 3 h.

Characterization Methods

rGO, 1-PB and 1-PB-rGO were dispersed in deionized water in concentration of 0.1 mg/mL for UV-Vis measurement. The solutions were then scanned from 200 to 400 nm using Thermo Scientific Genesys 10S UV-Vis. Fourier Transform Infrared Spectroscopy (FTIR) measurements were performed by pressing the materials with potassium bromide (KBr) to form pellets. The pellets were scanned in the frequency range of 400–4000 cm^{-1} using Perkin Elmer Spectrum 100.

Electrochemical Studies

The working electrodes were fabricated by mixing the material with carbon black and polyvinylidene fluoride (PVDF) in the ratio of 80:15:5. The mixture was then dispersed in N-methylpyrrolidone (NMP) and stirred overnight. The slurry was thrush coated onto nickel foam and dried in the oven at 80 °C for 12 h. To assemble the coin cell, two identical pieces of the fabricated electrode were sandwiched with a microglass fiber separator (25 μm, Filter Fioroni) and placed inside the coin cell (CR2032, MTI) with 6 M potassium hydroxide (KOH) as the electrolyte.

Cyclic voltammetry was performed at a potential window of −1–1 V at various scan rates ranging from 100 to 5 mV/s. Galvanostatic charge-discharge measurements were performed at different current densities for potential window of 0–1 V. Electrochemical impedance spectroscopy (EIS) was carried out at open circuit potential (OCP) for the frequency range of 500 kHz to 10 mHz. The experiments were conducted at room temperature using PARSTAT 2273.

Result and Discussion

Characterization of RGO and 1-PB-RGO

Figure 1 shows the UV-Vis spectra for rGO, 1-PB and 1-PB-rGO. The spectra for rGO exhibits an absorbance band at 266 nm which is the signature peak for rGO that has a maximum at 270 nm. The presence of this peak is due to the electronic $\pi - \pi^*$ transition of the C=C bond. It indicates that the electronic conjugation within the rGO sheets were restored upon reduction with hydrazine. Meanwhile, the 1-PB-rGO spectra exhibits three absorbance bands. An obvious band at 266 nm which represents the presence of rGO and a shoulder peak at 242 nm and a weak peak at 343 nm. The shoulder peak and weak peak are contributed by 1-PB [4].

To further confirm the successful functionalization of 1-PB on rGO, FTIR measurements were also conducted. Characteristic peaks for GO are observed in

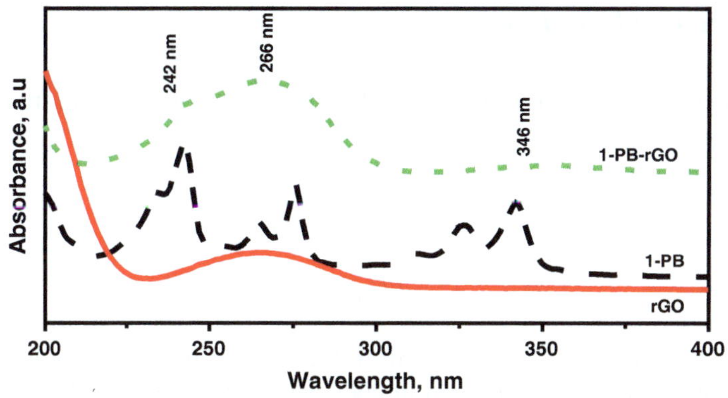

Fig. 1 UV-Vis spectra of rGO, 1-AP and 1-AP-rGO

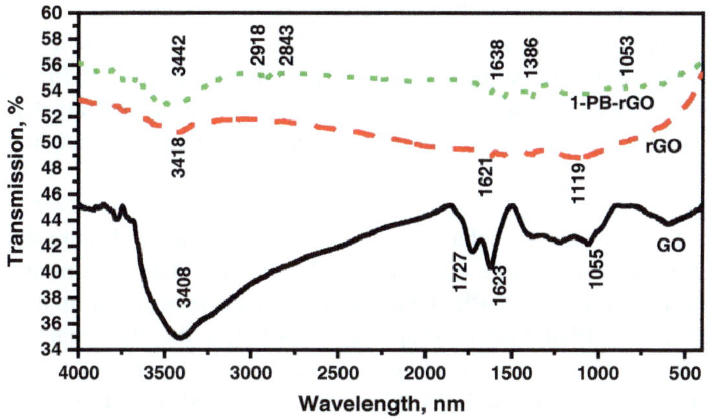

Fig. 2 FTIR spectra of GO, rGO and 1-AP-rGO

FTIR spectrum in Fig. 2 such as the peak due to broad and intense peak contributed by O–H groups at 3408 cm^{-1}, C=O group at 1727 cm^{-1} and C–O stretching at 1055 cm^{-1}. The intense peak contributed by O–H group have diminished significantly as a result of deoxygenation after reduction [6]. For the FTIR spectra of 1-PB-rGO, the signature peaks for rGO at 3442 and 1638 cm^{-1} is observed which is due to O–H group and C=O group respectively. A peak at 1053 cm^{-1} due to C–O strectching is detected too. The peaks contributed by the C–H stretching of alkane at 2918 and 2843 cm^{-1} and C–H rocking of alkane group at 1385 cm^{-1} are also observed. These peaks are most likely due to the functional groups that are present in 1-PB.

Electrochemical Studies

The electrochemical properties of rGO and 1-PB-rGO electrodes were studied using CV, CD and EIS. The specific capacitance for the materials were calculated from the discharge slope using the the following formula:

$$C_{sp}(F/g) = 4 \times C/m. \quad (1)$$

$$C = I/(dV/dt). \quad (2)$$

where C is the measured capacitance from the two-electrode system and m is the total mass of active materials present in both electrodes. The multiplier 4 is used to adjust the capacitance of the cell and the mass combined from the two electrodes to the capacitance and mass of a single electrode. dV/dt is calculated from the slope of the discharge curve [7].

Figure 3 shows the cyclic voltammogram of rGO and 1-PB-rGO at scan rate of 5 mV/s in 6 M potassium hydroxide (KOH) as the electrolyte. The cyclic voltammogram for rGO yielded an rectangular curve which suggest the idea capacitive behaviour of the rGO [8]. At the same scan rate, the CV curve for 1-PB-rGO is also rectangular in shape with no obvious redox peak. However, the larger area under the curve indicates that the 1-PB-rGO composite possess a better capacitive behaviour.

The galvanostatic charge-discharge study was conducted to observe the sustainability of the capacitive behaviour of the samples studied. Figure 4 shows the galvanostatic charge-discharge curve for rGO and 1-PB-rGO at current density of 0.1 A/g. For both rGO and 1-PB-rGO samples, it can be observed that the charge-discharge curve is triangular in shape which suggest that the charge discharge process is highly symmetrical, leading to ideal capacitive behaviour [9]. The specific capacitance for rGO and 1-PB-rGO were calculated from Eqs. 1 and 2 to be

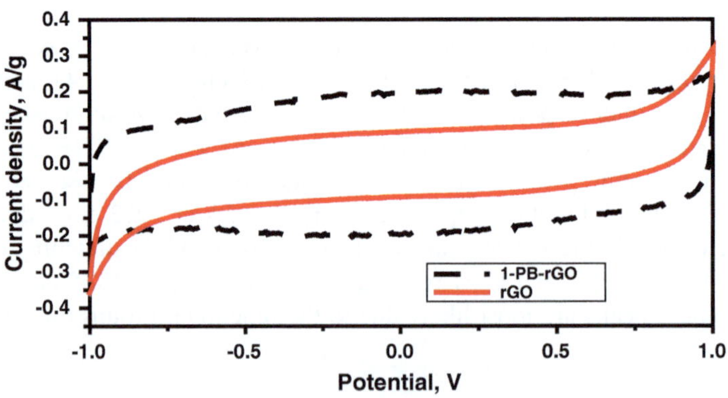

Fig. 3 Cyclic voltammetry (CV) curves for rGO and 1-PB-rGO at scan rate of 5 mV/s

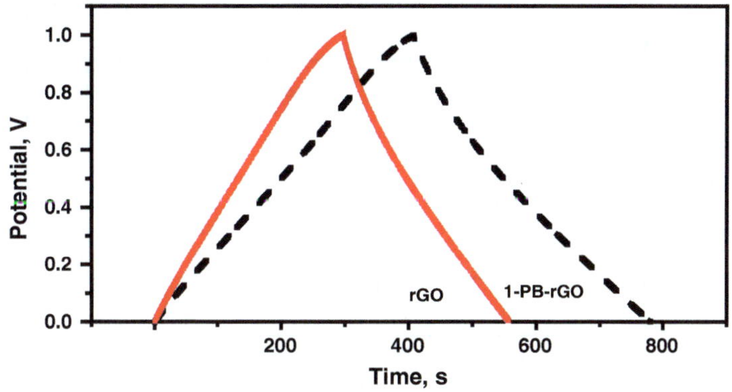

Fig. 4 Galvanostatic charge-discharge curve at current density of 0.1 A/g for rGO and 1-PB-rGO respectively

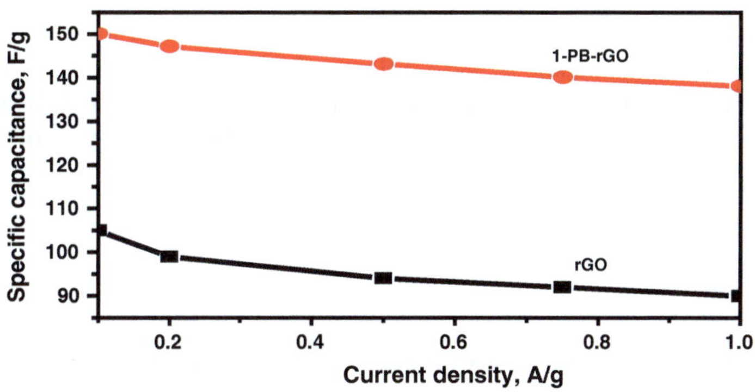

Fig. 5 Plot of specific capacitance of rGO and 1-PB-rGO electrode as a function of discharge current density

105 and 150 F/g (as shown in Fig. 5) respectively. The large enhancement of specific capacitance (approximately 45 %) observed is attributed to the intercalation of pyrene group between the rGO layers which prevented agglomeration of rGO. It is also attributed to the pseudocapacitance contributed by the –COOH functional group which is attached to the surface of the rGO (Fig. 6).

The electrochemical impedance spectroscopy (EIS) is a powerful tool used in the study of the ion transport at the interface and to probe the resistance limitations present in the supercapacitor fabricated. The intercept of the x-axis of the Nyquist plot at the high frequency end represents the solution resistance (R_s) as illustrated in Fig. 6 [10]. It can be observed from the inset illustration that the R_s is at ~0.6 ohm for rGO and ~0.5 ohm for 1-PB-rGO. This indicates that the functionalization of rGO with 1-PB improves the Rs of rGO.

This is usually followed by a semi-circular arch which is a representation of the interfacial charge transfer resistance (R_{ct}) and the interfacial capacitance (C_s) which

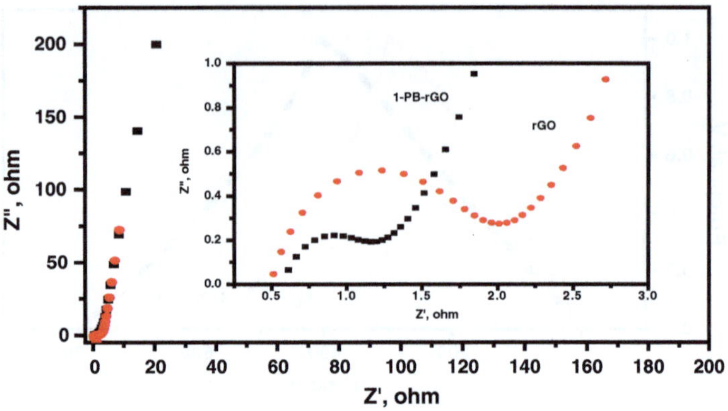

Fig. 6 Nyquist plot for 1-PB-rGO and rGO. Inset shows spectra at high frequency end

are connected at the electrode/electrolyte interface parallely. The Nyquist plots shows that R_{ct} for 1-PB-rGO is slightly larger than rGO which indicates that some of the pores could be blocked by the 1-PB particles [11]. The last part is the vertical line observed after the semi-circular arch. The slope for rGO and 1-PB-rGO is recorded to be $\sim 83°$ and $\sim 82°$ which shows that the supercapacitor functions closely to an ideal capacitor [10].

Conclusion

This work involves the non-covalent functionalization technique of rGO with 1-pyrenebutyric acid which is simple, cost effective and eco-friendly as it requires only simple sonication and no usage of strong oxidants or corrosive acids. The specific capacitance calculated to be 150 F/g which is an enhancement of 45 % compared to rGO. The enhancement is most probably due to the larger exposed surface area of the material as well as due to the attachment of –COOH functional group on the surface area of rGO. Therefore, 1-PB-rGO composite is an efficient choice of material to be use in the fabrication of electrode for energy storage application.

Acknowledgment KF Chong and co-workers would like to thank the funding of RDU 121212 and RDU 121213.

References

1. Ghosh S, An X, Shah R, Rawat D, Dave B, Kar S, Talapatra S (2012) Effect of 1-pyrenecarboxylic acid functionalization of graphene on its capacitive energy storage. J Phys Chem C 116:20688–20693
2. Xu Y, Bai H, Lu G, Li S, Shi G (2008) Flexible graphene films via the filtration of water soluble non-covalent functionalized graphene sheets. J Am Chem Soc 130:5856–5857

3. Luo L, Zhang Z, Ding Y, Deng D, Zhu X, Wang Z (2013) Label free electrochemical impedance genosensor based on 1-aminopyrene/graphene hybrids. Nanoscale 5:5833–5840
4. Li L, Zheng X, Wang J, Sun Q, Wun Q (2013) Solvent exfoliated and functionalized graphene with assistance of supercritical carbon dioxide. ACS Sustain Chem Eng 1:144–151
5. An X, Simmons T, Shah R, Wolfe C, Lewis KM, Washington M, Nayak SK, Talapatra S, Kar S (2010) Stable aqueous dispersions of noncovalently functionalized graphene from graphite and their multifunctional high-performance applications. Nano Lett 10:4295–4301
6. Choi EY, Han TH, Hong J, Kim JE, Lee SH, Kim HW, Kim SO (2010) Non-covalent functionalization of graphene with end functional polymers. J Mater Chem 20:1907–1912
7. Stoller MD, Ruoff RS (2013) Best practice methods for determining an electrode materials performance for ultracapacitor. Energy Environ Sci 3:1294–1301
8. Zhu Y, Murali S, Stoller MD, Velamakanni A, Piner RD, Ruoff RS (2010) Microwave assisted exfoliation and reduction of graphite oxide for ultracapacitors. Carbon 48:2106–2122
9. Mishra AK, Ramaprabhu S (2011) Functionalized graphene-based nanocomposites for supercapacitor application. J Phys Chem C 115:14006–14013
10. Hastak RS, Sivaraman P, Potphode DD, Shashidhara K, Samui AB (2012) All solid supercapacitor based on activated carbon and poly(2,5-benzimidazole) for high temperature application. Electrochim Acta 59:296–303
11. Su Y, Zhitomirsky I (2013) Eletrophoretic assembly of organic molecules and composites for electrochemical supercapacitors. J Colloid Interface Sci 392:247–255

Clarifying the Palm-Based Drilling Fluids Potentials

Hazlina Husin, Zulkafli Hassan, Azlinda Azizi, Mohd Fiqri Arman Mansor, Rosenadira Nasarauddin and Sharifah Nawirah Syed Ariffin

Abstract With the goal of increasing awareness of toxicity level and environmental impact of drilling fluid, and for providing continuous findings for a better drilling fluid formulation, an experimental work based on a rheological evaluation of drilling fluid was carried out. The use of palm-based and palm oil-based as an alternative base component in drilling fluid formulation have been explored in terms of sand content, pH, viscosity, yield point and gel strength. Standard laboratory formulation including distilled water, sodium hydroxide, bentonite, barite and 1,2-propylene glycol were used in this study. Tests conducted on palm-based drilling fluid with properties of 17.0 ppg and 80/20 oil-water ratio revealed that palm methyl ester C14 and palm kernel oil have sand content of less than 0.5 % vol. with excellent thermal resistance on high temperature for palm kernel oil based.

Keywords Palm oil · Palm methyl ester · Palm kernel oil · Drilling fluid · Rheological test

H. Husin (✉) · Z. Hassan · A. Azizi · M.F.A. Mansor · R. Nasarauddin · S.N. Syed Ariffin
Faculty of Chemical Engineering, Universiti Teknologi MARA, 40450 Shah Alam, Selangor Darul Ehsan, Malaysia
e-mail: hazlina858@salam.uitm.edu.my

Z. Hassan
e-mail: zulkafli1660@salam.uitm.edu.my

A. Azizi
e-mail: azlinda68@salam.uitm.edu.my

M.F.A. Mansor
e-mail: man37_jalil@yahoo.com.sg

R. Nasarauddin
e-mail: rosenadira@yahoo.com

S.N. Syed Ariffin
e-mail: shareefahnaweerah@yahoo.com

H. Husin
CoRe of Frontier Materials and Industry Applications, Universiti Teknologi MARA, 40450 Shah Alam, Selangor Darul Ehsan, Malaysia

M.A. Hashim (ed.), *ICGSCE 2014*, DOI 10.1007/978-981-287-505-1_23

Introduction

Drilling fluid or drilling mud is defined as any fluid used in a drilling operation. Drilling fluid plays a major role in drilling operation where it is pumped or circulated from the surface, down the drill string, through the bit and back to the surface [1]. The fluid is used to lift drill cuttings to the surface, control subsurface pressure, lubricate drill string, clean bottom-hole, aid formation evaluation process and protect formation. Unfortunately, the costs of drilling fluid constitute a significant fraction of the overall costs of drilling operation. In many cases, the cost ascribed to the drilling fluid is the costs associated with waste disposal and solids management. Thus, it is just as important to minimize costs associated with these two aspects as it is to ensure that the drilling fluid fulfils its primary functions.

Among the three types of drilling fluid; the oil-based, the water-based and the air/foam; oil-based drilling fluid has been the drilling fluid of choice for complex conditions such as high temperature, high angle, extended-reach well and hydratable shales. Also, oil-based drilling fluid produces low waste. However, a common base component used in the oil-based drilling fluid formulation, the diesel oil, is toxic and creates environmental problems when they are discharged [2]. According to the Effluent Limitation Guidelines (ELGs) [3, 4], a discharging law by Environmental Protection Agency (EPA), any waste produced from oil-based drilling fluid cannot be directly discharged on-site. ELGs prohibit releases of free oil, as detected by the static sheen test, from drilling fluid and drill cuttings discharges.

There have been continuous studies on the replacement of diesel oil by palm resources [5]. Study found that palm-oil has the lowest production cost from other major oils [6]. The used of palm-oil is also advocated because it is assumed to dramatically reduce CO_2 emissions [7] and may satisfy demanding technical criteria and strict environmental standards for offshore work [8]. This study aims to clarify the potentials of a few compounds from palm-oil resources at different temperatures, used as the base component in drilling fluid formulation.

Experimental

In this study, palm resources such as refined palm-oil, palm kernel oil and palm methyl esters were employed as the base fluid into standard formulation components of drilling fluid [9] as in Table 1 made of 17.0 ppg and 80/20 oil-water ratio. Palm methyl esters are made through palm oil transesterification process. The conversion of palm oil to palm methyl ester involves the use of methanol as a raw material and a basic catalyst. Conversion of palm oil compounds into smaller molecules of palm methyl esters allows them to behave similar to diesel fuel. Procedures for rheology investigation was based on API recommended standard (API-RP 13B-1 2009) [10]. The mixture was continuously stirred until a

Table 1 Standard formulation of drilling fluids [9]

Component	Quantity (g)
Distilled water	350
Calcium hydroxide	1.0
Bentonite	6.0
1,2-propylene glycol	0.7
Barite	130

homogeneous system obtained. The rheological properties of the mixture were determined.

Sand Test, 200 mesh sand screen was employed to trap the sand and other course particles in the drilling fluids. The volume of sand including the spaces between the grains is expressed as a percentage of the volume of the drilling fluid. The sand content of drilling fluid is the percentage (volume fraction) of particles of diameter larger than 74 μm.

Rheology Test, The FANN (Model 35A) viscometer was used in order to get viscosity, gel strength and yield point measurement. The FANN (Model 35A) viscometer is a coxial cylindrical rotational type, used to determine single or multiple-point viscosities. It has fixed speeds; 3 (GEL), 6, 100, 200, 300 and 600 RPM that are switch-selectable with the RPM knob. A freshly mixed sample was placed in the cup, the upper housing of the viscometer was tilted back, the cup was located under the sleeve (the pins on the bottom of the cup fit into holes in the base plate), and the upper housing was lowered to its manual position. Then, the knurled knob was turned between the rear support posts to raise or lower the rotor sleeve until it was immersed in the sample of the scribed line. After that, the sample was stirred for about 5 s at 600 RPM and then the RPM desired was selected for the best. The dial reading was waited to stabilize and then the dial reading and RPM were recorded.

Resistivity Test, 5 mL of the mixture was taken to undergo resistivity test using the FANN (Model 653 Resistivity Meter) Analog Resistivity Meter.

Result and Discussion

There are two types of oil that can be produced from palm resources. These are the palm oil which can be produced from the fibrous mesocarp of palm fruit and the lauric oil from the palm kernel. During the milling process, the palm fruit bunches are sterilised and stripped from the fruitlets. They are then digested and pressed to extract the palm oil. The nuts are separated from fibre in the press cake and cracked to obtain palm kernels which are crushed to obtain palm kernel oil and palm kernel cake. Fractionation process of palm oil and palm kernel oil produces the liquid stearin fraction and a solid stearin component. The fatty acid compositions of palm oil products are presented in Table 2. The results showed that palm oil has a

Table 2 Fatty acid composition of palm oil products, soy oil and coconut oil [11]

Fatty acid	Weight percentage			
	Palm oil	Palm kernel oil	Coconut oil	Soy oil
C6:0		0.3	0.2	
C8:0		4.4	8	
C10:0		3.7	7	
C12:0	0.2	48.3	48.2	
C14:0	1.1	15.6	18	
C16:0	44	7.8	8.5	6.5
C18:0	4.5	2	2.3	4.2
C18:1	39.2	15.1	5.7	28
C18:2	10.1	2.7	2.1	52.6
Iodine value	53.3	17.8	9.5	133

balanced ratio of saturated and unsaturated fatty acids while palm kernel oil has mainly saturated fatty acids which are broadly similar to the composition of coconut oil. When compared to soy oil, palm oil contains higher amount of saturated fatty acids. This means that it will be more stable and less prone to oxidation at high temperatures.

Table 3 shows the result for a sand content test on unconsolidated formation of palm-based drilling fluids using different types of base fluid. In order to achieve desired drilling fluid viscosity and drilling fluid weight, the sand content in the drilling fluid formulation should be controlled to be less than 0.5 % vol [12]. This is because the presence of high solid particles increases the drilling fluid viscosity and drilling fluid weight from desired values hence affects the drilling efficiency, which in turn increases maintenance costs. Furthermore, high sand content can lower the rate of penetration which in turn increases horsepower required to circulate the drilling fluid. According to Table 3, palm methyl esters C12 and C16 showed intolerable range of sand content. In contrast, palm methyl ester C14 and palm kernel oil showed a more reasonable range. Ismail and Khor studied palm methyl ester C13 and C15 and found that the preliminary result of these palm oil esters too has a potential to be used as based drilling fluid [2].

The study continues on the investigation of 2 sets of palm kernel oil-based formulation at different temperature; 25.6 °C (labelled as PKO-1) and 100.7 °C

Table 3 Percentage of sand content in drilling fluids with different base component

Base component	Sand content (% vol)
Palm kernel oil	0.50–0.80
Refined palm oil	3.50–5.00
Palm methyl ester, C12	1.00–6.00
Palm methyl ester, C14	0.25–1.50
Palm methyl ester, C16	3.50–4.25

(labelled as PKO-2). Investigation at high temperature aims to observe the probability of a thermal degradation process in the physicochemical structure of the fluid occurred at elevated temperature via rheological test. Furthermore, the measurement of rheological properties (such as viscosity) in drilling fluid is important in the development of rheological models at high temperature because these fluids have a very complex physicochemical structure which enables a non-Newtonian behaviour to be exhibited. Initially, the formulation of PKO-1 at room temperature was made for pH 9.79. Previous study has reported that the optimum range of drilling fluid pH is between 8.0 and 10.5 [13]. When temperature was increased, the pH of the palm kernel oil-based drilling fluid was found to be acidic at 100.7 °C (PKO-2). This indicates that at high temperature condition, palm kernel oil-based drilling fluid will require more alkaline components, such as soda ash or caustic soda, to elevate the pH. Alternatively, a chemical inhibitor with/without scavenging chemical may be needed. Commonly, the pH of drilling fluid is within an alkaline region so as to minimize the corrosion effect and corrosion issues on drilling equipment.

Rheology test is the basis for all analyses of wellbore hydraulics [14]. Rheological properties exert a considerable influence upon hydraulic horsepower. These include measurement of yield point, viscosity and gel strength (10′ and 10″). Measured using a viscometer [10], the results are vital in ensuring the performance of drilling fluid controls. The viscosity of PKO-2 sample showed lower reading than PKO-1. The plastic viscosity and apparent viscosity for PKO-2 is 13 cP and 14 cP respectively. This means that temperature has significant impact on the viscosities of palm kernel oil-based drilling fluid. Comparing the viscosities at these two temperatures, we found that gelation point was not observed at elevated temperature, in particular at 100.7 °C. When compared to previous studies viscosities of the investigated palm kernel oil-based drilling fluid is similar to that other reported viscosities for drilling fluid, within a range of 14–18 cP [15], and this reported value is also close to other commercial mineral oil-based drilling fluid [16]. This shows that the palm kernel oil compound used as the base component in the drilling fluid is compatible to the commercially systems in the sense of molecular stability at elevated temperature condition.

Yield point is defined as the initial resistance to flow. Drilling fluid with high yield point give better performance in carrying drill cuttings as compared to low yield point of the same drilling fluid [17]. Palm kernel oil-based drilling fluids (PKO-1 and PKO-2) demonstrated similar yield point value as 10.0 ppg palm methyl ester-based drilling fluids [2]. Yield point of drilling fluid can be increased by adding suitable gelling agent. Gel strength measured in 10 s–10 min period represents the gelling effect of drilling fluid in static condition. The gel building up property is vital in order to suspend the drill cuttings when drilling operations are temporarily ceased. Increment of the gel strengths (as shown in Table 4) with time confirms the thixotropic property of the palm kernel oil-based drilling fluid. The PKO systems show an acceptable gelling effect. Gel strength for commercial mineral oil-based [16] and palm methyl ester-based (C13 and C15) [2] drilling fluid was reported to be 9/14 and 9/12 respectively. Gelling effect can be improved by

Table 4 pH, rheology and resistivity tests

Sample	pH	Plastic viscosity (cP)	Apparent viscosity (cP)	Yield point (lb/100 ft^2)	Gel strength-10 s (lb/100 ft^2)	Gel strength-10 min (lb/100 ft^2)	Resistivity (Ohm)
PKO-1	9.79	59	62.5	7	9	12	12
PKO-2	5.69	13	14	2	9	16	12

increasing water content which behaves as pseudo particles and enhances suspension capability of the system [2].

Resistivity is a characteristic electrical property of a material and is equal to the electrical resistance of a 1 m^3 of the material to passage of a 1 A electric current perpendicular to two parallel faces [18]. Solid materials exhibit intermediate electrical properties depending on their physical and chemical properties. Control of the resistivity of a drilling fluid is desirable in order to get better evaluation of formation characteristics from an electric log. Resistivity of drilling fluid will be higher if water disperses well in oil phase. It means that the mixture will produce a good emulsion. In contrast, when water disperses hardly in oil phase the resistivity of drilling fluid will be lower since the mixture formed bad emulsion. The stability of emulsion of synthetic-base drilling fluid is reflected in the value of electrical stability. Since the 2 samples (PKO-1 and PKO-2) have the same formulation, there was no change observed. The resistivity of the drilling fluids observed was 12Ω for both low and high temperatures. This means that temperature has no effect on the emulsion property of the palm kernel oil-based system.

Conclusion

Up to date, drilling fluid manufacturers are moving towards a diverse range of formulation for a variety field condition in achieving their goal of efficient drilling operation. This includes the formulation from palm resources, in particular palm kernel oil. This study concluded that palm-based drilling fluid has a potential to be used as based drilling fluid. Besides palm kernel oil-based drilling fluid is robust at high temperature, usage of palm kernel oil will also allow manufacturers to develop value-added drilling fluid formulations that meet the strongly increasing market demand for environmental-friendly and sustainable sourcing of raw materials.

Acknowledgement The authors would like to thank Research Management Institute (RMI), Universiti Teknologi MARA for the Excellent Fund (600-RMI/ST/DANA 5/3/Dst (57/2009)) for funding the research project. Special grateful to the Ministry of Education, Malaysia (600-RMI/ERGS 5/3 (30/2013)) for providing financial support. The authors wish to thank the referee for making this a better paper.

References

1. Growcock F, Harvey T (2005) Drilling fluids processing handbook. Elsevier Inc., Oxford
2. Ismail AR, Khor SF (2001) Feasibility study of palm oil esters as based fluid in drilling operation. In: Regional conference for young scientists, Skudai, Malaysia, pp 23–28
3. Burke CJ, Veil JA (1995) Potential environmental benefits from regulatory consideration of synthetic drilling muds. In: Environmental assessment division. Argonne National Laboratory, Illinois
4. Veil JA, Burke CJ, Moses DO (1995) Synthetic drilling fluids—a pollution prevention opportunity for the oil and gas industry. In: Annual conference and exposition of the water environment federation, Argonne National Laboratory, Illinois, Miami Beach, Florida
5. Salleh MK, von Tapavicza S (2005) Palm oil derived esters—an environmentally safe drilling fluid. Oil Palm Ind Econ J (5)
6. Corley RHV (2009) How much palm oil do we need? Environ Sci Policy 1:134–139
7. Reijnders L, Huijbregts MAJ (2008) Palm oil and the emission of carbon-based greenhouse gases. J Clean Prod 16:477–482
8. Tapavicza SV (2005) Special report: vegetable esters make drilling fluids more environmentally friendly. In: Cognis Deutschland GmbH & Co. KG, Düsseldorf, Germany, 2005
9. Al-Bagoury M, Steele CD (2012) A new, alternative weight material for drilling fluids. In: IADC/SPE drilling conference and exhibition, San Diego, California, USA
10. American Petroleum Institute (2009) Recommended practice for field testing water-based drilling fluids. In: ANSI/API recommended practice RP 13B-1, 4th edn. American Petroleum Institute
11. Ahmad S (2000) Non-food uses of palm oil and palm kernel oil. MPOPC Palm Oil Inf Ser. MPOB, Kuala Lumpur, p 24
12. American Petroleum Institute (1996) Recommended practice for drilling fluid processing systems evaluation. In: API recommended practice 13C, 2nd edn. American Petroleum Institute
13. Amani M, Al-Jubouri M, Shadravan A (2012) Comparative study of using oil-based mud versus water-based mud in HPHT fields. Adv Pet Explor Dev 4:18–27
14. Azar JJ, Samuel GR (2007) Drilling engineering. PennWell Corporation, Oklahoma
15. Azizi A, Ibrahim MSN, Hamid KHK, Sauki A, Ghazali NA, Mohd TAT (2013) Agarwood waste as a new fluid loss control agent in water-based drilling fluid. Int J Sci Eng 5(2):101–105
16. Ismail AR (2001) Managing the environmental friendly drilling fluids in petroleum industries. In: The 2nd international conference on disaster management, Surabaya, Indonesia
17. Amani M, Al-Jubouri M, Shadravan A (2012) Comparative study of using oil-based mud versus water-based mud in HPHT fields. Adv Pet Explor Dev 4(2):18–27
18. A.S.S. Committee (2005) Drilling fluid processing handbook. Gulf Professional Publishing, Oxford

The Study of Temperature Profile and Syngas Flare in Co-gasification of Biomass Feedstock in Throated Downdraft Gasifier

Muddasser Inayat, Shaharin A. Sulaiman, Aini Abd Jamil, Fiseha M. Guangul and Samson M. Atnaw

Abstract Biomass gasification is a common technology, which converted solid biomass into gaseous fuel at high temperature reactions in the presence of gasification agent. In this paper, co-gasification of lignocellulosic biomass materials with oil palm fronds (OPF) in a downdraft gasifier is presented. The biomass feedstocks considered were sugar cane bagasse (SCB) and wood (*acacia mangium*). Only one material was co-gasified with OPF at a time, with blending ratios of 80:20, 50:50 and 20:80. The resulting temperature profiles in the reactor and the syngas flare duration were recorded. It was found that the blend of 80:20 wood and OPF gave the best result as it produced the longest steady flare duration (49.5 min). On the other hand, a significant bridging problem was observed in the co-gasification OPF and SCB, and thus implying the need for process improvement.

Keywords Co-gasification · Downdraft gasifier · Blending ratio · Temperature profile

M. Inayat (✉) · S.A. Sulaiman · A. Abd Jamil · F.M. Guangul
Department of Mechanical Engineering, Universiti Teknologi PETRONAS, 32610 Bandar Seri Iskandar, Perak Darul Ridzuan, Malaysia
e-mail: muddasser_engr@yahoo.com

S.A. Sulaiman
e-mail: shaharin@petronas.com.my

A. Abd Jamil
e-mail: ainiabdjamil@gmail.com

F.M. Guangul
e-mail: fiseha.mekonnen62@gmail.com

S.M. Atnaw
Faculty of Technology, Universiti Malaysia Pahang, Lebuhraya Tun Razak, 23600 Kuantan, Pahang, Malaysia
e-mail: mekbibsams@ump.edu.my

M.A. Hashim (ed.), *ICGSCE 2014*, DOI 10.1007/978-981-287-505-1_24

Introduction

Fossil fuel reserves are depleting due to growth in population and human activities. According to a forecast by the International Energy Agency (IEA), the demand for energy by 2050 is expected to grow by three folds as compared to the current consumption [1]. As a result, there is need to generate power from alternative sources such as biomass [2]. Malaysia is bestowed with a huge quantity of biomass resources due to its location near the equator with high precipitation rate and long sunny days throughout the year. This makes biomass the most prominent source of energy among other renewable sources in Malaysia. Forestry wood, oil palm industry, and rubber cultivation are the main sources of the biomass waste stream in Malaysia [3]. Although utilization of biomass as sources of renewable energy have a many benefits, it has also several issues such as interruption and discontinuity in feedstock supply [4]. The supply of some biomass materials can be affected by natural disasters such as flood and heavy rain. These aforementioned issues could be addressed through co-gasification.

The performance of gasification and quality of syngas depends upon numerous factors such as gasifying medium (air, steam or O_2), heating rate, temperature and equivalence ratio. In addition, the characteristics of the feedstock, i.e., the elemental composition, heating value, fixed carbon, volatile matters, moisture content and ash content would determine the performance of gasification [5]. In gasification, the temperature is a key factor that directly affects syngas composition [6]. High gasification temperature favors increase in gas yield due to a high decomposition of cellulose and hemicellulose. Steam reforming and water-gas reaction favor higher H_2 concentration and lower CH_4 concentration in the syngas for gasification temperature of between 750 °C and 800 °C. At a higher temperature between 850 °C and 900 °C, boudouard reaction favors higher CO concentration in the syngas [7].

The aim of this paper is to study the performance of co-gasifying different biomass feedstock, namely wood, oil palm fronds and sugar cane bagasse. The study involved measurements of the temperature profile in the reactor and nature of the flare resulted from the syngas produced.

Characterization of Feedstock

Oil palm fronds (OPF), wood and sugar cane bagasse (SCB) were selected as feedstock in the study. The oil palm fronds were collected from the FELCRA palm oil plantation in Bota Kanan, approximately 220 km to the Northwest of Kuala Lumpur. Wood (acacia mangium) was collected from the landscape area within the campus of Universiti Teknologi PETRONAS Malaysia. SCB was collected from small scale sugar cane juicer hawker nearby the campus area.

The bagasse was chopped to the length of 25–40 mm, while, OPF and wood were cut into small blocks with a maximum size of 25 mm. Feedstock was dried in

Table 1 Characterization results of wood, OPF and SCB

Biomass	Ultimate analysis (wt%)					Proximate analysis (wt% on dry basis)				HHV kJ/kg
	C	H	O[a]	N	S	MC	VM	FC	Ash	
Wood	43.54	3.59	51.70	1.00	0.16	4.25	88.07	10.77	1.16	17,526
OPF	42.57	5.71	51.02	0.42	0.29	6.15	80.55	16.43	3.02	17,045
SCB	42.02	4.01	51.95	1.28	0.74	20.67	65.30	11.35	2.69	17,486

[a]By difference basis. *MC* Moisture content; *VM* Volatile matters. *FC* Fixed carbon. *HHV* Higher heating values

an oven at 105 °C for 24 h. The maximum moisture content was 12 % on wet basis. Ground 250 μm sieved feedstock samples were used for characterization tests. A Leco CHNS-932 analyzer was used to determine carbon, hydrogen, nitrogen and sulfur contents of the feedstock according to ASTM D3176-89 standard [8]. The proximate analysis, for determination of moisture content, volatile matter, fixed carbon and ash, was performed according to ASTM E1131-98 standard test method by using STA 6000 TGA analyzer [9]. For measurements of higher heating value (HHV), a Leco AC-350 bomb calorimeter was used. The calorific value was measured through measurement of the heat release after combustion of the samples. Characterization results of wood, OPF and SCB are presented in Table 1.

Experimental Set-Up

The experimental set-up is shown in Fig. 1. For temperature measurements, thermocouples T_1–T_7 were mounted as shown in Fig. 1, and were recorded every 30 s by the data logger and computer. A pre-mixed blend of feedstock was fed into the gasifier through feeding holes. Ignition was started through the ignition hole by using scrap paper. After steady combustion was achieved, all lids were closed and air was supplied by blower at 400 L/min. The air flow rate was regulated and measured by using a valve and a rotameter. The syngas was ignited at the flare point. Char and ash were collected from the ash box and on the grate after the process was ended and were then weighed.

Results and Discussion

The duration of the process to fully consume 12.8 kg blend of OPF (80 %) and wood (20 %) was 90 min. The temperature variation with time for different zones inside the gasifier is shown in Fig. 2. The flare was obtained only after 34 min from the start time of the experiment, and the combustion zone temperature was recorded to be 504 °C. Continuous and stable flare was obtained for about 44 min. The maximum

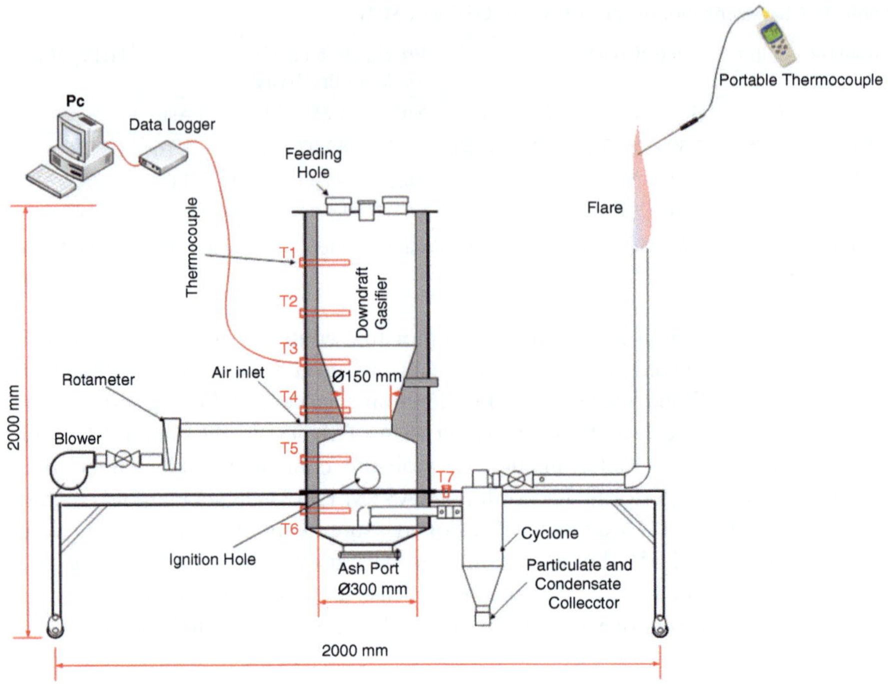

Fig. 1 Experimental set-up

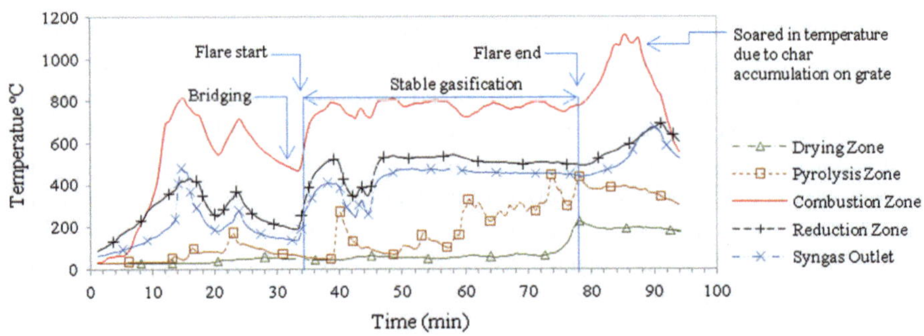

Fig. 2 Temperature profile for 80 % OPF and 20 % wood with time

temperature recorded at the flare point was 719 °C. A bright yellowish flare was obtained at the start of gasification due to the excess of hydrocarbon contents present in the syngas. However, the color of flare changed to light yellowish near the end of gasification. It was noted that, during the last 10 min of gasification, the temperature of the combustion zone increased to a peak value of 1112 °C. The significant increase in combustion zone temperature near the end of the operation is attributed to the

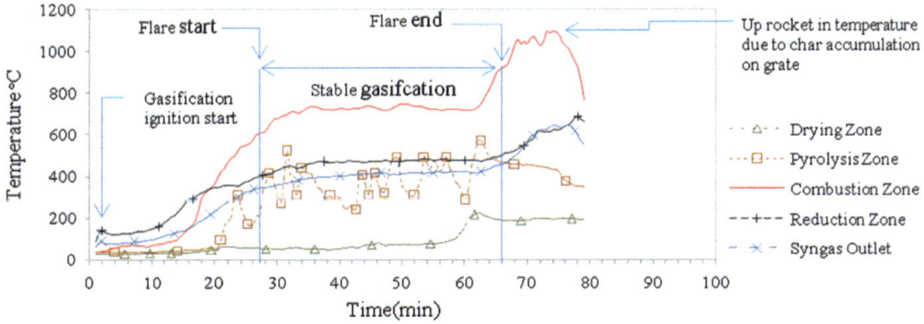

Fig. 3 Temperature profile for co-gasification of 50:50 OPF and wood blend with time

batch fed nature of the gasification process, whereby most part of the volatile matter components will be consumed during the steady operation duration, leaving combustion to be the only dominant operation for the last few minutes of the operation, causing the increase in temperature. The average temperatures in the combustion and reduction zones were 766 and 545 °C, respectively, during 34–78 min (marked by region between vertical lines) of gasification. The average temperature of syngas at the outlet was 323 °C. Similar to that reported by Guangul et al. [6] bridging was observed at start of experiment, probably due to the fibrous nature of OPF feedstock and this was overcome by stirring the feedstock inside the gasifier as soon as a drop in temperature in the reactor was detected.

The temperature variation for different zones with time for co-gasification of OPF (50 %) and wood (50 %) is shown in Fig. 3. The experiment took 79 min, which was shorter than that, for the mixture of 80 % OPF and 20 % wood. The syngas could ignite after 27 min from the start of the experiment and combustion zone temperature was reached 609 °C. The resulting flare was stable for 39 min. The maximum temperature of the flare at top was 744 °C. The flare was bright yellowish at the start of the process, changed to light yellow-bluish towards the end of the experiment. The average temperatures of the combustion and reduction zones were 732 and 469 °C, respectively during 27–78 min of gasification. The maximum combustion zone temperature was 1096 °C and sustained for 10.5 min of the experiment. The significant increase in combustion zone temperature near the end of the operation was similar to the Fig. 2. The average temperature of syngas outlet was 272 °C. Different that the experiment of OPF (80 %) and wood (20 %), no bridging problem was observed during the co-gasification of 50 % OPF and 50 % wood mixture.

The temperature variation for different zones of the gasifier with time for co-gasification of 20:80 OPF and wood is shown in Fig. 4. The experiment took 94 min, which was longer than that all blends. The flare could ignite after 28 min of experiment start and sustained for 49.5 min, average combustion zone temperature at instant was recorded as 741 °C. The maximum temperature recorded at the top of flare point was 719 °C. The colour of flare was bright yellow at the start of gasification, however, it changed to light yellowish near the end of gasification. The maximum combustion zone temperature was 1078 °C and sustained for 10 min near

Table 2 Pollutant emissions based on various energy sources [4, 14]

Energy sources	Percentage in electricity generation mix on 2012 (%)	Emission factor (kg/kWh)		
		CO_2	SO_2	NO_x
Coal	38.9	1.18	0.0139	0.0052
Crude oil	1.0	0.85	0.0164	0.0025
Gas	52.7	0.53	0.0005	0.0009
Hydro	7.3	0.00	0.0000	0.0000
others	0.2	0.00	0.0000	0.0000

$$\text{For } DR_d > SR_{\min}, ER_d = DR_d$$
$$AR = \sum_{d=1}^{Y} ER_d \tag{4}$$

where DR_d = solar radiation on day d, kWh/m^2; ER_d = effective solar radiation on day d, kWh/m^2; AR = annual effective solar radiation, kWh/m^2.

Table 2 shows the emission factors of different energy sources and the percentage of energy source in electricity generation mix on 2012. The emissions of pollutant from the combustion of fossil fuels for electricity generation are dependent on the characteristics of the energy source. Different source has different contents of carbon, sulphur and nitrogen or their compounds [1].

Benefit Analysis of SWH, The benefit of SWH was calculated based on the solar radiation in Malaysia which includes cost savings by replacing energy and pollution mitigation.

The cost saving by replacing energy can be estimated by the amount of energy generated by SWH under annual ER and the price of replaced energy. The replaced energy is electricity with unit price of Ringgit Malaysia (RM) 0.334/kWh [15]. The cost saving by replacing energy was calculated by Eqs. 5 and 6:

$$BE = AR \times A \times \gamma_{SWH} \tag{5}$$

$$CBE = BE \times P \tag{6}$$

where BE = total energy reduction, kWh; CBE = cost saving by replacing energy, RM; P = unit price of replaced energy, RM/kWh.

The cost saving from pollution mitigation can be estimated by reduction of GHGs and air pollutants emitted quantities once using SWH and the price of treating those pollutants. Pan et al. [13] reported that the prices for treating 1000 kg of CO_2, NO_x and SO_x are approximately RM 80, RM 2698.5 and RM 2624.2, respectively. The cost saving by pollution mitigation was calculated by Eqs. 7–8:

$$BP_{p,f} = BE \times EF_{p,f} \times PF_{p,f} \tag{7}$$

$$CBP_f = \sum\nolimits_p (BP_{p,f} \times TE_p) \tag{8}$$

where EF_{pf} = emission factor of pollutant p by energy source f, kg/kWh; PE_f = percentage of energy source f in mix; BP_{pf} = amount of GHGs and air pollutant p that reduces (e.g. CO_2, NO_x and SO_x) when SWH avoids using electricity produced by energy source f, kg; CBP_f = cost avoided to treat GHGs and air pollutants emissions when SWH avoids using electricity produced by energy source f, RM; TE_p = unit price for treating pollutant p, RM/kg. Both EF_{pf} and PE_p values are presented in Table 2.

Results and Discussion

The daily solar radiation data of study area was collected from National Aeronautics and Space Administration (NASA) climate internet database [20]. The variability of data was measured by coefficient of variation (CV), as in Eq. 1. If CV is low (<1), the data have low variability and vice versa. From the data analysis, the annual solar radiations at the study area have low CV (0.18). Thus, the daily solar radiations during the year have low variability which indicates that the applicability of SWH in Malaysia is not affected by seasonal solar radiation variation.

Benefit analysis, The benefit of using a SWH was analysed based on amount of energy that generated under solar radiation includes cost savings by replacing energy and pollution mitigation. The annual ER of the five study areas is roughly 60 % of its annual solar radiation, which means that only 60 % of the annual solar radiation is effectively used by SWH to produce hot water. This is due to Malaysia, which being hot and humid, has high tap water temperature throughout the year and resulted in low SR_{min}. Besides, the average daily solar radiation in Malaysia is 5.0 kWh/m^2, which is higher than SR_{min}. This indicated that most of the daily solar radiation in Malaysia exceeds SR_{min}, as shown in Fig. 1.

Cost saving by replacing energy, Cost saving by replacing energy was determined via Eq. 6 [13], based on the amount of energy generated under annual ER once SWH replace electricity with solar energy to produce hot water. Based on

Fig. 1 Daily solar radiation of study areas

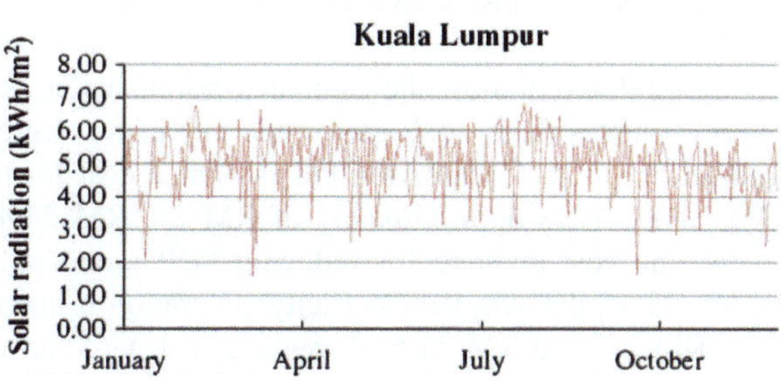

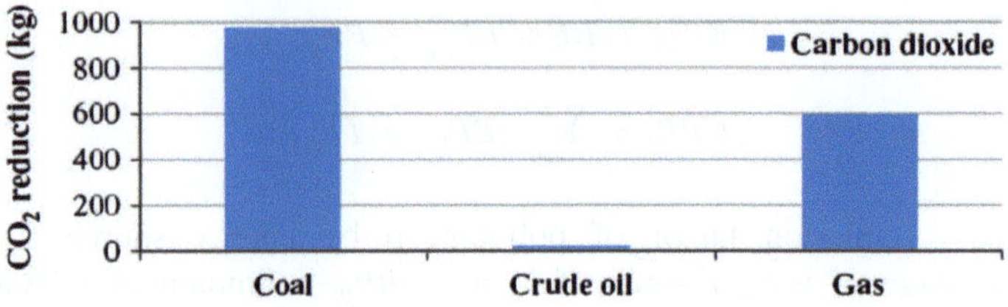

Fig. 2 Carbon dioxide reduction in KL region

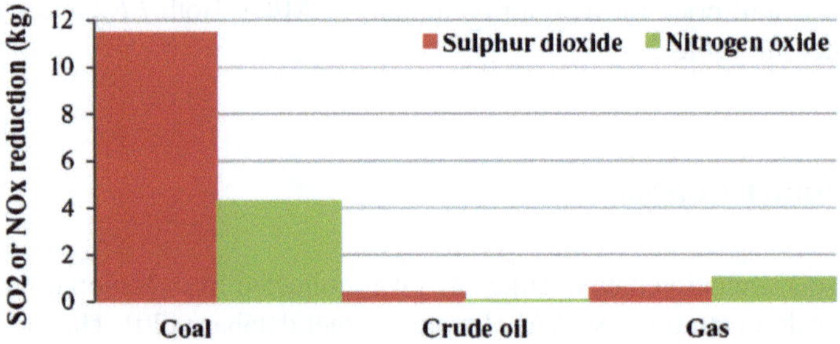

Fig. 3 Sulphur dioxide and nitrogen oxide reduction in KL region

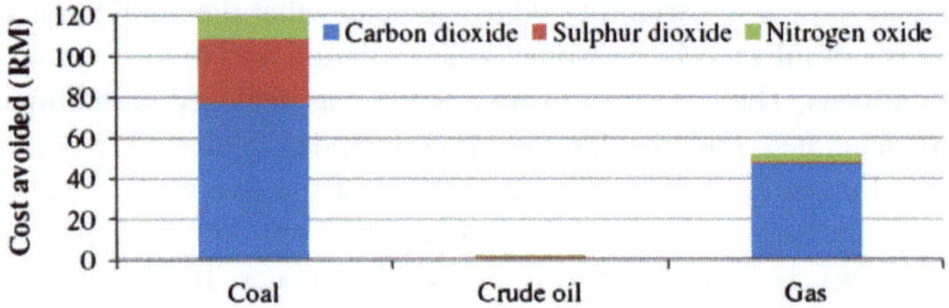

Fig. 4 Cost avoided for reducing GHGs and air pollutants emission in KL region

calculation, a typical SWH can save an average RM 708.3 of energy cost per year, which equivalent to RM 59 per month.

Cost saving by pollution mitigation, Cost saving by pollution mitigation, based on the reduction in amount of GHGs and air pollutants when using SWH and the price of treating those GHGs and air pollutants was estimated based on Eq. 8 and Table 2.

Figure 2 and 3 show the estimated pollutants emissions reduction from various energy sources (e.g., coal, gas, crude oil) in KL region, once the SWH employed instead of EWH. Considering the estimated cost for treating 1 metric ton of CO_2, NO_x and SO_2 as approximately of RM 80, RM 2698.5 and RM 2624.2, respectively, the cost avoided for reducing GHGs and pollutants emissions were estimated. Thus, the avoided electricity generation from coal, oil and gas can save RM 120, RM 2.5 and RM 51.5 of treating cost per year as shown in Fig. 4. Although

CO_2 has much lower unit treating cost than those of other air pollutants, however for the three main energy sources, the cost avoided for reducing CO2 was the highest. This is because the reduction in CO_2 emission is the highest among all air pollutants when replacing electricity with solar energy by the utilization of SWH.

Conclusion

In this study, financial evaluation of SWH in Malaysia was carried out which take into account the SWH's benefits. By applying SWH, the use of electricity for heating water can reduce, consequently lower the electricity bill and reduce the GHGs and pollutant emissions, produced by electricity generation of energy saving. It was found that the installation of a SWH can save an average 2120.70 kWh of electricity which equivalent to RM 708.3 of energy cost and avoid an average 1583 kg CO_2, 12 kg SO_2 and 5 kg NO_x emissions. It can be concluded that SWH is feasible to be used in Malaysia. However, To promote the widespread uses of SWH in Malaysia, the attentions from both government and public are required such as introduce policies by government that encourage people to use SWH. For example, government should provide subsidies to SWH's owner. Beside, a media seminar can also conduct to increase the awareness and understanding of public towards the benefits of SWH.

References

1. Saidur R, Masjuki HH, Jamaluddin MY, Ahmed S (2007) Energy and associated greenhouse gas emissions from household appliances in Malaysia. Energy Policy 35:1648–1657
2. Shafie SM, Mahlia TMI, Masjuki HH, Andriyana A (2011) Current energy usage and sustainable energy in Malaysia: a review. Renew Sustain Energy Rev 15:4370–4377
3. Ali R, Daut I, Taib S (2012) A review on existing and future energy sources for electrical power generation in Malaysia. Renew Sustain Energy 16:4047–4055
4. EPU (Economic Planning Unit) (2013) The Malaysian economy in figures at http://www.epu.gov.my/documents/10124/72ac36d7-fe5a-489b-a34c-a2cb2be073a6
5. Kaldellis JK, El-Samani K, Koronakis P (2005) Feasibility analysis of domestic solar water heating systems in Greece. Renew Energy 30:659–682
6. Kalogirou SA (2004) Environmental benefits of domestic solar energy systems. Energy Convers Manag 45:3075–3092
7. Al-Badi AH, Albadi MH (2012) Domestic solar water heating system in Oman: Current status and future prospects. Renew Sustain Energy Rev 16:5727–5731
8. Mekhilef S, Safari A, Mustaffa WES, Saidur R, Omar R, Younis MAA (2012) Solar energy in Malaysia: current state and prospects. Renew Sustain Energy Rev 16:386–396
9. Ibrahim A, Jin GL, Daghigh R, Salleh MH, Othman MY, Ruslan MH (2009) Hybrid photovoltaic thermal (PV/T) air and water based solar collectors suitable for building integrated applications. Am J Environ Sci 5(5):618–624
10. Chua SC, Oh TH (2012) Solar energy outlook in Malaysia. Renew Sustain Energy Rev 16:564–574

11. Ong HC, Mahlia TMI, Masjuki HH (2011) A review on energy scenario and sustainable energy in Malaysia. Renew Sustain Energy Rev 15:639–647
12. NASA (National Aeronautics and Space Administration) (2013) NASA prediction of worldwide energy resource (POWER) near real-time daily global radiation and meteorology. http://power.larc.nasa.gov/cgibin/cgiwrap/solar/timeseries.cgi
13. Pan TC, Kao JJ, Wong CP (2012) Effective solar radiation based benefit and cost analyses for solar water heater development in Taiwan. Renew Sustain Energy Rev 16:1874–1882
14. Mahlia TMI (2002) Emissions from electricity generation in Malaysia. Renew Energy 27:293–300
15. Tenaga Nasional Berhad (2013) Pricing and Tariff. http://www.tnb.com.my/residential/pricing-and-tariff.html

Part IV
Process Control, Modelling, Simulations, Data Mining

Introducing Pandanaceae Via IOS-Based Software Application

Hannis Fadzillah Mohsin, Ibtisam Abdul Wahab and Galoh Rashidah Haron

Abstract Pandanaceae is an application, based on the iPhone Operating System (iOS) that introduces the screwpines or *pokok pandan*, within a hand click. It describes both medicinal and traditional uses, chemical constituents, scientific references and geodata of this Asia Pacific plant species. For the purpose of the application prototype, the user is introduced to eight selected plants within the *Pandanus* genus of the Pandanaceae plant family. Fascinating pictorial views of the whole *Pandanus* plant and the major plant parts are not to be missed. From the users' perspective, they can select the image of one specific *Pandanus* species and the scientific information on the biological and pharmacological properties of that particular plant will be displayed. The researchers' years of experiences on sampling the *Pandanus* and the laboratorial work can be shared via this software, interactively. The molecular structures of the chemical ingredients of the *Pandanus* extracts are also shown. Consequently, *Pandanus* research profiling can be concluded; which touches on the scientists who are conducting experiments on this plant and their area of studies. Owing to the lack of information regarding Pandanaceae in a mobile application format, the concept is presently materialized.

Keywords Iphone · Pandanaceae

H.F. Mohsin (✉) · I. Abdul Wahab
Faculty of Pharmacy, UiTM, 42300 Puncak Alam, Selangor Darul Ehsan, Malaysia
e-mail: hannis@puncakalam.uitm.edu.my

I. Abdul Wahab
e-mail: ibtisam@puncakalam.uitm.edu.my

G.R. Haron
MIMOS Berhad, Technology Park Malaysia, 57000 Kuala Lumpur, Malaysia
e-mail: rashidah@mimos.my

M.A. Hashim (ed.), *ICGSCE 2014*, DOI 10.1007/978-981-287-505-1_26

Introduction

Pandanaceae is a plant family, which consists of more than six hundreds species. It is also selected as a product label, given to a software application, which is designed based on the iPhone Operating System (iOS). This application firstly introduces a limited number of the most common pandan species (also known as the screwpines) to the natural product lovers, within a hand click. It also describes both medicinal and traditional uses, chemical constituents, scientific references and geodata (geographical locations) of this Asia Pacific plant species. The user of this application is anyone who has the interest in learning such local plant. From a curiosity element, a user can join the group of researchers and share the knowledge. Without having a book, a digital library concerning pandan's facts and images are reachable. Comparable applications on other plant herbs could also be witnessed [1–3], and now, local researchers' years of experiences on sampling the *Pandanus* and experimental work in the laboratory [4] can be shared via this application. The usability of this prototype is currently described.

Experimental

The Pandanaceae application is developed based on the iOS, using Xcode software (version 5.0), on a MacBook Pro, a Macintosh portable computer. For the purpose of the application prototype, the user is introduced to eight selected plants within the same *Pandanus* genus of the Pandanaceae plant family (Table 1). The selection criteria are based on the most studied samples by local researcher [5, 6]. The *Pandanus* plants are available locally; it can be definitely witnessed in the botanical gardens, university campus and also planted as a landscaper along the roads. They are ranging from a miniature species to a larger one, which can grow up to fifteen meters in height. It is noteworthy that certain *Pandanus* are synonymous. For example, *Pandanus amaryllifolius* is also botanically recognized as *Pandanus odorus* [7].

Table 1 The selected *Pandanus* plant for the prototype

No.	*Pandanus* sample	Local name
1	*Pandanus pygmaeus*	Yellow pandan
2	*Pandanus sanderi*	Unknown
3	*Pandanus tectorius*	Unknown
4	*Pandanus amaryllifolius*	Fragrant pandan
5	*Pandanus dubius*	Mengkuang
6	*Pandanus conoideus*	Red pandan fruit
7	*Pandanus leram*	Nicobar breadfruit
8	*Pandanus odoratissimus*	Unknown
9	*Unidentified Pandanus*	Unknown

Fig. 1 The photograph of a *Pandanus* sample could be magnified, once the upmost left tile, is selected (*left*). The ninth tiles are dedicated for a collection of unknown *Pandanus* (*right*)

Result and Discussion

Fascinating pictorial views of the whole *Pandanus* plant and the major plant parts are successfully included (refer to Fig. 1). The screenshots display the characteristic four-fold leaves of this plant. In addition, the stilt roots are also the major botanical features for this species. Some of them display variegated leaves of white, yellow and green stripes. Furthermore, some photographs of unknown wild *Pandanus* are also included in this application (Fig. 1), in a hope to cultivate world web discussions for the purpose of the plant identification. From the users' perspective, they can select the image of one specific *Pandanus* species and the information on the biological and pharmacological properties of that particular plant will be displayed. The user will have a flow of information of the species, from the root up to the leaves of the species, interactively.

The molecular structures of the chemical ingredients of the *Pandanus* extracts are also shown (refer to Fig. 2). Miscellaneous information that can be reached includes the scientific evidences such as journals and other publications of the biopharmaceutical properties of *Pandanus*. Consequently, *Pandanus* research profiling can be concluded; which touches on the scientists who are conducting experiments on this plant and their area of studies. A wide range of investigation is involved, that includes *Pandanus*' conservation proposals, taxonomic research, ethnobotany and social studies, phytochemical and chemosynthetic work, mycological and zoological observations of *Pandanus*.

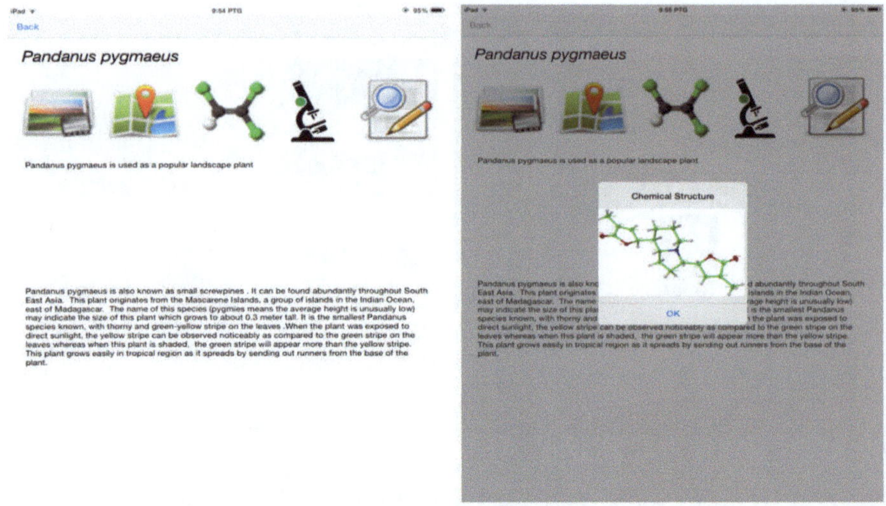

Fig. 2 The icons are arranged accordingly for a specific *Pandanus*, which begin with its photographic magnification, geodata, chemical constituents, microscopic observations and scientific publications related to that sample (*left*). Example of a molecular structure was shown (*right*)

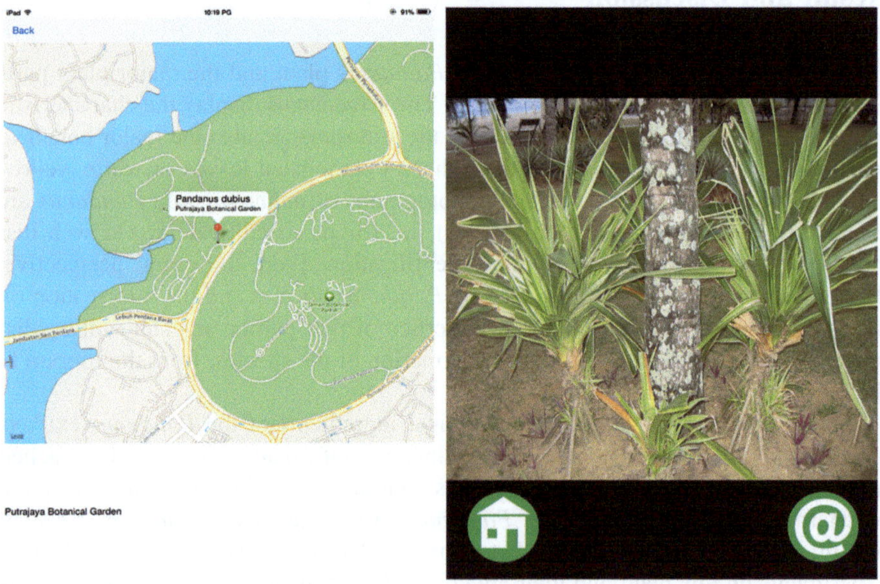

Fig. 3 The location of a *Pandanus* sample could be identified (*left*). The home button will guide the users back to the first window (*right*)

As shown in Fig. 3, the user could identify the location of a *Pandanus* sample, once the icon for the map is clicked, thus introducing the national botanical garden to the users. A window is also equipped with a home button, to guide users back to the display of the eight *Pandanus* tiles, after looking and differentiating with the unknown or wild *Pandanus*. In future, the user could also be able to turn for a landscape view of this application. The notes to each *Pandanus* species could be added, too. The user could also upload own captured *Pandanus* figures and interact with the researchers, in order to exchange information of discovered species. Other potential options for users would include the language change (either national language or English), without the internet access.

Conclusion

This application is designed for both iPhone and iPad. It is anticipated that this prototype should be easy to use and could offer the learnability of natural products to human. Built-in quizzes could also be proposed in order for users to quickly learn and distinguish between different *Pandanus* species. Later, to accommodate Android smart phone based users, this effort will be extended to Android operating system. Android users will also experience this pandan walk-about in their devices. Owing to the lack of information regarding Pandanaceae in a mobile application format, the concept is presently materialized. Continuous efforts could be planned, in order to digitalize the information, related to another hundreds of *Pandanus* recorded species. Thus, a comprehensive and fuller version of this application is recommended and physically, achievable. The prime motivation factor on pushing such technologies is aimed at positioning Malaysian academic and research institutions that contribute to the Apple store.

Acknowledgment This work is supported by Ministry of Education (Research Acculturation Grant Scheme) and Universiti Teknologi MARA (Principal Investigator Support Initiative Grant).

References

1. Alphablind Studio, Wild Berries & Herbs—Nature Mobile (2014) Information on https://itunes.apple.com/us/app/wild-berries-herbs-nature/id435903166?mt=8
2. Natator Publishing, Herbology On The Go (2014) Information on https://itunes.apple.com/us/app/herbology-on-the-go/id386509054?mt=8
3. Alphablind Studio, Medicinal Plants Pro—Nature Mobile By (2014) Information on https://itunes.apple.com/us/app/medicinal-plants-pro-nature/id499824923?mt=8
4. Mohsin HF (2012) Chemical constituents of Pandanaceae species. PhD thesis, Universiti Teknologi MARA, Malaysia
5. Armayni UA (2014) The isolation of alkaloids from *Pandanus* species. Master of Science thesis, Universiti Teknologi MARA, Malaysia (2014) (in print)

6. Md Yusof Y (2013) The hexane extraction of *Pandanus Conoideus*. Bachelor of Pharmacy Thesis, Universiti Teknologi MARA, Malaysia (2013)
7. Salim AA, Garson MJ, Craik DJ (2004) New alkaloids from *Pandanus amaryllifolius*. J Nat Prod 67(1):54–57

Investigation of Heat Treated Electrodeposited CoNiFe on Microstructure and Hardness

Nor Azrina Resali, Koay Mei Hyie, M.N. Berhan and C.M. Mardziah

Abstract In this research, heat treatment is the final finishing process applied on nanocrystalline CoNiFe to improve microstructure for good hardness property. Nanocrystalline CoNiFe has been synthesized using the electrodeposition method. This study investigated the effect of heat treatment at 500, 600, 700 and 800 °C on electrodeposited nanocrystalline CoNiFe. The heat treatment process was performed in the tube furnace with flowing Argon gas. In this paper, physical properties such as phase and crystallographic structure, surface morphology, grain size and hardness of nanocrystalline CoNiFe was studied by changing the heat treatment temperatures. The nanocrystalline CoNiFe phase revealed the Face Centered Cubic (FCC) and Body Centered Cubic (BCC) crystal structure. FESEM micrographs showed that the grain sizes of the coatings were in the range of 78.76–132 nm. Dendrite shape was found in the microstructure of nanocrystalline CoNiFe. The nanocrystalline CoNiFe prepared in heat treatment temperature of 700 °C, achieved the highest hardness of 449 HV. The surface roughness of nanocrystalline CoNiFe heated at 700 °C was found to be smaller than other temperatures.

Keywords CoNiFe · Heat treatment · Oxygen content · Dendrite · Hardness

N.A. Resali (✉) · K.M. Hyie · M.N. Berhan · C.M. Mardziah
Faculty of Mechanical Engineering, Universiti Teknologi MARA, 40450 Shah Alam, Selangor, Malaysia
e-mail: azrinaresali@yahoo.com

K.M. Hyie
e-mail: hyie1105@yahoo.com

M.N. Berhan
e-mail: Berhan@salam.uitm.edu.my

C.M. Mardziah
e-mail: mardziahcm@yahoo.com

M.A. Hashim (ed.), *ICGSCE 2014*, DOI 10.1007/978-981-287-505-1_27

Introduction

Engineering materials are heat treated under controlled sequence of heating and cooling to alter their physical and mechanical properties to meet desired engineering applications. Heat treatment is a combination of timed heating and cooling applied to a particular metal or alloy in the solid state to produce certain microstructure and desired mechanical properties [1].

Generally, the goal of the sintering process is to produce a coherent body with controlled microstructure, in some cases with controlled porosity [2]. Sintering phenomena can be classified into three stages. Initial stage of sintering corresponds to the situation when necks are forming and growing between grains as shown schematically in Fig. 1. At the end of this stage, the contact area increases up to 20 % with only a small densification. A marked decrease in the specific surface area of the compact occurs due to surface smoothing. The second stage of sintering is characterized by a more or less continuous network of pore channels along the grain edges. During this stage, the pore channel shrinks and grains grow. Grain growth can be more extensive in the final stage and difficulties are commonly encountered in the removal of porosity [3].

Electrodeposition is an electroplating process for producing a dense, uniform and adherent coating. This process offer many advantages and unique possibilities in the development of nanocrystalline materials with the low cost processing, rapidity, higher deposition rate and etc. [4].

A number of electrodeposited alloys are of interest due to various properties. Ternary alloys of CoNiFe are probably the most promising materials due to their excellent magnetic and physical properties such as superior corrosion resistance, low coercivity, and high resistivity. There are several factors including the crystallite size, the crystal orientation, and the alloy composition, that can affect the mechanical properties of these deposited films [5]. These properties also can be affected by the finishing the coating with heat treatment process.

During heat treatment, the process parameters such as temperature play an important role on properties of nanocrystalline CoNiFe. This is due to the grain boundary diffusion and volume diffusion that depends on temperature, the size,

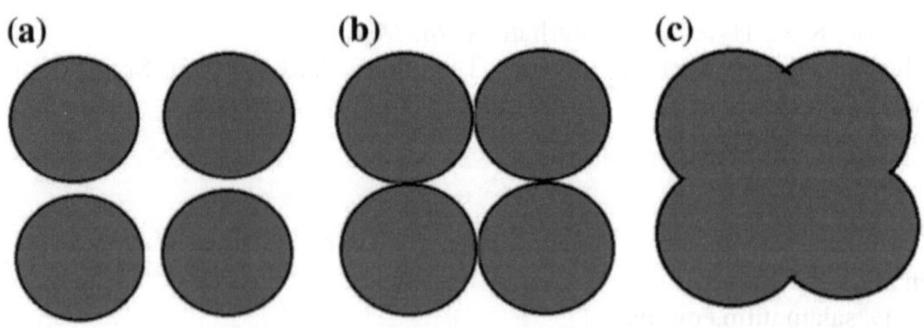

Fig. 1 Illustration of the sintering stage. **a** Formation of necks. **b** Evolution of necks and pores diffuse through and accumulate at grain boundaries. **c** Some pores are eliminated and closed pores appear

composition and distribution of grains of the material. Sintering process gives a huge effect on the microstructure, surface roughness and hardness. This is proven based on the previous researches [6, 7].

Al-Qura'n [7] studied the effect of heat treatment on the hardness of chromium-nickel steel. The best result of microhardness test was for the samples annealed at 740 °C for a period of 60 min. The authors concluded that heat treatment improves the quality in terms of microstructures and surface roughness and thus produced the highest hardness. The applied heat treatment in the final stage of electrodeposition nickel (EN) coating changes its properties significantly [6]. Mechanical properties of EN coatings such as hardness and wear resistance are enhanced by heat treatment. When the sintering temperature is increased, the grain size increases thus make the pore size to decrease [8]. Besides, it is found that the hardness will increase as the sintering temperature is increased. In contrast, the hardness will decrease if the grains contain an oxide ion at high temperature.

Remarkable increasing of surface hardness of CoNiFe coating up to 200–300 HV by elecrodeposition was reported [9, 10]. Among those surface modification techniques, electrodeposition is the simplest method with low set-up cost. However, low adhesion of CoNiFe presents the drawback of this route. The subsequent heat treatment can potentially improve the adhesion of CoNiFe [11]. Furthermore, the intermetallic formed during heating can further enhance the hardness of the CoNiFe. During heat treatment, the process parameters such as holding time and temperature play an important role on phase formation of CoNiFe.

In the present work, nanocrystalline CoNiFe were heat-treated under different temperatures to study the effect of heat treatment on the microstructure, grain size and hardness.

Experimental

A layer of CoNiFe was applied on a stainless steel substrate by the electrodeposition technique, using graphite as an anode [9, 10]. Figure 2 shows the schematic of electrodeposition process consists of four elements which are power supply, sample, graphite and solution. The aqueous bath consists of $CoSO_4$, $NiSO_4$, $FeSO_4$, H_3BO_3, and Saccharin. In this experiment, electrodeposition was conducted at 50 °C using a direct current (DC) condition at 0.133 A/cm^2. The holding time for the deposition was 15 min. 15 min for deposition have ample time to produce fully coating. It is believed that the coating have sufficient time for the grains to distribute and rearrange themselves in the microstructure. If the deposition time increases, some of the grains tend to fused, combining with each other and producing larger grains.

All samples were heat treated in a tube furnace under an argon atmosphere at 500, 600, 700 and 800 °C. The soaking time and heating rate in this experiment were fixed at 45 min and 8 °C/minute, respectively. After the heat treatment cycle, the samples were cooled in the furnace.

Following the electrodeposition and heat treatment steps, the samples were qualitatively characterized by glancing incident angle X-ray diffractometer (XRD) at the incident angle of 5°. Microstructures of the heat treated samples were observed by a Field emission scanning electron microscope (FESEM). Hardness of the samples was measured with a Vicker's microhardness using 100 g loading.

Results and Discussion

Physical structure, phase and crystallite sizes, The physical structure, phase and crystallite sizes of nanocrystalline CoNiFe were characterized by XRD analysis. XRD patterns of all the nanocrystalline CoNiFe were measured from 2θ angle of 30° to 100° with Cu Kα radiation. XRD patterns of nanocrystalline CoNiFe on different heat treating temperatures are shown in Fig. 2. All nanocrystalline CoNiFe showed the similar XRD patterns.

The reflection patterns of these nanocrystalline CoNiFe reveal the characteristics peaks of CoNi and FeNi. The nanocrystalline have a combination of face-centered cubic (fcc) structure and body centered cubic (bcc) phase.

After heat treatment at 500 °C, some small peak of FeNi (111) can be observed with low relative intensities. The peak with low intensities indicate that the volume fraction of intermetallic phase (FeNi) formed by heat treatment under these conditions are low. By increasing heat treatment temperature to 700 °C, higher relative intensity peak of FeNi can be observed. This implies that the volume fraction of the intermetallic formed by heat treatment at 700 °C is significantly higher than that formed at 500 and 600 °C [12].

There were no significant changes in the XRD patterns at heat treating temperatures below 700 °C, indicating that no phase transition occurred within this

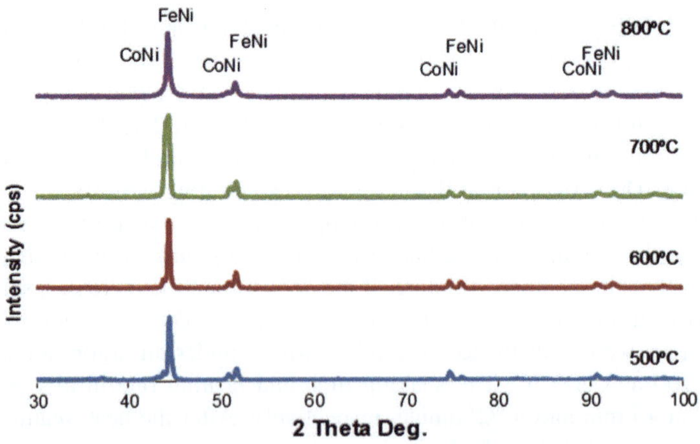

Fig. 2 XRD patterns of nanocrystalline CoNiFe at different heat treating temperatures

temperature range. When the heat treatment temperature was increased to 800 °C, the structure became crystalline, and new peaks corresponding to crystalline fcc CoNi (111) appeared. This behavior can be attributed to the crystallization of pure nickel followed by the precipitation of CoNi from the supersaturated FeNi solid solution [13].

As the heat treatment was increased, the (111) peak becomes narrow and intense while other peaks also appeared because of the crystallization. In addition, the relationship between the FWHM and grain size indicate that the grain size might have increased with increasing temperature.

FESEM observations, The comparison between the surface morphologies of nanocrystalline CoNiFe at different heat treating temperatures is presented in Fig. 3. The average grain sizes of the samples 500, 600, 700 and 800 °C were about 78, 105, 119 and 132 nm, respectively. The increase of grain size may be ascribed to the grain growth and the precipitation of intermetallic compound resulted from heat treatment.

At low temperature, the grain is decreased and the grains began to adhere and grew together. When the temperature has been increased, grain growth began to occur. Grain boundaries and a metastable pore phase were established. When the grain growth continued, the cross-sectional area of the pore phase was decreased. Then, the pore phase becomes discontinuous, and cause the closed pores remained

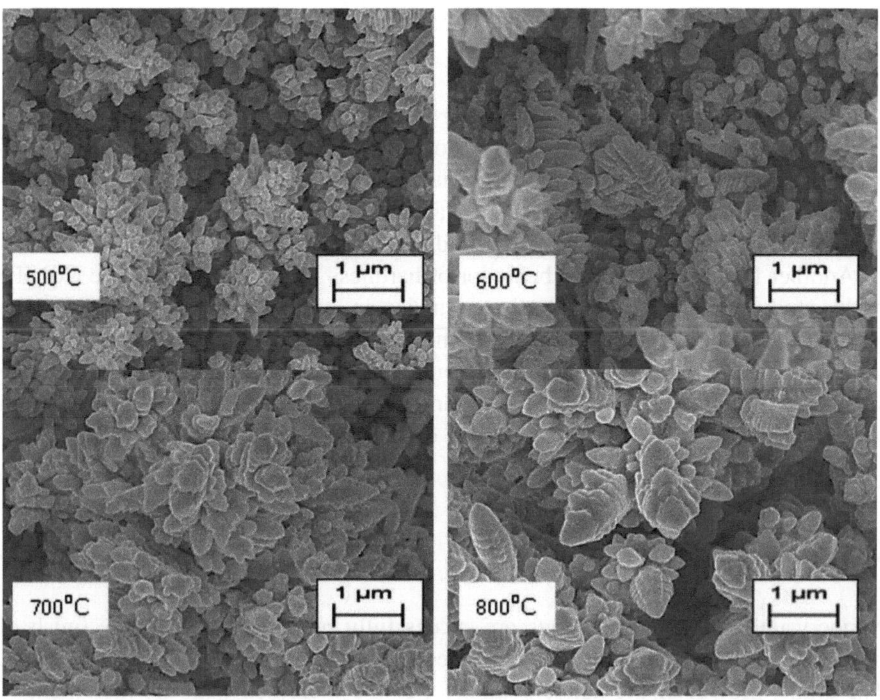

Fig. 3 Surface morphologies of nanocrystalline CoNiFe at different temperatures

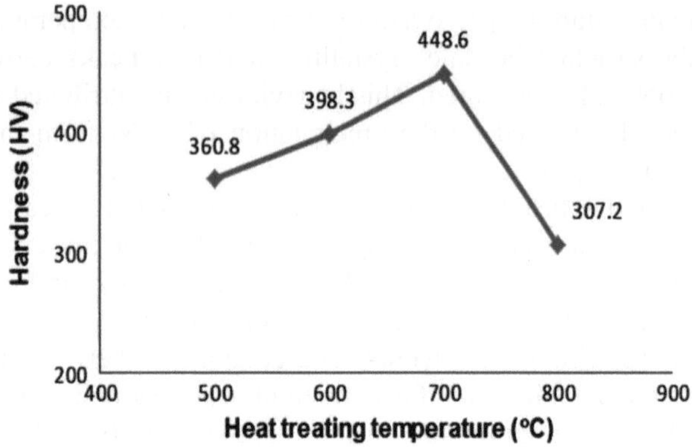

Fig. 4 The average hardness for nanocrystalline CoNiFe at different heat treating temperatures

at the grain boundaries. Complete removal of the remaining pore and leading to a completely dense material [14].

Nanocrystalline CoNiFe heated at 700 °C was denser and less voids if compared with other samples. A greater number of grain boundaries with the highest proportion of atoms inside the boundaries created an extremely high volume fraction of grain boundaries in the CoNiFe microstructure [9, 10]. Therefore, nanocrystalline CoNiFe were believed to develop a dense and less voids phase.

Hardness, Fig. 4 shows the average hardness of nanocrystalline CoNiFe at different heat treating temperatures. The hardness was increased with increasing heat treating temperature. It reached the maximum hardness at 700 °C and drop drastically after that.

The increase in hardness was associated with a structural change in the coating. At higher temperatures (>700 °C), the coating began to soften as the grains conglomerated and reducing the number of hardening sites. This process also removes FeNi (111) from the coating, producing a separate phase of soft nickel within the matrix and further reducing the bulk hardness [15].

Another factor affecting the behavior of hardness of the nanocrystalline CoNiFe in this study may be attributed to the grain size and porosity effect. The smaller the grain size leads to higher hardness of the electrodeposits. However, higher heat treating temperature (>700 °C) was found to significantly reduce the coating strength. As can be noticed that the hardness increases and reaches the maximum value of about 449 HV at 700 °C which coincides with less voids.

Conclusion

Nanocrystalline CoNiFe coatings are successfully prepared with surface finish of heat treatment. The optimum temperature of nanocrystalline CoNiFe was 700 °C. The grains growth of the nanocrystalline CoNiFe increased with the increment of heat treatment temperature. Voids phase area was decreased when the grain growth occurs. The hardness of nanocrystalline CoNiFe increased with the reduction of void in microstructure. The maximum hardness is 449 HV at 700 °C.

Acknowledgment This work has received financial support by Fundamental Research Grant Scheme (FRGS) [600-RMI/FRGS 5/3 (19/2013)] and Principal Investigator Support Initiative (PSI) [600-RMI/DANA 5/3/PSI (155/2013)] from Research Management Institute (RMI) of Universiti Teknologi MARA (UiTM). The author also would like to thanks to Ministry of Education Malaysia for financial support.

References

1. Fadare DA, Fadara TG, Akanbi OY (2011) Effect of heat treatment on mechanical properties and microstructure of NST 37-2 steel. J Miner Mater Charact Eng 10(3):299–308
2. Issariyapat A, Swangsak P, Boonyongmaneerat Y, Visuttipitukul P (2011) Effects of heat treatment on the interfacial structure of Nickel-Aluminum coating composites. Adv Mater Res 154:1462–1467
3. Kang S-JL (2005) Handbook of sintering: densification, grain growth and microstructure. Elsevier Butterworth Heineman, Amsterdam
4. Bakonyi I, Toth-Kada E, Tarnoczi T, Varga L K, Cziraki A, Gerocs I and Fogarassy B., Structure and properties of fine-grained electrodeposited nickel, Nanostructure Material. 3 (2993) 155
5. Fortas G, Sam S, Fekih Z, Gabouze N (2009) Electrodeposition of CoNiFe alloys on n-type silicon. Mater Sci Forum 609:207–212
6. Al-Qura'n F (2009) Effect of heat treatment on the microstructure and hardness of Chromium-Nickel steel. Contemp Eng Sci 8:355–359
7. Liu WL, Hsieh SH, Chen WJ, Hsu YC (2009) Growth behavior of electroless Ni–Co–P deposits on Fe. Appl Surf Sci 255:3880–3883
8. Al Qura'n F (2009) Effect of heat treatment on the microstructure and hardness of Chromium–Nickel steel. Contemp Eng Sci 2(8):355–359
9. Hyie KM, Resali NA, Abdullah WNR (2012) Study of alloys addition to the electrodeposited nanocrystalline cobalt. Adv Mater Res 486:108–113
10. Resali NA, Hyie KM, Abdullah WNR, Ghani MAA, Kalam A (2013) The effect of bath pH on the phase formation of ternary Co-Ni-Fe nano-coatings. Appl Mech Mater 391:9–13
11. Tamil Arasu P, Dhanasekaran R, Senthil Kumar P, Srinivasan N (2013) Effect of hardness and microstructure on En 353 steel by heat treatment, research inventy. Int J Eng Sci 2(11):01–05
12. Bordia RK, Camacho-Montes H (2012) Sintering: fundamentals and practice. The American Ceramic Society, Published 2012 by John Wiley & Sons, Inc.
13. Slokar L, Matković T, Matković P (2011) Alloying and heat treatment effects on the microstructure and hardness of biomedical titanium alloys. University of Zagreb Faculty of Metallurgy, Aleja narodnih heroja 3, 44103 Sisak, Croatia
14. Kofstad P (1998) High temperature corrosion. Elsevier Applied Science Publishers LTD, 52 Vanderbilt Avenue, New York, USA
15. Zhou J, Soboyejo WO (2004) Compression–compression fatigue of open cell aluminum foams: macro-/micro- mechanisms and the effects of heat treatment. Mater Sci Eng A-Struct Mater 369(1):23–35 (Journal: Materials Science and Engineering A-structural Materials Properties Microstructure and Processing)

Computational Fluid Dynamics Modeling of Mercury Emissions in Pulverized Coal Combustion

Nor Fadilah Mohamad, Noor Hidayu Abdul Rani, Sherif Abdulbari Ali and Sharifah Aishah Syed A. Kadir

Abstract Mercury is among the main pollutants in the atmosphere. Mercury can be released to the atmosphere by human activities including coal combustion, cement production, incineration and others. Coal combustion is primarily contributed to mercury emissions. This study is aimed at research and implementation of mercury model in computational fluid dynamics modeling to predict the distribution of mercury when coal is combusted in the furnace. The mercury model was implemented in CFD Fluent that consists of three models including One Step Model, Two Step Model and Detail Wilcox Model. The post process modeling is used for mercury emissions prediction since the low amounts of mercury has negligible impact on coal combustion. The Detailed Wilcox mercury model was applied and the result obtained was compared with measured from the plant.

Keywords Mercury · Emissions · Combustion · Modeling

Introduction

Mercury emissions are one of the most problematic environmental concern in the world because of its toxicity and bioaccumulation within an aquatic food chain [1]. Mercury is very harmful and can damage nerve systems of humans and lead to neurological syndrome. Mercury has become an issue after the tragedy of methyl

---

N.F. Mohamad (✉) · N.H. Abdul Rani · S.A. Ali · S.A. Syed A. Kadir
Faculty of Chemical Engineering, Universiti Teknologi MARA, 40450 Shah Alam, Malaysia
e-mail: fala5494@yahoo.com

N.H. Abdul Rani
e-mail: hidayu_rani@yahoo.com

S.A. Ali
e-mail: abdulbari@salam.uitm.edu.my

S.A. Syed A. Kadir
e-mail: drsharifah@salam.uitm.edu.my

M.A. Hashim (ed.), *ICGSCE 2014*, DOI 10.1007/978-981-287-505-1_28

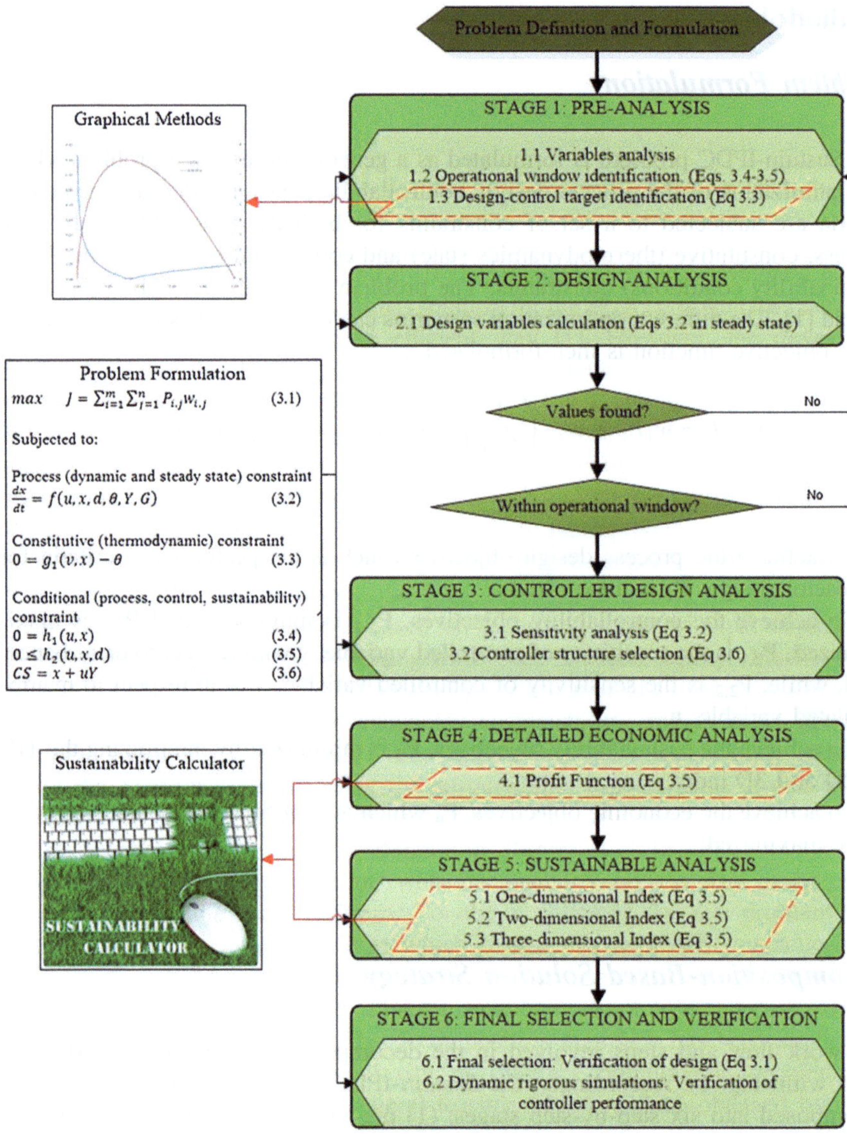

Fig. 1 Sustainable integrated process design and control methodology for reactor systems

Result and Discussion

Process Description

The multi-step consecutive reaction of cyclohexanone production provides an interesting challenge for the *Sustain*-IPDC application. The results will show the capability of the *Sustain*-IPDC to handle or solve complex processes as well as the simple process. Next, the following situation is considered in the *CSTR*. The production of the desired product, cyclohexanone is to be done from the reaction of the only reactant in the reactor, cyclohexanol. In spite of cyclohexanone, hydrogen is also produced as a by-product from the first reaction.

Problem Formulation

The *Sustain*-IPDC problem for the process described above is defined in terms of a performance objective (with respect to design, controlled, sustainable and detailed economic), and the three set of constraints (process, constitutive and conditional). It can be referred in Hamid [1]. For this case, each component used is presented as *a*, *b*, c and *d* which respect to cyclohexanol, cyclohexanone, high boiler, and hydrogen respectively.

Stage 1: Pre-analysis. The objective of this stage is to define the operational window and set the targets for the design-controller solution by using attainable region diagram [4]. Based on the plotted region, the maximum point at the diagram (Point A) is chosen as the design target together with the other two points (alternative design for verification purposes), (refer with: Fig. 2).
Stage 2: Design analysis, In this stage, all the targets selected in the previous stage then undergo the validation process for finding its design values by calculating its residence time and the design variables such as volume and height of the *CSTR* (refer with: Table 1).

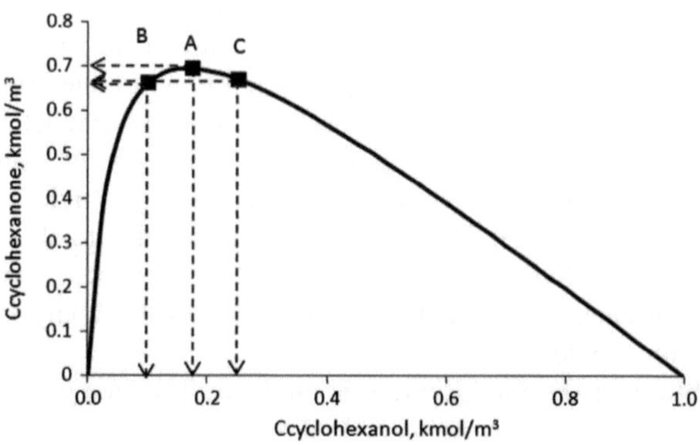

Fig. 2 Attainable region diagram for C_a with respect to C_b at 394.26 K

Table 1 Values of residence time with the corresponding process-controlled and design-manipulated variables at different *CSTR* designs

Targets	T (hr)	Process-controlled		Manipulated	Design	
		C_a (kmol/m^3)	C_b (kmol/m^3)	F (m^3/hr)	V (m^3)	h_r (m)
A	0.52	0.175	0.6940	10.0	5.15	2.27
B	0.98	0.100	0.6616	10.0	9.84	4.33
C	0.33	0.250	0.6696	10.0	3.28	1.44

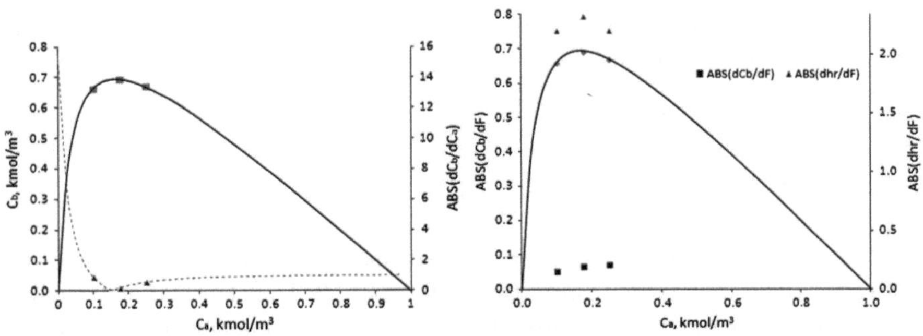

Fig. 3 Attainable region diagram for the desired concentration of b with respect to reactant, C_a, and its corresponding derivatives of C_b with respect to disturbance C_a at feed (*left*) and attainable region diagram for the desired concentration of b with respect to reactant, C_a, and its corresponding derivatives of C_b and h_r with respect to manipulated variable *F* (*right*)

Stage 3: Controller design analysis, At this stage, the controllability performance of each of the feasible candidates is evaluated and validated for the selection of controller structure. There are two criteria that need to be analyzed which are sensitivity of controlled variable y with respect to disturbance d, *dy*/*d*d and sensitivity of controlled variable y with respect to the manipulated variable u, *dy*/*d*u. The results clearly show that design A has the best controller design compared to designs B and C (Fig. 3).

Stage 4: Sustainability analysis, At this stage, the sustainability index for three-dimensional (3D) sustainability analysis is based on Martins et al. [2]. Here, the indexes used are energy intensity, material intensity, potential chemical risk and potential environmental impact. Energy and material used give negative impact to the environment (waste generated), positive impact to economy (value creation) and give both impact to society (higher living standard and future generation deprive due to depletion). While the other two indexes could have a poor working environment (environment), financial loss due to shutdown (economic) and illness (society). In this case, those four indexes are located inside a 3D index that comply all aspects in sustainability. The result for this analysis shown in table below (refer with: Table 2).

Stage 5: Detailed Economic analysis, The analysis is continued by identifying the profit and total cost of each design candidates. Based on the analysis, design at

Table 2 Values of all index in 3D sustainability analysis for each candidates

Candidates	Index	Potential environmental impact	Potential chemical impact	Material intensity	Non renewable energy intensity
A	Full index	0.5913	57.4522	1.1700	0.0017
	Scale index	1.0000	0.9506	0.8752	0.8947
	3D index	0.9301			
B	Full index	0.5888	60.4383	1.3369	0.0019
	Scale index	0.9958	1.0000	1.0000	1.0000
	3D index	0.9989			
C	Full index	0.5580	54.2111	1.1040	0.0016
	Scale index	0.9437	0.8970	0.8258	0.8421
	3D index	0.8771			

Point B is the best in economic analysis because of its high profit. This is due to the large design volume of *CSTR* at this point which leads to high production of the products. Then, it follows by design and Points A and C, respectively. The *CSTR* designs will be analyzed by using the profit function in order to analyze the design that will provide the maximum profit. There are four criteria used to calculate the profit function, which are profit of product, cost of material, depreciation cost and operating cost. The economic analysis is calculated by assuming that the *CSTR* is operating in 5 years and 340 days per year. From the results, design at Point B has the highest profit compared to designs at Points A and C (Table 3).

Stage 6: Final selection and verification. The objective of this stage is to select the best candidate by analyzing the value of the multi-objective function. The multi-objective function is calculated by summing up the multi-objective function value. From the tabulated results in Table 4, it is clearly seen that the multi-objective function, *J* for *CSTR* design A is higher compared to other designs.

Table 3 Economic analysis for each candidates

Candidates	I_1 (\$/yr)	I_2 (\$/yr)	C_R (\$/yr)		C_D (\$/yr)	C_O (\$/yr)
			Purchase	Installed		
A	69,402.284	46,254.6	28,901.1	2890.1	5298.5	0.0
B	126,306.297	88,304.1	37,432.2	3743.2	6862.6	0.0
C	42,610.647	29,434.7	24,121.1	2412.1	4422.2	0.0

Table 4 Final result of multi-objective function

	$P_{1,1}$	$P_{2,1}$	$P_{2,2}$	P_3	$P_{4,1}$	
A	0.694	0.178	2.398	0.930	17,849.2	
B	0.662	0.874	2.257	0.999	31,139.6	
C	0.670	0.552	2.276	0.877	8753.7	
	$P_{1,1s}$	$P_{2,1s}$	$P_{2,2s}$	$P_{3,1s}$	$P_{4,1s}$	J
A	1.000	0.203	1.000	0.550	0.573	**10.39**
B	0.953	1.000	0.941	1.000	1.000	5.89
C	0.965	0.631	0.949	0.337	0.281	7.79

The best candidate is highlighted in **bold**

Therefore, it is verified that the *CSTR* design at the maximum point of attainable region is an optimal solution for *Sustain*-IPDC of a cyclohexanone production in *CSTR* which satisfies the design, control, sustainability and economic criteria.

Conclusion

In this work, a systematic model-based methodology has been developed for sustainable integrated process design and control (*Sustain*-IPDC) for a continuous-stirred tank reactor system (*CSTR*) by using a 3D sustainability index. This methodology has been applied and verified for a single *CSTR* system. The main contribution of this work that has been obtained from this work is that, the methodology applied a step-by-step procedure which allows a systematic analysis being done at each stage. Every step of the design methodology is clear with respect to calculations/analysis and generic in terms of application range which makes the application of the methodology is quite easy.

Acknowledgment The financial support from Zamalah Universiti Teknologi Malaysia is highly acknowledged.

References

1. Hamid MKA (2011) Model-based integrated process design and controller design of chemical processes. Ph. D. Thesis, Technical University of Denmark
2. Martins A, Mata T, Costa C, Sikdar S (2007) Framework for sustainability metrics. Ind Eng Chem Res 46:2962–2973
3. Karunanithi APT, Achenie LEK, Gani R (2006) A computer-aided molecular design framework for crystallization solvent design. Chem Eng Sci 61:1247–1260
4. Hamid MKA, Sin G, Gani R (2010) Integration of process design and controller design for chemical processes using model-based methodology. Comput Chem Eng 34:683–699

Three-Dimensional Sustainability Analysis for Integrated Process Design and Control for Separation of Benzene and Toluene Mixture

Mohamad Zulkhairi Nordin, Mohamad Dzulfadzli Jais, Mohamad Rizza Othman and Mohd Kamaruddin Abd Hamid

Abstract The objective of this paper is to present the development of a sustainable integrated process design and control methodology for a distillation column system. The sustainable integrated process design and control methodology for a distillation column system is developed to ensure the distillation column design is more cost efficient, controllable and sustainable to meet the product quality. The sustainable integrated process design and control problem for a distillation column system is typically formulated as a mathematical programming (optimization with constraints) problem, and solved by decomposing it to six sequential hierarchical sub-problems: (i) pre-analysis, (ii) design analysis, (iii) controller design analysis, (iv) sustainability analysis, (v) detailed economics analysis and (vi) final selection and verification. In the pre-analysis sub-problem, the concept of driving force is used to locate the optimal design-control solution targets, which are defined at the maximum point of the driving force diagram. The sustainability analysis sub-problem was calculated by using the three-dimensional sustainability index. The results through case study of benzene-toluene separation process shows the proposed methodology is capable to find the optimal solution that satisfies design, control, sustainability and economic criteria in a simple and efficient way.

M.Z. Nordin · M.D. Jais · M.K. Abd Hamid (✉)
Process Systems Engineering Centre (PROSPECT), Faculty of Chemical Engineering, Universiti Teknologi Malaysia, 81310, UTM Johor Bahru, Johor, Malaysia
e-mail: kamaruddin@cheme.utm.my

M.Z. Nordin
e-mail: mzulkhairi90@gmail.com

M.D. Jais
e-mail: mdzulfadzli3@gmail.com

M.R. Othman
Process System Engineering Group (PSERG), Faculty of Chemical and Natural Resources Engineering, Universiti Malaysia Pahang, Lebuhraya Tun Razak, 26300, Gambang, Kuantan, Pahang, Malaysia
e-mail: rizza@ump.edu.my

M.A. Hashim (ed.), *ICGSCE 2014*, DOI 10.1007/978-981-287-505-1_30

Keywords Optimal · Integrated · Design · Control · Sustainability

Introduction

Integrated Process Design and Control (IPDC) methodology was developed and able to obtain an optimal solution for the IPDC problem [1]. The IPDC methodology has shown that distillation column design at the highest point of driving force (DF) is the best in terms of design and controllability. However, the developed methodology for the IPDC did not consider sustainability aspect in the early of design stage. Distillation column design can be further improved by including sustainability aspect within the developed IPDC methodology to ensure that the design is controllable, as well as sustainable.

Sustainability can be defined as the development that meets the needs of the present without compromising the future generations to meet their own need. Basically, sustainability is made up from three principles which are environmental, economic and social. From the three principles of sustainability, three different sustainability indexes can be obtained which are one-dimensional (1-D) index, two-dimensional (2-D) index and three-dimensional (3-D) index as shown in Fig. 1 [2–4]. Each index is represented with its own metrics; use to assess the sustainability performance of a process or a system, in order to evaluate the progress toward enhancing the sustainability [5]. Furthermore, indicator also can be used to assist decision makers in evaluating the alternatives. This paper will highlight the implementation of sustainability criteria by using the 3-D index in the IPDC methodology known as Sustainable Integrated Process Design and Control (*Sustain*-IPDC).

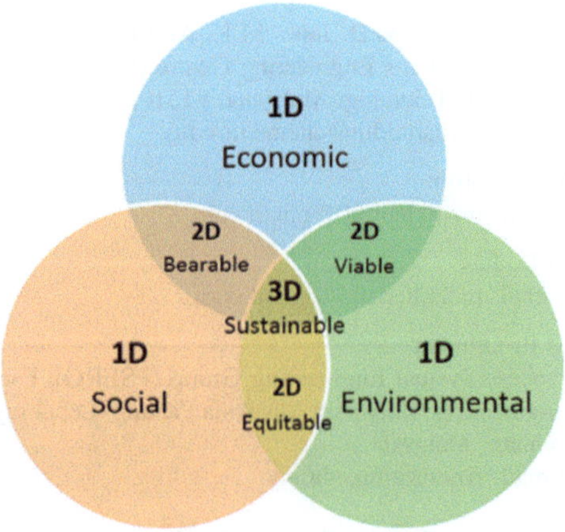

Fig. 1 Venn diagram of sustainability criteria

Methodology

The *Sustain*-IPDC problem is formulated as a generic optimization problem in which a performance objective in terms of design, control, sustainability and cost is optimized subject to a set of constraints as Eqs. 1–6.

$$\max J = \sum_{i=1}^{m} \sum_{j=1}^{n} P_{i,j} w_{i,j} \tag{1}$$

Subjected to:
Process (dynamic and/or steady state) constraint

$$\frac{dx}{dt} = f(u, x, d, \theta, Y, G) \tag{2}$$

Constitutive (thermodynamic) constraint

$$0 = g_1(v, x) - \theta \tag{3}$$

Conditional (process, controllability, sustainability) constraint

$$0 = h_1(u, x) \tag{4}$$

$$0 \le h_2(u, x, d) \tag{5}$$

$$CS = x + uY \tag{6}$$

whereas:
x set of process controlled variables
u set of manipulated (design) variables
d set of disturbance variables
θ set of constitutive variables
v set of chemical system variables
G set of independent variables
i category of objective function
j specific term of each category
Y set of binary decision variables
w_i weight factor of each objective term
P_i objective term

Sustain-IPDC problem for a distillation column system is formulated as a mixed-integer dynamic optimization (MIDO). The design with the maximum value of J is indicated as the best distillation design. The scenarios can be generated as follows:

- To achieve the process design objective, P_1 which is the performance criteria for a distillation design is maximized.
- To achieve the controller design objectives, $P_{2,1}$ is minimized and $P_{2,2}$ is maximized. $P_{2,1}$ is the sensitivity of controlled variable, y with respect to disturbance, d. While, $P_{2,2}$ is the sensitivity of controlled variables, y with respect to manipulated variables, u.
- To achieve the sustainability objectives, $P_{3,j}$ is minimized. There are four indexes used to achieve the sustainability objective which are $P_{3,1}$, $P_{3,2}$, $P_{3,3}$ and $P_{3,4}$. These indexes represent the material intensity, energy intensity, potential chemical risk and potential environmental impact respectively.
- To achieve the economic objectives, P_4 which is net profit function are need to be maximized.

Then the multi-objective function can be formulated as Eq. 7 (same weight factor, $w_i = 1$):

$$\max J = P_1 + \frac{1}{P_{2,1}} + P_{2,2} + \frac{1}{P_{3,1}} + \frac{1}{P_{3,2}} + \frac{1}{P_{3,3}} + \frac{1}{P_{3,4}} + P_4 \tag{7}$$

The problem is solved by decomposing it into six stages: (i) pre-analysis, (ii) design analysis, (iii) controller design analysis, (iv) sustainability analysis, (v) detailed economics analysis and (vi) final selection and verification, as shown in Fig. 2. In this way, the solution of the decomposed set of sub-problem is equivalent

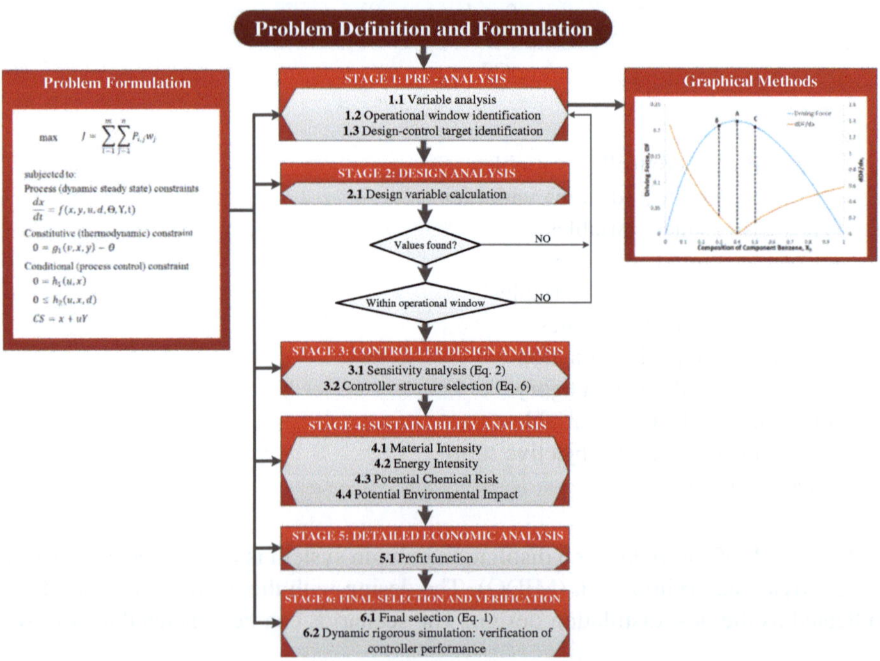

Fig. 2 Sustainable integrated process design and control methodology for distillation column

to the original problem. As each sub-problem is being solved, a large number of feasible solution within the search space is identified and hence eliminated. Therefore, while the problem complexity may increase with every subsequence stage, the number of feasible solution is reduced at every stage.

Application of Methodology for Distillation Column Systems

The application of the *Sustain*-IPDC is illustrated by the separation process of Benzene and Toluene.

Stage 1: Pre-analysis. The objective of this stage is to define the operational window where the optimal solution is located. The operational window is identified based on bottom and top product quality. Step-by-step algorithm for a simple distillation by Bek-Pederson and Gani is implemented in this step [2]. The driving force (DF) diagram is developed by using the Eq. 8, where y_i and x_i are phase composition of component i and α_{ij} is the relative separability factor for binary pair.

$$DF = y_i - x_i = \frac{x_i \alpha_{ij}}{1 + x_i(\alpha_{ij} - 1)} - x_i \tag{8}$$

DF diagram for Benzene-Toluene system at P = 6 atm is drawn as shown in Fig. 3. The target for optimal sustainable process-control design solution for distillation column is identified at the maximum point of driving force (Point A). Two other points are identified as alternative designs (Points B and C) to compare with Point A.

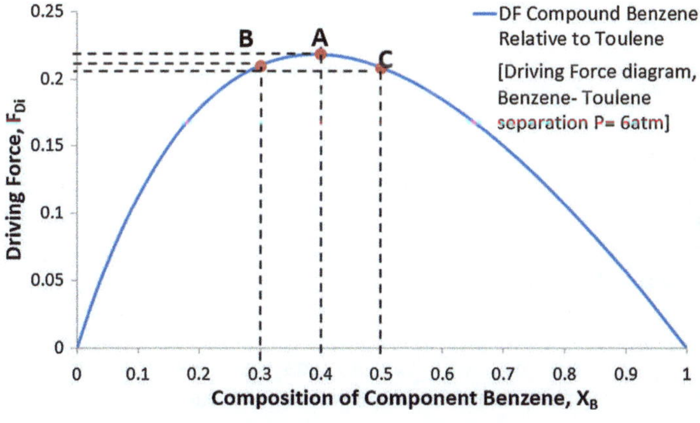

Fig. 3 Driving force diagram for separation of Benzene-Toluene by a distillation column

Table 1 Values of design variable for separation Benzene-Toluene for distillation column design

Point	N. Stage	N. Feed	RBmin	RRmin	RB	RR	D (kmol/hr)	Q_C (kW)	Q_R (kW)
A	20	12	1.70	2.58	2.04	3.09	8.061	217.4	244.9
B	20	14	1.38	2.88	1.66	3.46	8.061	222.6	249.9
C	20	10	2.05	2.33	2.46	2.79	8.061	244.0	272.2

Stage 2: Design Analysis. All targets identified in Stage 1 is validated by finding the acceptable values of controlled and manipulated variables. Each points are analyzed by applying the methodology proposed by Bek-Pederson and Gani [6]. Then, from the results, other design-process variables of distillation column are obtained by using the Aspen HYSYS steady-state process simulator. The results are summarized in Table 1. It is noted that design A corresponds to the lowest energy consumption.

Stage 3: Controller Design Analysis. At this stage, the controllability performance of each of the feasible candidates is evaluated and validated for the selection of controller structure. There are two criteria that need to be analyzed which are sensitivity of controlled variables **y** with respect to disturbances **d** and sensitivity of controlled variables **y** with respect to the manipulated variables **u**. The details procedure for controller design analysis are shown in Hamid's work [7]. Figure 4 shows the plot of derivative DF with the respect to x_B (disturbance). A smaller derivative means that the process sensitivity is lower, hence, from process control of view, distillation column design A is less sensitive to the effect of disturbance thus make it more robust in maintaining its controlled variable in the presence of disturbances. The controller structure is selected by calculating the derivatives values of the potential controlled

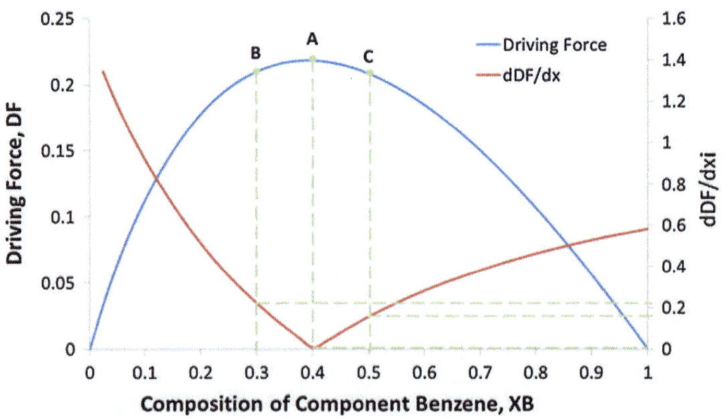

Fig. 4 Plot of driving force and derivative of driving force with respect to composition as a function of composition for composition of Benzene

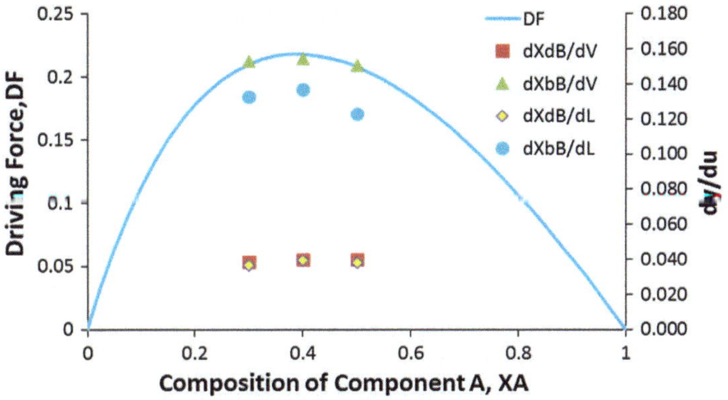

Fig. 5 Driving force diagram for separation of Benzene-Toluene by distillation with corresponding derivatives of x_{dB}, x_{dT} x_{bB} and x_{bT} respect to L and V

variables (x_{dB},x_{dT}, x_{bB}, x_{bT}) with respect to manipulated variables (V and L), in a constant step size. The value of the derivatives are plotted in Fig. 5. Higher gain in process control means the process will require smaller control action to maintain the controlled variable at the optimal set point value. So, it can be concluded that to maintain the purity of the product in the distillation column can be done by directly manipulating the value of V and L.

Stage 4: Sustainability Analysis. The purpose of this stage is to analyze the optimal design of distillation column in terms of sustainability. The sustainability of each distillation design is assessed by using 3D metrics. There are four indexes of 3D metrics used to assess the sustainability analysis which are material intensity (MI), energy intensity (EI), potential chemical risk (PCR) and potential environmental impact (PEI) [4]. In order to fulfill the sustainability criteria, it must have low impacts to the economic losses as well as environmental impacts. Material and energy consumption will be measured to satisfy sustainability objective functions. The values of each sustainability metric are tabulated in Table 2. It can be seen that Point A has the lowest index compared with other points. Thus, it proves that the distillation design at the maximum point of driving force will have the best objective function in term of sustainability.

Stage 5: Detailed Economic Analysis. The distillation designs will be analyzed using profit function in order to analyze the design that will provide the maximum profit [8]. There are four criteria used to calculate the profit function, which are profit of product, cost of material, depreciation cost and operating

Table 2 The sustainability analysis values at different distillation column design

	MI	EI	PCR	PEI
Point A	0.9938	2.6393	0.0001	0.0012
Point B	0.9938	2.6955	0.0001	0.0012
Point C	0.9938	2.9477	0.0001	0.0012

Table 3 The detaile economic analysis values at different distillation design

Cost	Product	Raw material	Depreciation	Operation	Profit, ∅
Point A	108,533,304	81,600,000	1,872,649	1,712,683	23,347,971
Point B	108,533,304	81,600,000	1,896,790	1,749,208	23,287,304
Point C	108,533,304	81,600,000	1,900,070	1,912,927	23,120,306

Table 4 The values of multi-objective function calculation

	$P_{1,1}$	$P_{2,1}$	$P_{2,2}$	$P_{3,1}$	$P_{3,2}$	$P_{3,2}$	$P_{3,2}$	$P_{4,1}$	
A	0.2180	0.0065	0.0720	0.9938	2.6393	0.0001	0.0012	23,347,971	
B	0.2098	0.2205	0.0680	0.9938	2.6955	0.0001	0.0012	23,287,304	
C	0.2082	0.1558	0.0690	0.9938	2.9477	0.0001	0.0012	23,120,306	
	$P_{1,1s}$	$P_{2,1s}$	$P_{2,2s}$	$P_{3,1s}$	$P_{3,1s}$			$P_{4,1s}$	J
A	1.00	0.03	1.00	1.00	0.89	1.00	1.00	1.00	**40.46**
B	0.96	1.00	0.99	1.00	0.91	1.00	1.00	0.99	8.04
C	0.96	0.71	0.98	1.00	1.00	1.00	1.00	0.99	8.34

The best candidate is in **bold**

cost. The economic analysis is calculated by assuming that the distillation column is operating in 5 years and 340 days per year. The result of the calculation is summarized in Table 3. From the results, distillation design at Point A has the highest profit and this verified that the best design in terms of economic criteria.

Stage 6: Final Selection and Verification. The objective of this stage is to select the best candidate by analyzing the value of the multi-objective function. The multi-objective function is calculated by summing up the objective function value using Eq. 7. Each objective value is normalized with respect to the highest value since the range of the objective function can be different. From the result tabulated in Table 4, it is clearly seen that the multi-objective function, *J* for distillation column design A is the highest compared to other designs due to it has the best value for design, controller design, sustainability and economic. Therefore, it is verified that the distillation column design at the maximum point of driving force is an optimal solution for *Sustain*-IPDC of a Benzene-Toluene separation process which satisfies the design, controller design, sustainability and economic criteria.

Conclusion

As a conclusion, a systematic-based methodology has been developed for Sustain-IPDC for a distillation column system. This methodology has been applied and verified for a single distillation column. The presented case study of Benzene-

Toluene separation process has shown that the optimal design is at the highest point of driving force, with respect to design, control, sustainability and economic criteria. The optimal solutions were obtained in an efficient and systematic way.

Acknowledgment The financial support from Universiti Teknologi Malaysia (RUGS Tier 1 Q.J130000.2509.07H39) and Ministry of Education of Malaysia are highly acknowledged.

References

1. Hamid MKA, Sin G, Gani R (2010) Integration of process design and controller design for chemical processes using model-based methodology. Comput Chem Eng 34:683–699
2. Shadiya OO, Hogh KA (2013) Sustainability evaluator: tool for evaluating process sustainability. Environ Prog Sustain Energy 32:749–761
3. Uhlman BW, Sailing P (2010) Measuring and communicating sustainability through eco-efficiency analysis. Chem Eng Process 106:17–26
4. Martins AA, Mata TM, Costa CAV, Sikdar SK (2007) Framework for sustainability metrics. Ind Eng Chem Res 46:2962–2973
5. Sikdar SK (2003) Sustainable development and sustainability metrics. AIChE J 49(8):1928–1932
6. Bek-Pedersen E, Gani R (2004) Design and synthesis of distillation systems using a driving force based approach. Chem Eng Process 43:251–262
7. Hamid MKA (2011) Model-based integrated process design and controller design of chemical processes, DTU PhD Thesis, ISBN: 978-87-92481-39-9
8. Li CS, Zhang XP, Zhang SJ, Suzuki KZ (2009) Environmentally conscious design of chemical processes and products: multi-optimization method. Chem Eng Res Des 87:233–243 IChemE

Nucleation Kinetics of Carbamazepine-Saccharin (CBZ-SAC) Co-crystal

Khairool Azizul Mohammad, Syarifah Abd Rahim and Mohd Rushdi Abu Bakar

Abstract In this study the effect of the heating/cooling rate of various CBZ-SAC mixture in ethanol was investigated in order to deduce the metastable zone width (MSZW) and the nucleation order. Nývlt's equation involving the maximum temperature difference ΔT_{max} between dissolution and crystallization temperatures of CBZ-SAC co-crystals with the heating/cooling rate r was applied for the study of nucleation kinetics by the polythermal method. A step towards the determination of dissolution temperature of the crystals has been improved. The results showed that dissolution temperature for CBZ-SAC co-crystal mixture increased with the increase in heating rate, CBZ-concentration and the ratio of SAC to CBZ in ethanol solution. The analysis revealed that a nucleation order for CBZ-SAC co-crystals increases with the SAC to CBZ ratio and is in the range of 1.58–6.31.

Keywords Crystallization · Co-crystal · Carbamazepine · Metastable zone width · Nucleation kinetics

---

K.A. Mohammad · S. Abd Rahim (✉)
Faculty of Chemical and Natural Resources Engineering, University Malaysia Pahang, 26300 Pahang, Malaysia
e-mail: syarifah@ump.edu.my

K.A. Mohammad
e-mail: kazizul@ump.edu.my

M.R. Abu Bakar
Department of Pharmaceutical Technology, Kulliyyah of Pharmacy, International Islamic University Malaysia, Bandar Indera Mahkota, 25200 Kuantan, Pahang, Malaysia
e-mail: rushdi@iium.edu.my

M.A. Hashim (ed.), *ICGSCE 2014*, DOI 10.1007/978-981-287-505-1_31

Introduction

Crystallization from solution has been studied intensively and widely used in many industries and applications. The main purpose of crystallization is to separate and purify solids in a suitable solvent which lead to the control of the shape and size of final crystalline products. This technique provides an end compound with a high degree of purity that has many desirable properties [1]. A broad knowledge of the crystallization process is also essential to prevent the undesired forming of crystals in industrial applications such as a tendency for crystal forming in biodiesel, which can deteriorate the performance of fuel combustion [2].

There are two steps in forming a crystal/solid from a solution: (i) nucleation, which the creation of minute nuclei in a supersaturated solution (ii) crystal growth that transform the stable nuclei into solid crystals [1]. The crystallization occurs in a metastable region, which exists in a supersaturated solution. Metastable zone width (MSZW) is located between supersaturation and saturation limit of a system. The optimization of a crystallization process and the quality of the forming crystal can be simply controlled when the crystallization runs within the MSZW [2, 3]. The factors that can affect the MSZW are cooling/heating rates, solution temperature, impurities or seeds, stirring rate and solvent [2, 4–6]. Therefore, it is particularly important to identify these parameters precisely for accessing the MSZW. An analysis of the change in physical quantities such as particle count number, turbidity, intensity (absorbance or transmittance), and electrical conductivity were used in the research works as an indicator of the nucleation event [7]. Among these, the turbidity measurement is an inexpensive method which several chemical processes are using it [8].

The maximum super-cooling temperature ΔT_{max} described as the difference between the dissolution temperature T_{diss} and the crystallization temperature T_{crys} at which a detectable crystallisation commences: $\Delta T_{max} = T_{diss} - T_{crys}$ [5]. Figure 1 illustrates a saturation–supersaturation diagram of the crystals.

A solution of concentration C_1 with a temperature T_a, above dissolution temperature T_{diss1} was cooled at a constant cooling rate 'R' to a temperature 'T_b'. First crystals were detected at the temperature T_{crys1} in the solution. The solution was

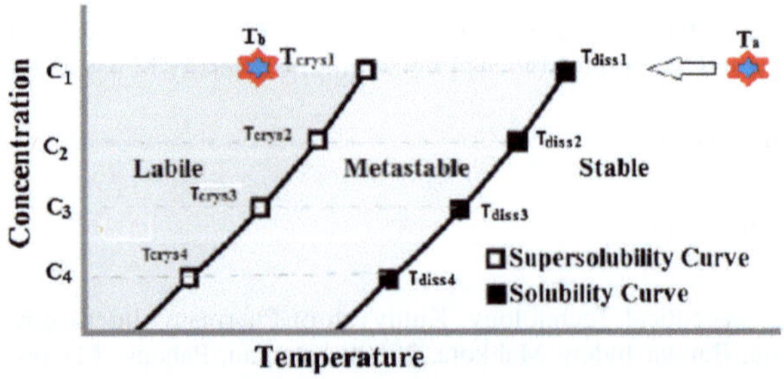

Fig. 1 Solubility and supersolubility curves of a solid from solution (figure modified from [1])

then heated at a constant rate to a temperature T_a. The crystals were fully dissolved at the temperature T_{diss1}. Therefore, the maximum super-cooling ΔT_{max} at the concentration C_1 was the difference between the dissolution temperature T_{diss1} and the crystallization temperature T_{crys1}, i.e. $\Delta T_{max}(C_1) = T_{diss1} - T_{crys1}$. There were saturation concentrations C_2 and C_3 of solute in the solutions corresponding to dissolution temperature T_{diss2} and T_{diss3}, respectively.

Figure 1 illustrates (i) the stable zone, which is unsaturated, and crystallization cannot occur in this region; (ii) the metastable zone, which is supersaturated, and spontaneous crystallization is unlikely to occur, but crystallization can occur by adding a seed; (iii) the labile zone, which is non-stable zone and spontaneous crystallization occurs mostly to in this zone [1].

Experimental

In this study, 250 ml vacuum jacketed vessel equipped with a stirrer run with a speed of 300 rpm, a turbidity probe, and a valve for draining purpose. An analysis software was connected for the sophisticated automation for control and analysis of the operations. The turbidity sensor was calibrated before each experiment as 100 % for a suspension of CBZ- SAC co-crystals and 0 % for a clear solution. Different mol ratio (3.5, 3.0 and 2.5) of SAC to CBZ with different concentrations of ethanolic solutions containing 19.14, 17.96, 17.01 and 15.83 mg/ml CBZ in 200 ml ethanol were made. Four different heating and cooling rates (0.25, 0.5, 0.75, and 1 °C/min) were applied in this crystallization experiments.

The initial suspension (3.5 SAC/CBZ ratio and 19.14 mg/ml CBZ in 200 ml ethanol) was heated to 60 °C, which is higher than dissolution temperature. As can be seen in Fig. 2, the stirred mixture was kept at 60 °C for 40 min to make sure that the solid was fully dissolved. The mixture solution was afterward cooled at a rate 0.25 °C/min to a temperature T_2, which is 2–3 °C lower than the crystallization temperature T_{crys1}. The temperature of the medium at 20 % turbidity serves as crystallization temperature T_{crys1} of the co-crystal. The solution was again heated to 60 °C and left for 40 min to minimize memory effects and get the complete dissolution of solid. The steps were repeated with a new cooling rate (0.5, 0.75, and 1 °C/min) to obtain the crystallization temperature (T_{crys2}, T_{crys3}, T_{crys4}) of the co-crystal and then with the different ratio of SAC to CBZ.

There was some modification in the stage to determine the dissolution temperature T_{diss} by comparing with previous researches [6, 9, 10] as illustrated in Fig. 3. The solution was dissolved at 60 °C for 40 min and cooled with a fixed rate of 0.75 °C/min to a temperature T_b, which is 2–3 °C lower than the crystallization temperature T_{crys3}. The temperature T_b was kept constant for 40 min between point b and c. The solution was then heated with the different rate of 0.25, 0.5, 0.75, and 1.0 °C/min. Different heating rates resulted in different dissolution temperatures (T_{diss1}, T_{diss2}, T_{diss3}, and T_{diss4}). Solution at point e, f and g follows the same

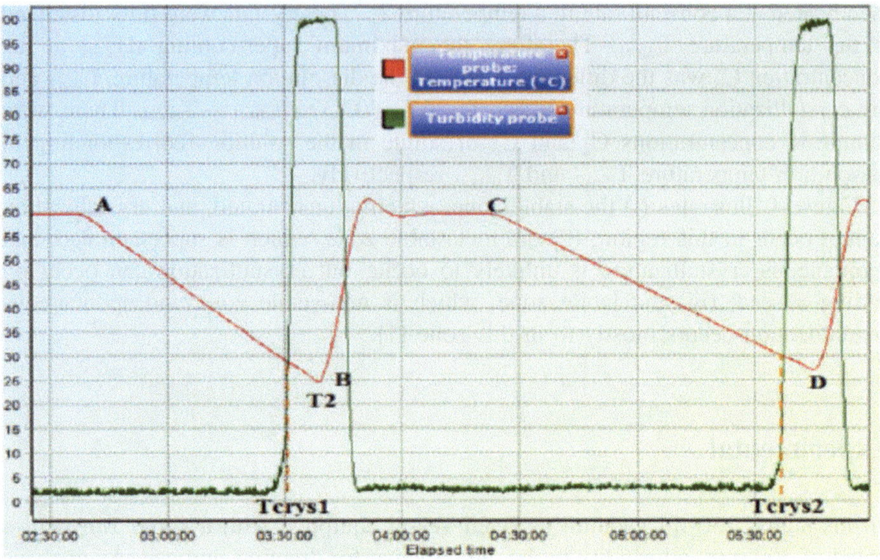

Fig. 2 Method for determining the crystallization temperature from the solution. *Lines AB* and *CD* are with cooling rate of 0.75 and 0.5 °C/min respectively

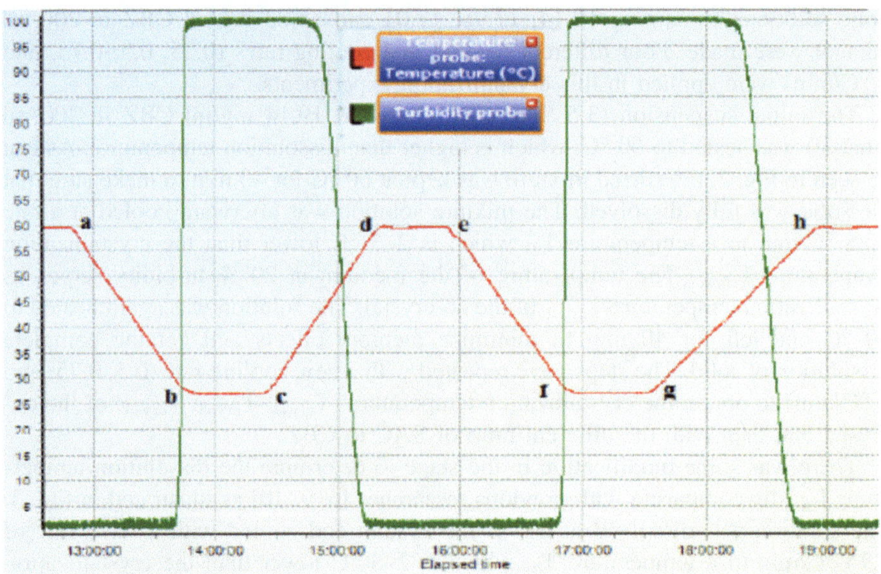

Fig. 3 New approached method for determining the dissolution temperature of the crystal. *Lines abc* and *efg* have the same cooling rate and isothermal temperature. *Lines cd* and *gh* are with heating rate of 0.75 and 0.5 °C/min respectively

profiles as point a, b, and c that resulted in the similar properties of co-crystal at a point c and d. Therefore, the proper comparison can be done for the effect of different heating rate on MSZW.

Result and Discussion

The data were processed to obtain the solubility–supersolubility curves and nucleation. The effect of heating and cooling rate R on T_{diss} and T_{crys} at different concentrations and different ratio of SAC/CBZ can be plotted in polynomial form ($T = ar^2 + ba + c$). The c-value is the interception of the curve at zero heating/cooling rates. This value can be taken as the equilibrium dissolution temperature T_{dissO} and the equilibrium crystallization temperature T_{crysO} [6]. The results for T_{dissO} and T_{crysO} are shown in Table 1. The graph of temperature (T_{diss} and T_{crys}) versus cooling/heating rate for the CBZ concentration of 19.14 mg/ml and different ratio of SAC to CBZ is plotted as shown in Fig. 4. The results show that T_{diss} for CBZ-SAC co-crystal increases with the increase of the heating rate and the ratio of SAC to CBZ in ethanol solution. The graph also represents that T_{crys} for CBZ-SAC co-crystal decrease with increasing heating rate and the ratio of SAC to CBZ in ethanol solution.

The effect of CBZ concentrations on MSZW is shown in Fig. 5. This figure shows that as the CBZ-concentration increases, T_{diss} increases and T_{crys} decreases with heating/cooling rate. These results are in accordance with previous works

Table 1 T_{crysO}, T_{dissO}, and dC/dT_{dissO} value of CBZ-SAC in ethanol

SAC/CBZ mol ratio	CBZ concentration, C (mg/ml)								dC/dT_{dissO}
	19.14		17.96		17.01		15.83		
	T_{crysO} (°C)	T_{dissO} (°C)	T_{crysO} (°C)	T_{dissO} (°C)	T_{crysO} (°C)	T_{dissO} (°C)	T_{crysO} (°C)	T_{dissO} (°C)	
3.5	45.98	55.56	40.15	54.00	39.06	52.05	35.32	43.82	0.252
3.0	44.91	53.49	35.00	50.50	31.50	50.07	32.15	39.70	0.212
2.5	35.62	51.22	30.28	49.75	27.81	47.00	26.61	38.74	0.236

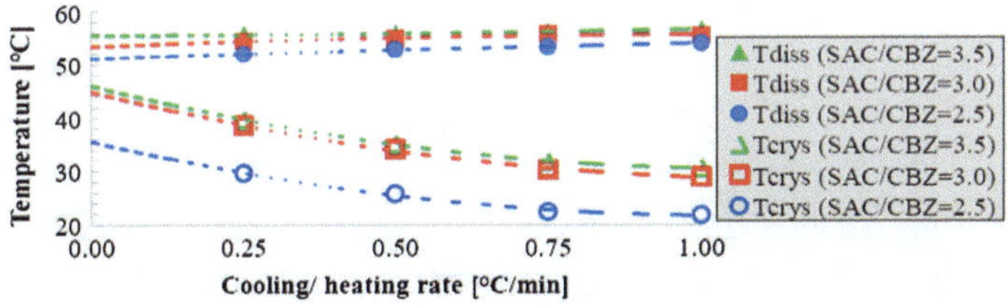

Fig. 4 MSZW for CBZ concentration of 19.14 mg/ml and different ratio of SAC to CBZ

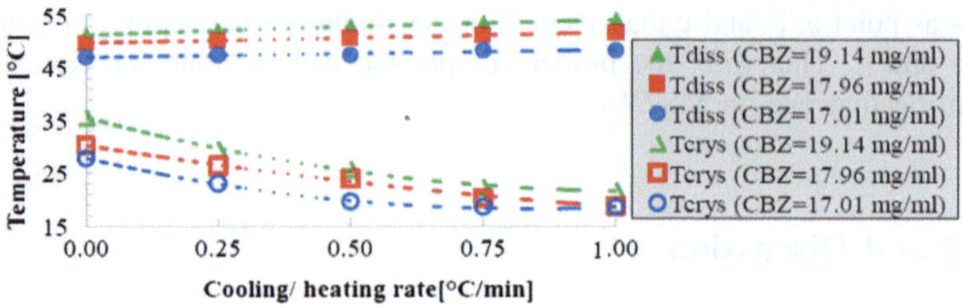

Fig. 5 MSZW for ratio (SAC/CBZ) of 2.5 and different values of CBZ concentration

Table 2 MSZW of CBZ-SAC co-crystals

SAC/CBZ mol ratio	MSZW (°C)			
	CBZ concentration, C (mg/ml)			
	19.14	17.96	17.01	15.83
3.5	9.57	13.42	12.99	8.50
3.0	8.58	15.07	18.57	7.55
2.5	15.60	19.48	19.19	12.13

[5, 10]. From the T_{crys0} and T_{diss0} results in Table 1, the graph concentration of CBZ, C against T_{diss0} was plotted and the slope of the graph dC/dT_{diss0} was obtained.

Table 2 shows the MSZW of CBZ-SAC co-crystals. MSZW is measured by the difference between T_{diss0} and T_{crys0} at equilibrium (zero rates) position. The lowest concentration of CBZ i.e., 15.83 mg/ml shows the smallest MSZW for all (SAC/CBZ) mol ratios. This trend is in accordance with previous work [5]. Moreover, for all CBZ concentration, an increase in (SAC/CBZ) mol ratio resulted in a decrease in MSZW; thus crystallization is easier to occur.

Using Nyvlt's approach, the nucleation rate J is defined as $J = k(\Delta C_{max})^m$, where k is the mass nucleation rate constant, m is the nucleation order, and ΔC_{max} is the absolute supersaturation (kgm^{-3}). ΔC is equal to $(dC/dT)\Delta T$. As known previously cooling rate R is equal to $\Delta T/\Delta t$. J also can be defined as $J = (dC/dT)R$. By substitutes all these equation, the following expression is obtained: $\ln(R) = (m-1)\ln(dC/dT) + \ln(k) + m\ln(\Delta T)$. By plotting ln(R) versus ln(ΔT), the value of m and k can be calculated. Value m is the slope of the graph, and the interception b is equal to $(m - 1)\ln(dC/dT) + \ln(k)$. Figure 6 shows the graphs of

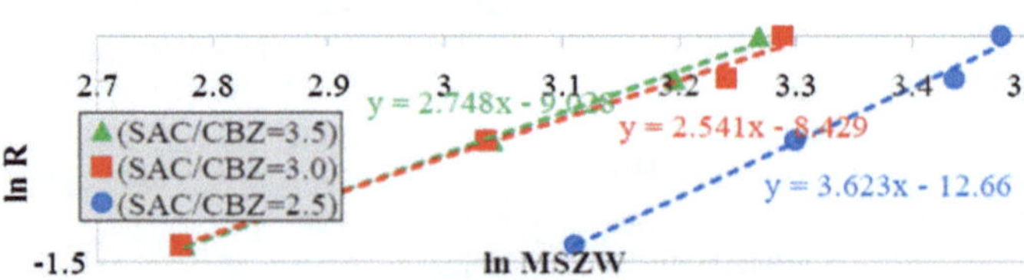

Fig. 6 Ln R versus ln MSZW for CBZ concentration of 19.14 mg/ml

Table 3 Nucleation order, m and the y-interception value of ln R versus ln MSZW, b

SAC/CBZ mol ratio	CBZ concentration, C (mg/ml)							
	19.14		17.96		17.01		15.83	
	m	b	m	b	m	b	m	b
3.5	2.75	−9.03	3.51	−11.64	2.61	−9.14	1.59	−5.23
3.0	2.54	−8.43	3.46	−11.88	4.82	−16.77	2.19	−6.83
2.5	3.62	−12.66	4.05	−14.14	6.31	−21.57	3.87	−11.85

Table 4 Nucleation kinetic constant of CBZ-SAC in ethanol

SAC/CBZ mol ratio	Nucleation kinetic constant, k			
	CBZ concentration, C (mg/ml)			
	19.14	17.96	17.01	15.83
3.5	1.33E−03	2.79E−04	9.88E−04	1.20E−02
3.0	2.39E−03	3.18E−04	1.97E−05	6.92E−03
2.5	1.41E−04	5.95E−05	9.23E−07	4.52E−04

ln(R) versus ln MSZW for CBZ concentration of 19.14 mg/ml and different ratio of (SAC/CBZ). Values of m and b are shown in Table 3 and the calculated nucleation kinetics k is given in Table 4. The analysis reveals that the nucleation order m for CBZ-SAC co-crystals increases with the SAC to CBZ ratio and have the average value of 3.44. This value is acceptable as reported in the literature that, the values of m are usually in the range of 1.65–4.9 for organic compounds [11].

Conclusion

In conclusion, this newly improved method is an accurate method for determination of dissolution temperature of the co-crystals. In this method, the supersaturated solution is assumed to have the same property and structure of the crystals due to the same profile of the previous cooling during nucleation event. Therefore, the proper comparison can be done for the effect of different heating rate on MSZW. It can also be concluded that the cooling/heating rate can strongly affect the MSZW of a CBZ-SAC mixture in ethanol solution. The MSZW is wider for a higher heating/ cooling rate. Moreover, the mol ratio of SAC to CBZ and the concentration of CBZ gave the tremendous effect on dissolution and crystallization temperature of the CBZ-SAC mixture in ethanol solution. Another advantage of this technique is that the machine can be pre-set according to desired profile, thus saving the human labour.

References

1. Mullin JW (2001) Crystallization. Butterworth-Heinemann, London
2. Mota FL, Teychéne S, Biscans B (2014) Measurement of the nucleation and growth kinetics of some middle distillate fuels and their blends with a model biodiesel fuel. Ind Eng Chem Res 53(7):2811–2819
3. Kannan V, Brahadeeswaran S (2013) Investigations on influence of coloration on growth of high quality hydrazonium l-tartrate single crystal. J Cryst Growth 374:71–78
4. Kulkarni SA, Kadam SS, Meekes H, Stankiewicz AI, ter Horst JH (2013) Crystal nucleation kinetics from induction times and metastable zone widths. Cryst Growth Des 13(6):2435–2440
5. Qian Y, Lu G, Sun Y, Song X, Yu J (2014) Metastable zone width of $SrCl_2 \cdot 6H_2O$ during cooling crystallization. Cryst Res Technol 49(1):78–83
6. Corzo DMC, Borissova A, Hammond RB, Kashchiev D, Roberts KJ, Lewtas K, More I (2014) Nucleation mechanism and kinetics from the analysis of polythermal crystallisation data: methyl stearate from kerosene solutions. CrystEngComm 16(6):974–991
7. Kubota N (2008) A new interpretation of metastable zone widths measured for unseeded solutions. J Cryst Growth 310(3):629–634
8. Bernardo A, Giulietti M (2010) Modeling of crystal growth and nucleation rates for pentaerythritol batch crystallization. Chem Eng Res Des 88(10):1356–1364
9. Bernardes CE, da Piedade MEM (2010) A fully automatic apparatus for thermal analysis of crystallization from solution and metastable zone width determinations. J Therm Anal Calorim 100(2):493–500
10. Bonnin-Paris J, Bostyn S, Havet J-L, Fauduet H (2011) Determination of the metastable zone width of glycine aqueous solutions for batch crystallizations. Chem Eng Commun 198 (8):1004–1017
11. de Arunmozhi G, de M Gomes E (2004) Metastability and crystal growth kinetics on L-arginine phosphate. Cryst Res Technol 39(1):34–39

Part V
Advanced Materials and Nanotechnologies

Activation of Titanium Dioxide Under Visible-Light by Metal and Non-metal Doping

Krishnan Jagannathan, Sikirman Arman and Nerissa Mohamad Elvana

Abstract Surface modification of the titanium dioxide, TiO_2 by doping with nitrogen or iron in order to produce visible-light active photocatalyst was investigated. Solgel method was used in preparing the modified TiO_2. Tetra titanium isopropoxide, ammonium nitrate and ferric nitrate were used as precursors with the dopant concentration of 1 % and calcination temperature of 600 °C. The prepared photocatalyst samples were characterized by XRD, BET and UV-vis DRS in order to study their physical properties. The results from XRD demonstrated that all the photocatalysts were anatase phase with high surface area. The reduction of band gap energy was shown by adding a non-metal dopant into TiO_2 lattice. Meanwhile, FTIR results revealed the presence of N and Fe dopant in the prepared photocatalysts. The properties of the photocatalyst affected its effectivity which was tested by performing a standard batch photocatalytic degradation experiment with methylene blue as a model pollutant under visible light. The result showed that N-doped TiO_2 and Fe-doped TiO_2 photocatalyst degraded the methylene blue within five hours of irradiation time yielding photodegradation efficiency of 72.3 and 61.98 %, respectively.

Keywords Doping · Titanium dioxide · Visible light · Solgel · Photocatalytic degradation · Nitrogen · Iron

K. Jagannathan (✉) · S. Arman · N.M. Elvana
Fakulti Kejuruteraan Kimia, Universiti Teknologi MARA, 40450 Shah Alam, Selangor, Malaysia
e-mail: jagannathan@salam.uitm.edu.my

S. Arman
e-mail: 2012873978@isiswa.uitm.edu.my

N.M. Elvana
e-mail: elvana68@yahoo.com

M.A. Hashim (ed.), *ICGSCE 2014*, DOI 10.1007/978-981-287-505-1_32

Introduction

Photocatalysis is one of the highly effective applications in destructing organic pollutants to harmless end products at ambient conditions using light and photocatalyst [1]. Titanium dioxide (TiO_2) is the commonly used photocatalyst in many environmental applications such as decolorizing of dyes, degradation of organic pollutants, water purification and hydrogen production from water splitting. It is due to its desirable characteristics such as economical, environmental friendly, structural stability and high degradation rate [2, 3]. However, pure TiO_2 Titanium dioxide can be used only under UV light because of its high band gap energy (3.2 eV). The application of UV is highly energy intensive and hence it makes the process expensive. Besides, in the solar light spectrum, the UV light exists only 3–5 % compared to visible-light (45 %) thereby limiting the usage of sun light. In order to activate TiO_2 under visible-light ($400 < \lambda < 500$ nm), it is necessary to modify its surface by doping it with certain elements.nitrogen (N) and iron (Fe) were selected as dopants in this study because they both could prevent the undesirable recombination of the electron-hole pair, otherwise which would decrease the performance of the photodegradation [4].

Iron doped TiO_2 under visible-light showed better performance in degrading the toluene gases as compared to pure TiO_2 under visible-light [5]. The presence of iron dopant would incorporate into TiO_2 lattice, which made it photo-response to visible-light and thus increased the photocatalytic activity. Meanwhile, nitrogen dopant induced reduction of the band gap between conduction and valence band [6]. Besides, the nitrogen dopant absorbed the visible-light to generate electron-hole pair and thereby finally increasing the photocatalytic activity. In this study, nitrogen and iron doped TiO_2 samples were prepared by using solgel method and their surface properties were compared. Finally, the effectivity of prepared photocatalysts was tested by performing a standard batch photocatalytic degradation experiments under visible-light with methylene blue as a model pollutant.

Materials and Method

Preparation of Photocatalyst, N doped TiO_2 was prepared using Solgel method which had been described in our previous study [2]. Titanium (IV) isopropoxide (TTIP) (Merck, USA), and Ammonium nitrate (NH_4NO_3) (Merck, USA) were used as starting materials in preparing the non-metal doped photocatalyst. These prepared photocatalyst formed were labeled as N-TiO_2. For metal doped TiO_2, the photocatalyst prepared were done by changing the precursors to Ferric nitrate (Fe $(NO_3)_3$) (Friendemann Schmidt, Australia) as the source for iron dopant and labeled as Fe-TiO_2.

Characterization of Photocatalyst, The photocatalysts were analyzed using X-ray diffractometer (XRD, Rigaku Ultime IV) with Cu Kα radiation sources and operating at a scanning rate of 8 °/s in a 2θ range of 20–80° to determine the phase of the crystal formed. The photocatalysts were then analyzed using AUTOSORB-1 Quanta Chrome Instruments (USA) at −196 °C to obtain the BET (Brunauer, Emmett, and Teller) surface area. Fourier Transform Infrared (FTIR) Spectroscoy (Perkin Elmer Spectrum One) was used to determine the specific absorption peak wave number in the range of 4000–500 cm^{-1} with 4 cm^{-1} of resolution. Finally, the band gap of photocatalysts surface was determined using UV-vis Spectrophotometer (Lambda 35, Perkin Elmer) where $BaSO_4$ was used as a reflectance standard in UV-vis diffuse reflectance method.

Photocatalytic Degradation, The performance of the photocatalyst was determined by measuring the removal efficiency of Methylene Blue (MB) under visible light. The batch experiments were conducted only for a period of five hours of irradiation time to compare the relative performance of the prepared photocatalysts. A 25 W Phillips Compact Fluorescence Lamp emitting at an average rate of 2.448 W/m^2 (555 nm) was used as the visible-light source. A 0.05 g of N-TiO_2 photocatalyst was added into 50 ml of MB solution having a concentration of 10 ppm. The MB solution was stirred for 30 min in the dark condition to ensure equilibrium adsorption prior to photodegradation. Then, the photocatalytic degradation experiment was run under visible-light meanwhile the samples were collected every 30 min for 5 h for the analysis of its residual concentrations. To ensure the photocatalyst separated from the MB solution, it was centrifuged for 5 min at 10,000 rpm. The decolorized MB was analyzed using UV-vis Spectrophotometer at 660 nm. The experiment was repeated for Fe-TiO_2 photocatalyst.

Results and Discussion

XRD, Fig. 1 shows the XRD patterns of the photocatalysts. Fe-TiO_2 and N-TiO_2 photocatalysts approximately fitted with the JCPDS file number (PDF no. 21-1272) where the samples were of anatase phase of TiO_2 with the characteristic features and diffraction peaks at 25.281° (101), 37.80° (004), 48.05° (200), 53.89° (105), 55.06° (211), 62.688° (204), 68.76° (116), and 75.029° (215). The photocatalysts were found in anatase phase due to the appropriate combinations of dopant concentration and calcination temperature used in the preparation process of photocatalysts as discussed in our previous study [2, 3]. The main peak patterns for both Fe-TiO_2 and N-TiO_2 were found at 25.361° and 25.379°, respectively. Both metal and non-metal dopants showed a shift to the right where iron dopants got shifted more compared to non-metal dopant. This was due to the formation of crystallographic point defects by the substitution of titania ion, Ti^{4+} by iron ion, Fe^{3+} owing to their similarities in ion radii [7] where most of the iron ions possibly got into TiO_2 Titanium dioxide structure at interstices or the ions might be occupied at the titanium lattice sites to form a solution with iron-titanium solids [8]. Meanwhile, the existence of the nitrogen dopant did not change the basic crystalline structure of TiO_2 but had intermingled within TiO_2 surface [7].

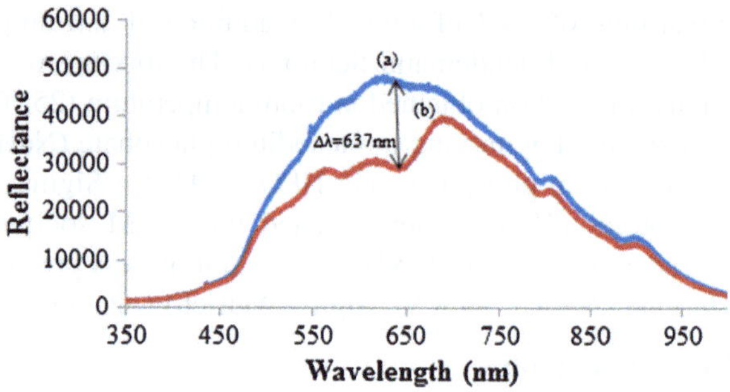

Fig. 1 Reflectance spectra of the immobilized Co^{2+} ion (*a*) and its complex with 50 ppm NH_3 (*b*) at pH 13

of 630 nm. When Co^{2+} ion-immobilized silica pellet reacted with NH_3, it changed to blue colour and gave maximum reflectance at 683 nm. However, the reflectance intensity was diminished when the bright pink-coloured Co^{2+} ion silica pellet was turned to dark blue. Figure 1 depicts the reflectance spectra of immobilized Co^{2+} ion and its NH_3 complex at pH 13.

The chemical reaction taken place between Co^{2+} ion and NH_3 can be described with Eq. 1:

$$CO^{2+}{}_{(aq)} + 6NH_{3(aq)} \rightleftharpoons \left[CO(NH_3)_6\right]^{2+}{}_{(aq)} \quad (1)$$

In general, NH_3 acts as both a base and a ligand. When an aliquot of NH_3 is introduced towards immobilized Co^{2+} ion, NH_3 molecules replace water ligands attached to the pink hexaaquacobalt(II) $[Co(H_2O)_6]^{2+}$ ion and that dark blue colours of immobilized hexamminecobaltate(II) ($[Co(NH_3)_6]^{2+}$) complex are formed. The formation of dark blue immobilized $[Co(NH_3)]_6^{2+}$ complex absorbed incident light transmitted by the bifurcated optical fiber and attenuated the reflectance signal over the visible wavelength. The immobilized Co^{2+} ion, on the other hand, demonstrated higher reflectance intensity as bright colours reflect better than dark. Because maximum reflectance difference between immobilized reagent and complex was observed at 637 nm, therefore, this wavelength was used as working wavelength for subsequent experiments.

pH Effect

Basically, NH_3 in water exist in equilibrium with NH_4^+ ion as shown in Eq. 2:

$$NH_4{}^+ + H_2O \rightleftharpoons NH_3 + H_3O^+ \quad (2)$$

The equilibrium is extremely dependant on the pH where high pH favours the formation of NH_3 and vice versa [10]. Figure 2 shows that the optimum response for NH_3 pellet sensor was obtained at pH 13 and 637 nm, whereby the largest relative reflectance acquired between immobilized reagent and complex.

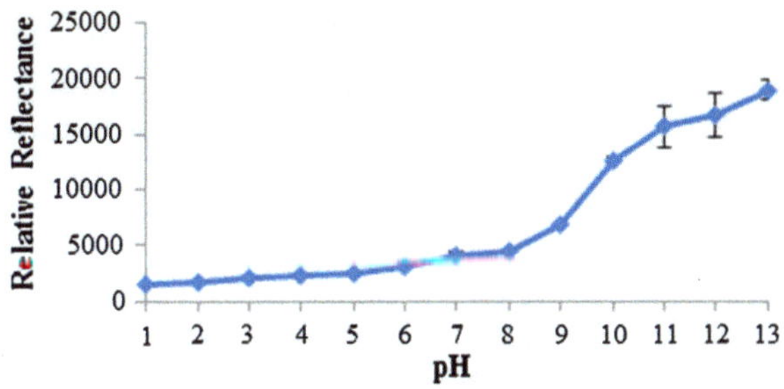

Fig. 2 The effect of pH on the immobilized $[Co(NH_3)_6]^{2+}$ complex at 637 nm using 50 ppm NH_3 at pH 13

From pH 1 to pH 6, the acidic conditions favouring the protonation of NH_3, thereby the formation of blue-coloured $[Co(NH_3)_6]^{2+}$ complex was not discernible, and rendered higher reflectance signal which was relatively similar to that of the immobilized Co^{2+} ion with no NH_3 was added. As the pH increased from pH 8 and onwards, the blue colouration of silica pellet was getting obviously developed. The deprotonation of analyte was high in basic conditions especially from pH 11.6 to pH 12.5, and the blue colour became darkest at pH 13 due to the presence of excess NH_3 in basic conditions, which promoted the formation of immobilized $[Co(NH_3)_6]^{2+}$ complex [11]. Thus, optimum pH 13 was applied for further optimization analyses.

NH_3 Detection

The silica pellet sensor was tested with a range of NH_3 concentrations. The reflectance signal increased linearly with increasing NH_3 concentration and gradually levelling off at a steady-state relative reflectance from 500 ppm NH_3 onwards (Fig. 3).

The calibration curve of relative reflectance versus NH_3 concentration was found linear over the range of 18–100 ppm NH_3 with a correlation coefficient (R^2) of 0.9866. The limit of detection (LOD) of the NH_3 pellet sensor was calculated to be 15 ppm NH_3. The pellet sensor was found taking 3 min for full colour development upon addition of NH_3 solution. Therefore, 3 min reaction time was fixed for every optical NH_3 detection. The measurement for identical addition of analyte was reproducible at 0.5 % RSD (n = 3), and the sensor repeatability was obtained at <5.0 % RSD (n = 5) indicated that the NH_3 pellet sensor can be satisfactorily reused for *in situ* environmental monitoring of NH_3 pollutant. The NH_3 optode retained almost 100 % of its initial reflectance response over 6 months' storage period at ambient temperature indicative of a highly stable immobilized reagent. Based on the linear range obtained, it can be deduced that this silica-based pellet sensor can be used for continuous monitoring of NH_3 in sewage, domestic waste water, landfill leachate, etc. whereby the typical NH_3 levels in those waters are normally between

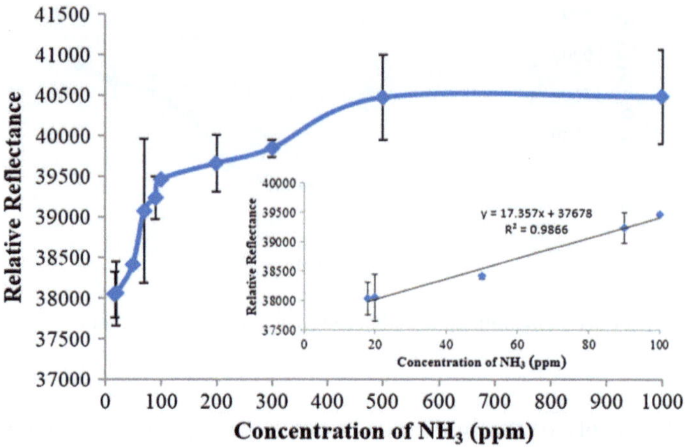

Fig. 3 The response curve of the pellet sensor generated using NH_3 concentrations from 18 to 100 ppm at pH 13. The inset shows the dynamic linear NH_3 concentration range for the silica-based pellet sensing material

10 and 40 ppm. However, it can be excessive up to 100 ppm or more in the highly polluted waters e.g. river waters.

Optical sensor for NH_3 detection in aqueous environment has recently been developed by Varghese et al. [12] based on evanescent wave fiber optic sensor with dual wavelength probing for NH_3 detection in water at ppb levels. Pre-treatment of the water sample (i.e. dilution) becomes necessary when the sensor is used for on-site analysis of sewage for sample concentration >17 ppm, and that direct determination of NH_3 in envronmental samples become impposible. Optical NH_3 sensor for higly polluted water has also been fabricated by Tan et al. [13] with broader dynamic linear range (~1000–5000 ppm NH_3) using reflectance approach. However, it took some 6 min long for assay of NH_3 in water. This suggests the potential of the proposed pellet-based sensor for in situ analysis of NH_3 in aqueous media, which may be performed in a quantitative or semi-quantitative manner based on pellet colour change.

Interference

Interference study is crucial to determine the pellet sensor performance for optical sensing of NH_3 in solution. Based on the results shown in Table 1, alkali, alkaline earth and SO_4^{2-} ions do not pose significant interference to the determination of 50 ppm NH_3 using the developed silica pellet sensor. The presence of those elements in the concentration higher than NH_3 is usually not possible in water samples [14].

Table 1 The pellet sensor response to various interfering agents at different concentration ratios

Interfering ion	Pellet sensor response at different interfering ion concentration to NH_s concentration ratios				
	1:0	1:1	1:10	1:100	1:1000
Na^+	38412.61 ± 41.69	38313.90 ± 353.85	31157.45 ± 354.09[a]		
K^+	38412.61 ± 41.69	38301.46 ± 268.26	38321.12 ± 346.82	38402.04 ± 43.19	33238.46 ± 5.53[a]
Mg^{2+}	38412.61 ± 41.69	38333.95 ± 340.64	38317.78 ± 83.88	33063.76 ± 118.79[a]	
Ca^{2+}	38412.61 ± 41.69	38426.01 ± 26.12	38417.17 ± 163.53	31245.89 ± 329.83[a]	
SO_4^{2-}	38412.61 ± 41.69	38423.42 ± 48.62	38404.83 ± 122.15	38333.04 ± 48.15	30133.04 ± 32.89[a]

[a] *t*–statistic exceeded *t*–critical value

Conclusion

The proposed pellet-based optosensor has the potentiality for visual determination of NH_3 in environmental water samples. The small size of the silica microparticles presented no significant barrier for diffusion of NH_3. The large interfacial contact area of the immobilized Co^{2+} ion on the microsilica with NH_3 allows mass transfer of NH_3 molecule to the microparticles' surfaces to effect the immobilized charge-transfer complex formation, thus faster response time and broader dynamic range were achieved. The proposed NH_3 pellet sensor involved fast detection procedure, which merely requires dispensing a small aliquot of sample over the sensor surface. It is also an economical sensing device since no substrate is required to contain the sensing matrix, whereby the stand-alone reagent phase in its finished form is free from physical support. The simplicty-based sensing method allows NH_3 determination for a large number of environmental samples to be carried out in a short period of time.

Acknowledgment We gratefully acknowledge Universiti Kebangsaan Malaysia for the Young Researchers Encouragement Grant (GGPM-2014-035) and Malaysian Ministry of Higher Education for Fundamental Research Grant Scheme (FRGS/2/2014/SG01/UKM/02/1) and Exploratory Research Grant Scheme (ERGS-RDU120604).

References

1. Appl M (1999) Ammonia: principles and industrial practice. Wiley, New York
2. Hasan HA, Abdullah SRS, Kamarudin SK, Kofli NT (2011) Problems of ammonia and manganese in Malaysian drinking water treatments. World Appl Sci 12(10):1890–1896
3. Shaposhnik A, Ryabtsev S, Zviagin A, Korchagina S, Meshkova N, Shaposhnik D, Vasiliev A (2011) Selective detection of ammonia and its derivatives using MOX-sensor and microreactor. Procedia Eng 25:1097–1100
4. World Health Organization (2011) Guidelines for drinking-water quality, 4th edn. World Health Organization, Geneva
5. Ni JQ, Heber AJ (2001) Sampling and measurement of ammonia concentration at animal facilities-A review. American society of agricultural engineers annual international meeting paper no. 01-4090
6. Aomura Y, Kobayashi Y, Miyazawa Y, Shimizu H (2010) Application of copper sulfate pentahydrate as an ammonia removal reagent for the determination of trace impurities in ammonia gas chromatography. J Chromatogr A 1217:1838–1844
7. Nakano N, Sugata K, Nagashima K (1995) Development of a monitoring tape for ammonia gas in air by fluorescence detection. Anal Chim Acta 302(23):201–205
8. Masserini RT, Fanning KA (2000) A sensor package for the simultaneous determination of nanomolar concentration of nitrite, nitrate, ammonia in seawater by fluorescence detector. Mar Chem 68(4):323–333
9. Rastegarzadeh S, Razei V (2007) An optical sensor for zinc determination based on zincon as sensing reagent. Sens Actuators B 129(1):327–331
10. Staden JF, Taljaard RE (1997) Determination of ammonia in water and industrial effluent streams with the indophenol blue method using sequential injection analysis. Anal Chim Acta 344:281–289

11. Pranaitytė B, Jermak S, Naujalis E, Padarauskas A (2007) Capillary electrophoretic determination of ammonia using headspace single-drop microextraction. Microchem J 86:48–52
12. Varghese BP, Pillai AB, Naduvil MK (2013) Fiber optic sensor for the detection of ammonia, phosphate and iron in water. J Optics 42(2):78–82
13. Tan LL, Ahmad M, Lee YH (2012) A novel optical ammonia sensor based on reflectance measurements for highly polluted and coloured water. Sens Actuators B: Chem 171–173:994–1000
14. Azrina A, Khoo HE, Idris MA, Amin I, Razman MR (2011) Major inorganic elements in tap water samples in Penisular Malaysia. Malays J Nutr 17(2):271–276

Low Noise and Properties of Double Layer Concrete Paving Blocks

Euniza Jusli, Hasanan Md. Nor, Ramadhansyah Putra Jaya, Zaiton Haron, Musli Nizam Yahya, M. Azman and Lee Kuan Lim

Abstract Double layer rubberized concrete block (DL-RCPB) was introduced to effectively make use of waste tyre and to improve the performance of concrete pavement block. This study was carried out to justify the effectiveness of using waste tyre rubber as aggregate. In this study, there were eight series of concrete mix with 10 and 20 % waste tyre rubber replacement level and different thickness of facing layer; 10, 20, 30 and 40 mm. The compressive strength was reduced when the percentage of waste tyre rubber was increased. Remarkably, all types of DL-RCPB with rubber percentage up to 20 % achieved minimum required compressive strength of 45 MPa. Moreover, DL-RCPB (20 %) with 40 mm top layer was proven

E. Jusli · H. Md. Nor (✉) · R.P. Jaya · L.K. Lim
Department of Geothechnics and Transportation, Faculty of Civil Engineering,
Universiti Teknologi Malaysia, 81310 Skudai, Johor, Malaysia
e-mail: hasanan@utm.my

E. Jusli
e-mail: eunizajusli@gmail.com

R.P. Jaya
e-mail: ramadhansyah@utm.my

L.K. Lim
e-mail: kevinleekuanlim@gmail.com

Z. Haron
Department of Structure and Materials, Faculty of Civil Engineering,
Universiti Teknologi Malaysia, 81310 Skudai, Johor, Malaysia
e-mail: zaitonharon@utm.my

M.N. Yahya
Department of Engineering Mechanics, Faculty of Mechanical and Manufacturing
Engineering, Universiti Tun Hussein Onn, Parit Raja, 86400 Batu Pahat,
Johor, Malaysia
e-mail: musli@uthm.edu.my

M. Azman
Department of Engineering, Razak School of Engineering and Advanced Technology,
Universiti Teknologi Malaysia, International Campus, Jalan Semarak,
54100 Kuala Lumpur, Malaysia
e-mail: az_man@ic.utm.my

M.A. Hashim (ed.), *ICGSCE 2014*, DOI 10.1007/978-981-287-505-1_34

to have the most effective sound absorption concrete mix design with up to 36 % of sound adsorption recorded. Reduction in block density was found when the percentages of rubber were increased. Porosity up to 17 % was recorded for DL-RCPB (20 %). The rough surface of rubber particles tends to attract air on the rubbers' surface and the trapped air tends to produce voids once the concrete hardened. Porosity of DL-RCPB was increased when the rubber was added to concrete mix, thereby decreasing the density of DL-RCPBs. Low density and high porosity increased the intensity of sound absorption and relatively gives positive impact on the development of low noise CBP as long as the compressive strength was sufficient.

Keywords Double layer · Rubberized · Concrete · Paving block · Tyre rubber

Introduction

In general, concrete paving block (CPB) is more dependable as compared to other pavement materials due to its better attribute in many aspects. Compressive strength, abrasion, skidding resistance, weathering durability, serving period, aesthetic feature and flexibility in terms of installation and removal are example of aspects that CPB has showed in remarkable performances [1]. Although previous studies on concrete paving blocks has improved many aspects in the pavement industry, but there are still some undesirable effects generated. Among the undesirable effects created are noise pollution and poor riding quality. Many researches have been done on the physical and mechanical properties of different types and design of concrete paving block [2–4]. These initiatives were taken in order to widen the usage and improve the performance of CPB. However, it seems that there were limited researches focusing on the development of low noise CPB. Waste tyre was well known for its high durability and elasticity properties that make it suitable materials substituent in concrete to improve sound and energy absorbability. Apparently, recycling waste tyre rubber as concrete materials seems to be an ideal solution towards treating this non-biodegradable waste. Therefore to effectively make use of the waste tyre rubber and increase the performance of CPB, double layer rubberized concrete paving blocks (DL-RCPBs) was introduced in a hope to tackle the insufficiencies [5, 6]. DL-RCPBs were designed in order to reduce the traffic noise due to the interaction between vehicle's tyre and pavement. Waste tyre rubber was used as a medium to absorb more sound and vibration in concrete, in returns to produce low noise CPB. The effect of layer and rubber content on several physical and mechanical properties of double layer rubberized concrete blocks was evaluated in this study. The properties that comprised in study include compressive strength, sound absorption, density, porosity and skid resistance. Moreover, the results found can also be used as indicators or references for future study as limited researches are been done on rubberized concrete with double layer characteristic.

Materials

Cement, Ordinary Portland Cement (TASEK cement) was used throughout the experiment. This type of cement was compiled to ASTM C150 Type I [7]. The cement are composed of SiO_2, Al_2O_3, Fe_2O_3, CaO, MgO and SO_3, respectively [8].

Aggregate, In this study, local natural crushed granite and natural river sand were used as coarse and fine aggregate. These aggregates are compiled the requirements stated in ASTM C136-06 [9]. Figure 1 illustrate the size distribution of coarse aggregates; aggregates fall in the fixed acceptance ranges. The aggregates used are already in saturated dry condition thus no further drying process is required.

Sand, For CPB concrete mixture, natural river sand complies with the requirement in ASTM C136-06 [9] was used. The size distribution of sand particles was presented in Fig. 2. The sand used was left for air drying for several days before it been used for the mixing.

Rubber Aggregate, There are two types of rubber aggregate; fine and coarse aggregates as shown in Fig. 3 has been adopted in this study mainly, each type of aggregate was classified according to its size. Fine tyre rubber with size ranging from 1 to 4 mm was used for sand replacement. Meanwhile, 5–8 mm of tyre rubber was utilized as coarse aggregate replacement. These rubber aggregates were supplied by Yong Fong Industry, Malaysia. Tyre rubber are composed of 48 % styrene-butadience rubber (SBR), 47 % carbon black, 1.9 % extender oil, 1.1 % zinc oxide, 0.8 %, sulfur, 0.7 % accelerator and 0.5 % streáric acid [10].

Superplasticizer, To maintain the feasibility of the concrete mixtures, a Glenium C380 superplasticizer was used throughout this study. This superplasticizer is chloride-free and complies with the requirements of the ASTM C494-13 [11].

Mix Proportion, Control mix with mix proportion of cement: sand: aggregate: superplasticizer; 1:1.5:1.7:0.003 and 0.47 water cement ratio (w/c) was used. The control mix was designed to carry a characteristic strength of 50 MPa at 28 days.

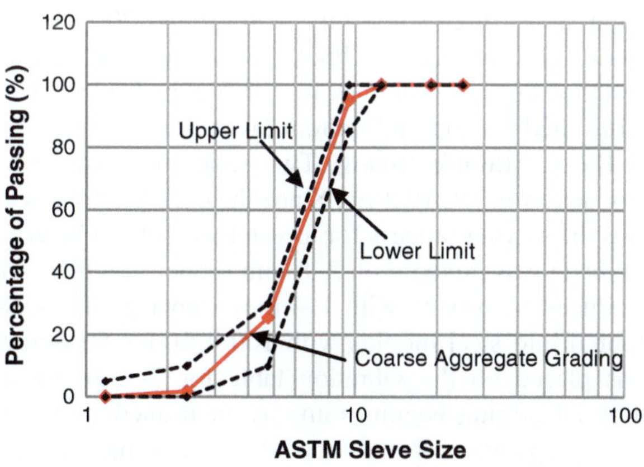

Fig. 1 Sieve analysis result for coarse aggregate

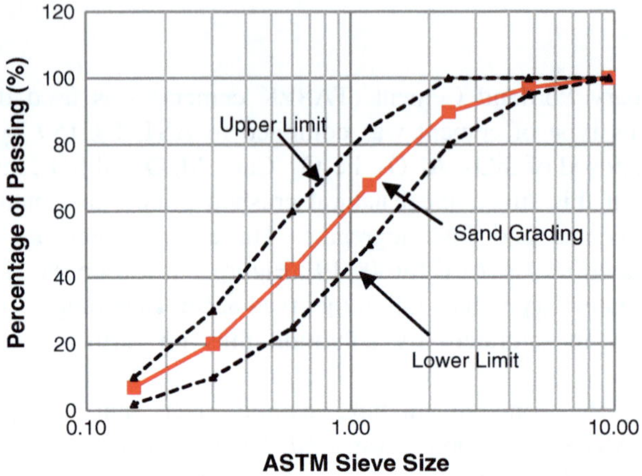

Fig. 2 Sieve analysis result for sand

Fig. 3 Waste tyre rubber

Eight series of concrete mix with 10 and 20 % of waste rubber aggregate substitutions and different thickness of facing layer was investigated in this study. The designed concrete samples was shown in Table 1. Coarse and fine tyre rubber was substituted for top and bottom layer, respectively.

Block Manufacturing, Two series of concrete mixture were produced in the production of DL-RCPBs. Since the concrete blocks consist of two concrete layers, thus the mixing process for top and the bottom layer was done in sequence. Series I (layer 1) was mixed with 5–8 mm rubber granules, whereas series II (Layer 2) was produced by mixing with 1–4 mm rubber granules. Series II concrete mixture was poured into steel moulds with 100 × 200 × 80 mm dimension. Then, steel mould was placed on the vibration table for 5 s. At the meantime, the top layer raw material mixing begun. Same as mentioned steps were used for bottom concrete layer preparation. Removal of the excess material and flattening the block surface

Table 1 Concrete mix design

Mix symbol	Mix proportion		Cement content (kg/m^3)	Water/cement ratio	Rubber content (%)
	Series I (C:A:S)	Series II (C:A:S)	Series I&II	Series I&II	Series I&II
CCPB	1:1.7:1.5	1:1.7:1.5	489	0.47	0
DL-RCPB (10 %)	1:1.5: 1.5	1:1.7:1.35	489	0.47	10
DL-RCPB (20 %)	1:1.35:1.5	1:1.7:1.2	489	0.47	20
DL-RCPB (30 %)	1:1.2:1.5	1:1.7:1.05	489	0.47	30
DL-RCPB (40 %)	1:1.0:1.5	1:1.7:0.9	489	0.47	40

C:A:S Cement: Aggregate: Sand
CCPB Control concrete paving block
DL-RCPB Double layer rubberized concrete paving block

was done after completing vibration process. The features of DL-RCPB were shown in Fig. 4.

All blocks were covered with moist hessians for a period of 24-h before demoulding. Air curing method was adopted as all the blocks were kept in laboratory under room temperature of 25 °C and 65 % relative humidity. The preparation method of cylinder samples with 99.5 mm diameter and 80 mm height for sound absorption test was similar as blocks.

Testing Methods, Sound absorption, compression test, density and porosity were carried out to investigate the properties of hardened DL-RCPB with different layer thickness. Sound absorption test was carried out using impedance tube method; a tube, two microphones and a digital frequency analysis system. In this test, an array of complex data obtained from the measured transfer function of wave frequencies would be processed by computer program. The absorption coefficient of the related material would then be generated later by the program in excel data form. Cylindrical sample with diameter of 99.5 and 80 mm height were tested according to ASTM E1050 [12]. The compression test was carried out using

Fig. 4 Double layer rubberized concrete block

MATEST compression machine with 3000 kN capacity and pace rate of 2.5 kN/s. Average of three samples each for 7 and 28 days strength with different mix designs were tested. According to BS 6717 Part 1 [13], 3 mm plywood was used applied on top and bottom of block surface. Density and prosity properties of DL-RCPBs were carried out according to the standard procedures stated in ASTM C642 [14].

Results and Discussion

Sound Absorption, The ability of DL-RCPBs to absorb sound was measured by the sound absorption coefficient (α). Figures 5 and 6 indicate an improvement of acoustic performance as the percentage of rubber substitution and thickness of top layer were increased. Apparently, similar graphs' pattern was observed for both graphs in Figs. 5 and 6. It was found that DL-RCPBs for 10 and 20 % rubber content have better sound absorption properties as compared to CCPB which only achieve 0.05 maximum α-value. From the results shown in Fig. 5, DL-RCPB (10 %) with 40 mm layer thickness was noticed to have the highest α-value of 0.253 at 700 Hz. This implied that the thickness of the top layer or amount of coarse aggregate has an influent on the acoustic performance. Besides, similar result pattern was shown by 20 % rubber content cylinders. Tyre rubber content of DL-RCPB (20 %) shown in Fig. 6 recorded the sound absorption coefficient of 0.364 which higher than 0.253 shown by the highest of DL-RCPB (10 %). Indeed, the damping of sound and reduction in dynamic energy were promoted when the rubber content increased in advance.

Compressive strength, The results of compression test for DL-RCPB with 10 and 20 % rubber substitution were illustrated in Fig. 7. Each value represents the average result of three samples. In general, the compressive strength of DL-RCPB decreases with the increase of rubber tyre substitution and thickness of facing layer.

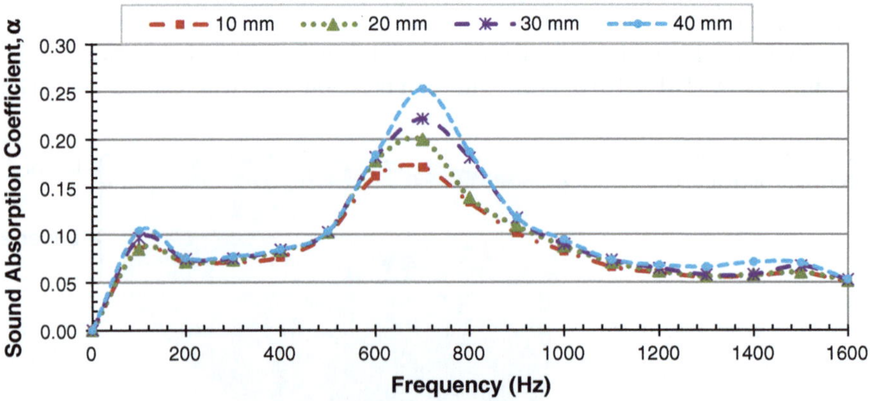

Fig. 5 Sound absorption of DL-RCPBs (10 %)

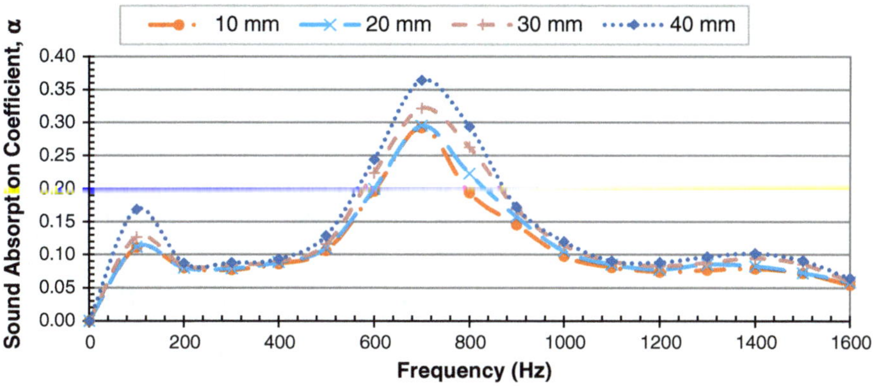

Fig. 6 Sound absorption of DL-RCPBs (20 %)

Fig. 7 Compressive strength of double layer rubberized concrete paving blocks

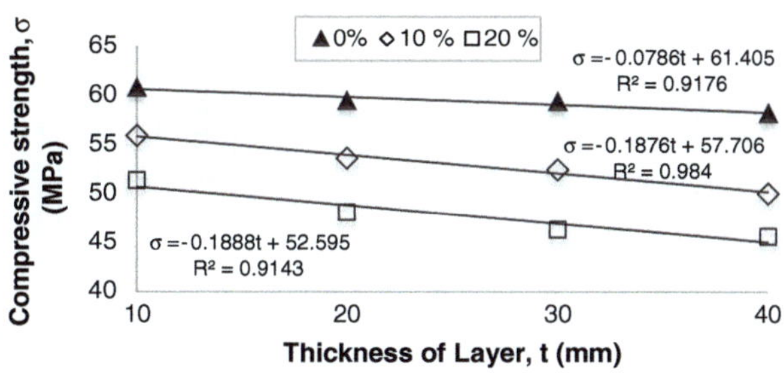

Compressive strength of DL-RCPB (20 %) with 40 mm thick of facing layer achieved the minimum requirement of 45 MPa. Besides, the results for DL-RCPB (20 %) indicate that the highest compressive strength recorded was 51.37 MPa. From the results, it shows progressive reduction of strength as the thickness of facing layer increase. This may due to less adhesion between tyre rubber and cement paste. For DL-RCPBs (10 %), the entire concrete block satisfied with minimum compressive strength requirement of 45 MPa. However, for DL-RCPB (20 %), only DL-RCPB with 10 mm top layer that satisfies the minimum strength requirement.

Density, The density results for DL-RCPB (10 %) and DL-RCPB (20 %) were presented in Fig. 8. Referring to Fig. 8, the density for both percentages of rubber substituted blocks decreased progressively as the thickness of the top layer increased. DL-RCPB (20 %) with 40 mm thickness of facing layer recorded the lowest density of 2218.30 kg/m^3 this value is 3 % lower than the control block. The density recorded for control block was 2298.51 kg/m^3. This phenomenon happened because of different complexity in block and the presence of the low density substitution material.

Porosity, A steady increment of porosity pattern was obtained with the increase of top layer thickness for both percentages of tyre rubber contents as shown in

Fig. 8 Density of DL-RCPBs

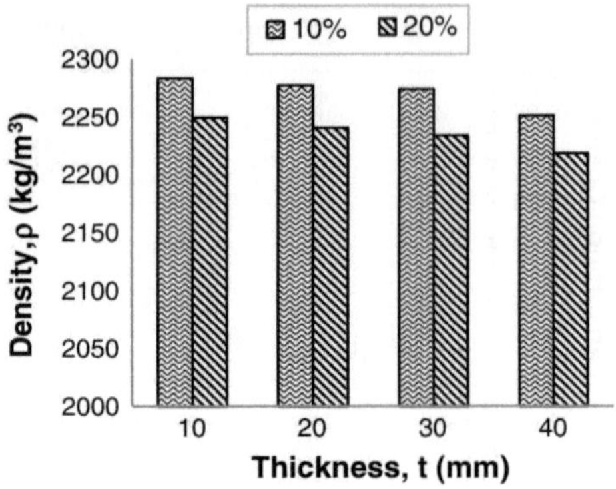

Fig. 9 Porosity of DL-RCPBs

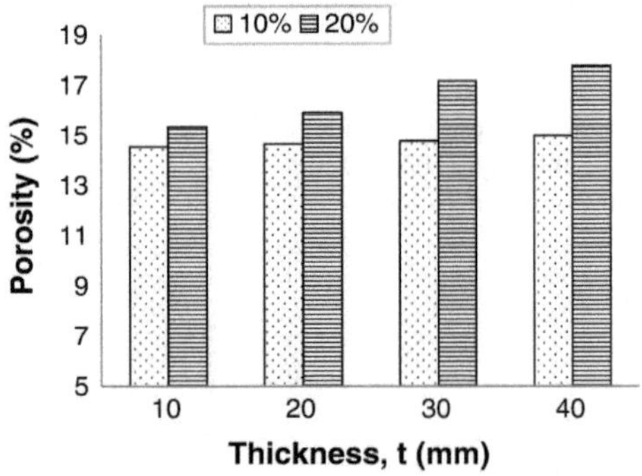

Fig. 9. Based on the results, the porosity of different mix designs was ranged from 14.54 to 15.00 % for 7 days, whereas at 28 days the porosity ranged from 15.36 to 17.78 %. From there the lowest porosity value was recorded for the control mix, whereas the highest porosity was displayed by DL-RCPB (20 %) with 40 mm facing layer thickness. Non-polar nature of the tyre rubber material was the reason behind this phenomenon. Regarding to that, air particle would easily attracted around the rubber aggregate surface during mixing process causing a weaker bond between non-polar particles and polar particle. According to Sandberg and Ejsmont [14], the air pumping effect was effectively reduces by porosity, thereby reducing the tyre-pavement interaction noise.

Conclusion

The conclusions that can be drawn as follows:

(i) Compressive strength of the DL-RCPB tends to decrease when more percentage of tyre rubber was substituted.

(ii) Concrete blocks with 10 and 20 % tyre rubber content regardless the thickness of facing layer was found to comply with minimum requirement of 45 MPa.
(iii) Sound absorption was increased when the percentage of tyre rubber substitution and the amount of coarse rubber aggregates was increased. DL-RCPB (20 %) with 40 mm facing layer thickness recorded up to 36 % of sound absorption improvement when compared with the control block.
(iv) Density indicated a reduction when the percentage of the tyre rubber was increased. Meanwhile, porosity experienced a rise when higher percentage of the tyre rubber was substituted.
(v) Porosity increased due to the physical characteristics of rubber particle, which has rough surface that tend to entrap air to their surface.

Acknowledgment The authors would like to express their gratitude toward Ministry of Higher Education of Malaysia (MOHE) and Universiti Teknologi Malaysia for the financial support by providing research grant RUG 03H47 for this study.

References

1. Van der Vlist AA (1980) The development of CBP in Netherlands. In: First international conference on concrete block paving, pp 14–22
2. Ling TC, Nor HM, Lim SK (2010) Using recycled waste tyres in concrete paving blocks. In: Proceeding of the Institution of Civil Engineers, pp 37–45
3. Poon CS, Chan D (2006) Paving blocks made with recycled concrete aggregate and crushed clay brick. Constr Build Mater 20:569–577
4. Azman M, Nor HM, Hainin MR, Yaacob H, Ismail CR, Nur Hafizah AK (2013) The effect of groove-underside shaped concrete block on pavement permanent deformation. Jurnal Teknologi 61(3):7–14
5. Euniza J, Nor HM, Ramadhansyah PJ, Zaiton H (2014) Effect of using waste tyre rubber on the properties of double layer rubberized concrete paving blocks. Jurnal Teknologi 71(3):1–5
6. Euniza J, Nor HM, Ramadhansyah PJ, Zaiton H (2014) Double layer concrete paving blocks using waste tyre rubber as aggregate replacement. Appl Mech Mater 554:128–132
7. ASTM C150-12 Standard specification for Portland cement, American society for testing and materials. Information on http://www.tasekcement.com/index/cement_facts/cement_quality_.html
8. ASTM C136-06:2011 Standard test method for sieve analysis of fine and coarse aggregate, Am Soc Testing Mater
9. Euniza J, Nor HM, Ramadhansyah PJ, Zaiton H (2014) Chemical Properties of Waste Tyre Rubber Granules. Adv Mater Res 911:77–81
10. ASTM C494:2013, Standard specification for chemical admixtures for concrete. Am Soc Testing Mater
11. ASTM E1050-10.2011 Standard test method for impedance and absorption of acoustical materials using a tube, two microphones and a digital frequency analysis system. Am Soc Testing Mater
12. BS 6717-1:1993 Precast concrete paving blocks-Part 1 : specification for paving blocks, British Standard
13. ASTM C642-06 Standard test method for density, absorption, and voids in hardened concrete. Am Soc Testing Mater
14. Sandberg U, Ejsmont JA (2002) Tyre/road noise reference book. Informex, Kisa

Synthesis of Modified Covalent Organic Framework-1 (COF-1) and Its Characterizations

Muhammad Falaq Muhammad Faisal, Ahmad Rafizan Mohamad Daud and Kamariah Noor Ismail

Abstract Three samples of covalent organic framework-1 (COF-1) assigned as COF-1 (S1), COF-1 (S2) and COF-1 (S3) were prepared by varying the initial mass of 1, 4-benzene diboronic acid (BDBA) used. The samples were characterized using Nitrogen adsorption for porosity analysis and XRD for crystallinity analysis. All samples exhibit Type IV isotherm indicating mesoporous materials. The BET surface area value showed an increasing trend with increasing mass of BDBA used. The highest BET surface area was recorded by COF-1 (S3) with a value of 107.9 m^2/g. This sample has the highest amount of micropore volume of 66 % and the lowest average pore diameter of 11.05 nm. XRD patterns revealed sharp peaks indicating a crystalline structure and the peak positions matched well with other works available in literature. The difference in intensities however was due to the exposure of the samples to atmosphere resulted in the re-formation of COF-1 structure onto BDBA structure.

Keywords Covalent organic framework-1 · Adsorption isotherm · Porosity · X-ray diffraction · Crystallinity

Introduction

Due to the ever improving industrial developments that necessitate usage of nanoporous materials with higher loading capacity and enhanced physico-chemical characteristics, scientists and engineers have focused their effort in developing

M.F. Muhammad Faisal · A.R. Mohamad Daud (✉) · K.N. Ismail
Faculty of Chemical Engineering, Universiti Teknologi MARA, 40450 Shah Alam, Selangor, Malaysia
e-mail: ahmad2057@salam.uitm.edu.my

M.F. Muhammad Faisal
e-mail: falaqfaisal@gmail.com

K.N. Ismail
e-mail: knoor@salam.uitm.edu.my

M.A. Hashim (ed.), *ICGSCE 2014*, DOI 10.1007/978-981-287-505-1_35

alternative nanoporous materials that could be applied for industrial purposes such as fuel storage [1], catalysis [2], optical sensors [3] and separations [4]. In recent years, nanoporous Covalent Organic Frameworks (COFs) [5–7] have been synthesized by Yaghi group to act as a possible material for hydrogen gas storage. COFs was developed to improve the characteristics of the previously developed Metal Organic Frameworks (MOFs) by the same research group [8]. The metal content which adds extra weight in MOFs is a drawback [9] and did not significantly influence its storage capabilities [10].

The first developed COFs, named COF-1, is synthesized through the molecular dehydration of 1, 4-benzene diboronic acid (BDBA) and is lightweight, thermally stable and low in density [5]. Though still in its early stage, COFs have proven to have potential for gas storage applications such as hydrogen, carbon dioxide and methane. Several studies revealed that COFs structure is in the microporous range and high in crystallinity such as COF-LZU1 [11], COF-18 [12] and COF-8 [6].

The aim of this study was to synthesize three modified samples of COF-1 using different mass of BDBA. All samples were characterized according to porosity and crystalline structure using nitrogen adsorption and X-Ray Diffraction (XRD) analyses respectively.

Experimental

Synthesis of COF-1, COF-1 was synthesized in accordance with the general procedures developed by [5] and methodology adopted from [13]. An ampule glass tube (Wheaton, I.D. = 22.5 mm) was filled with 25 mg of 1,4-benzene diboronic acid (Aldrich, ≥95.0 %, USA) and 1 ml of a 1:1 (volumetric ratio) solution of mesitylene:dioxane (Acros, 99 %, Belgium). The tube was connected to a steel tube with the other end connected to a vacuum pump. The tube was then flashed frozen in liquid nitrogen bath, evacuated to an internal pressure of 20 Pa and flame sealed using a blow torch. The length of the tube was reduced to 120 mm upon sealing and heated in an oven for 72 h at 100 °C. The resultant fine white powder was then filtered and washed with 15 ml acetone (Merck, ACS reagent, Germany). Similar procedure was applied where for the second and third samples, the initial mass of BDBA used was increased to 50 and 100 mg. The volume of mesitylene:dioxane solution used was 2 ml for 50 mg of BDBA and 4 ml for 100 mg of BDBA.

Characterizations of COF-1, Porosity analysis was obtained using Quantachrome Autosorb-1 micropore surface area analyzer. A liquid nitrogen bath held at 77 K was used to maintain the temperature during analysis and was continuously monitored using a computerized system. Prior to nitrogen adsorption, 0.2 grams of the sample was degassed at 120 °C for 12 h to remove any impurities and trapped gas molecules inside the pores.

X-ray diffraction (XRD) patterns were collected on a Rigaku diffractometer running at 40 kV, 20 mA using a Cu K_α ($\lambda = 0.1543$ nm) radiation source, with a scan speed of 5°/min and a 2θ step size of 0.02° over a range of 3–90°.

Results and Discussion

Isotherm Characteristics, Fig. 1 shows the nitrogen gas adsorption/desorption isotherm profile for all the synthesized COF-1 at 77 K. All the isotherm patterns exhibited a well-defined flat line which increases slightly with a sudden surge at the end. Based on the classification set by the IUPAC system, this isotherm exhibited Type IV which is in the mesoporous material range. Adsorption occurred at a relative pressure of 0.1 with a slightly increase uptake as the relative pressure was increased indicating the absence of many highly microporous materials (pores smaller than 2 nm).

The initial part of the isotherm followed the isotherm of a Type II material which is a macroporous material that attributes to monolayer-multilayer adsorption.

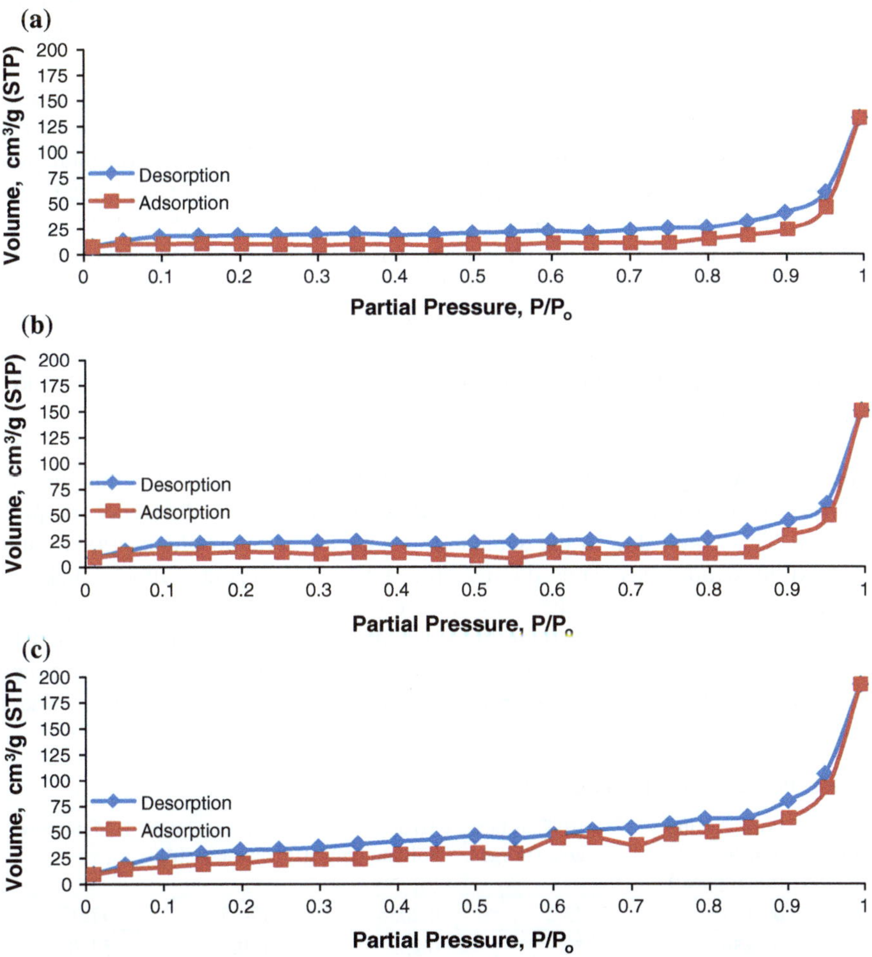

Fig. 1 Adsorption isotherm profile for **a** COF-1 (S1), **b** COF-1 (S2), **c** COF-1 (S3)

The scarcity of micropores inside the materials could be a drawback in its application for gas adsorption. The effectiveness of a storage gas material highly relies on the accessibility of the micropore inside the material to the small gas particles, which referred as *micropore filling* [14]. This finding also explains the lower value of BET surface area obtained for all samples of COF-1 with the highest achieved BET surface area value was COF-1 (S3) with 107.9 m^{-2}/g. The sudden enhanced uptake of N_2 at a relative pressure after 0.8 suggests the presence of small quantities of macropores (pores larger than 2 nm).

The isotherm also exhibits a hysteresis loop between adsorption and desorption lines, indicates the possible presence of many mesoporous structures within the synthesized COF-1. The hysteresis loop indicates the capillary condensation that is taking place in the mesopores [14]. So, it can be concluded that all the structures of COF-1 exhibit the Type IV isotherm which indicates the presence of mesoporous structures instead of Type I (microporous structures) which was reported by [5].

Porosity Characteristics, Table 1 presents the pore characteristics of all the samples. The highest BET surface area obtained is sample COF-1 (S3) with 107.9 m^2/g followed by COF-1 (S2) with 52 m^2/g and COF-1 (S1) with 43.2 m^2/g. Sample COF-1 (S3) also has highest amount of micropores volume which is 0.1980 cm^3/g which accounted to 66 % of the total pore volume. Its total pore volume is the highest at 0.2981 cm^3/g and also had the lowest average pore diameter of 11.05 nm as compared to COF-1 (S1) and COF-1 (S2). The average pore diameters for all COF-1 samples were in the range of 11.05–19.05 nm. The lower the average pore diameter resulted in a higher number of micropores available, thus a higher value of surface area were achieved. It can be further concluded that porosity for all the samples synthesized in this research can be classified as mesopores instead of micropores due to the large average pore diameter and low volume of micropores.

XRD Analysis, Fig. 2 shows the XRD patterns of all samples of COF-1. All samples showed good crystallinity where sharp peaks occurred while the 2θ positions were matched well with the works by [5] albeit difference in peak intensity. This could due to the sample exposure to the atmosphere prior to analyzing. If the sample was exposed to the atmosphere for a period of time, which was around ten days in this work there is a change in the crystallinity of the sample and the XRD pattern does not provide the same peaks intensity as that presented by Cote et al. [5]. Li and Yang [15] investigated this phenomenon by exposing their synthesized COF-1 to the atmosphere for seven days where the effect was a

Table 1 Pore characteristics of COF-1

COF-1 sample	Pore volume (cm^3/g) (micropores)	BET surface area (m^2/g)	Total pore volume (cm^3/g)	Average pore diameter (Å)	Micropore (%)
S1	0.1193	43.2	0.2058	190.5	57.97
S2	0.1256	52	0.2331	179.4	53.88
S3	0.1980	107.9	0.2981	110.5	66.42

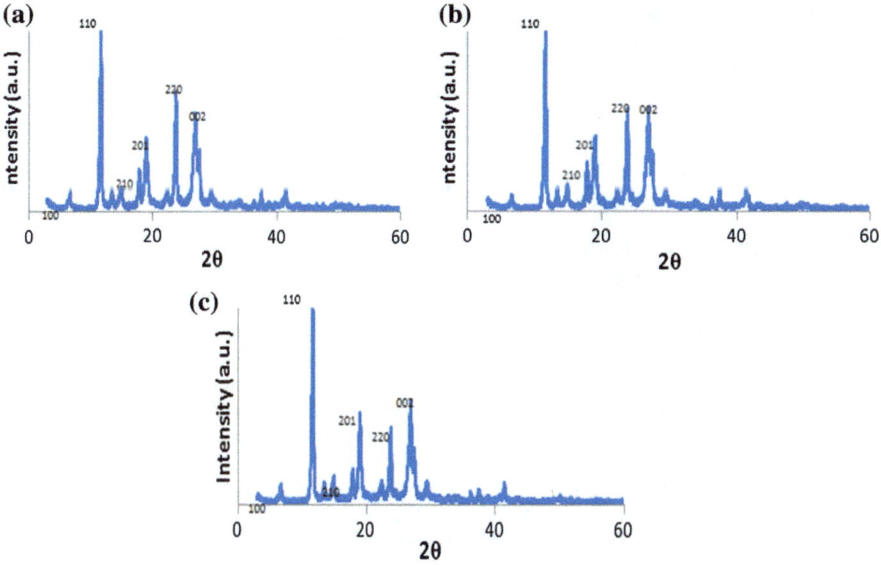

Fig. 2 XRD patterns for **a** COF-1 (S1), **b** COF-1 (S2), **c** COF-1 (S3)

different patterns formed, which correlates well with what was exhibited by the COF-1 synthesized in this work. Peaks at 12°, 19° and 27° indicated the prominent structure of COF-1, while peaks at 18°, 23° and 28° revealed the reformation of BDBA structure. Overall, there are no significant differences in XRD patterns for all COF-1 samples when compared with respect to initial mass of BDBA.

Conclusions

Modified COF-1 samples were prepared by varying the initial mass of reactant used. Adsorption isotherm profile and XRD analyses were used to evaluate characteristics of the modified COF-1. Overall, through the isotherm patterns obtained, the pores developed in all COF-1 samples prepared in this work were in the mesoporous range of Type IV isotherm. Sample COF-1 (S3) which used 100 mg of BDBA has the highest BET surface area of 107.9 m^2/g. This sample has also the highest amount of micropore volume and the lowest average pore diameter. The XRD analysis show good crystallinity structure but different in peak intensities. This indicates reformation of crystalline COF-1 structure into BDBA due to exposure to atmosphere. Considering that COF-1 synthesis research is very much at early stage, findings from this study deemed beneficial in improving the preparation method of COF-1 for future studies.

Acknowledgment The authors were grateful to the Ministry of Higher Education, Malaysia and Universiti Teknologi MARA for providing grants 600-RMI/ST/FRGS 5/3/Fst(101/2010) and 600-RMI/DANA 5/3/RIF(131/2012).

References

1. Benard P, Chahine R, Chandonia PA, Cossement D, Dorval-Douville G, Lafi L et al (2007) Comparison of hydrogen adsorption on nanaoporous materials. J Alloys Compd 446–447:380–384
2. Pisduangdaw S, Panpranot J, Methastidsook C, Chaisuk C, Faungnawakil K, Praserthdam P et al (2009) Characteristics and catalytic properties of Pt-Sn/Al2O3 nanoparticles synthesized by one-step flame spray pyrolysis in the dehydrogenation of propane. Appl Catal A 370:1–6
3. Fernandez-Sanchez JF, Cannas R, Spichiger S, Steiger R, Spichiger-Keller UE (2006) Novel nanostructured materials to develop oxygen-sensitive films for optical sensors. Anal Chim Acta 566:271–282
4. Ackley MW, Rege SU, Saxena H (2003) Application of natural zeolites in the purification and separation of gases. Microporous Mesoporous Mater 61:25–42
5. Cote AP, Benin AI, Ockwig NW, O'Keeffe M, Matzger AJ, Yaghi OM (2005) Porous, crystalline, covalent organic frameworks. Science 310:1166–1170
6. Cote AP, El-Kaderi HM, Furukawa H, Hunt JR, Yaghi OM (2007) Reticular synthesis of microporous and mesoporous 2D covalent organic frameworks. J Am Chem Soc 129:12914–12915
7. El-Kaderi HM, Hunt JR, Mendoza-Cortés JL, Cote AP, Taylor RE, O'Keeffe M, Yaghi OM (2007) Designed synthesis of 3D covalent organic frameworks. Science 316:268–272
8. Yaghi OM, Li G, Li H (1995) Selective binding and removal of guests in a microporous metal-organic framework. Nature 378:703–706
9. Klontzas E, Tylianakis E, Froudakis GE (2008) Hydrogen storage in 3D covalent organic frameworks. A multiscale theoretical investigation. J Phys Chem C 112:9095–9098
10. Assfour B, Seifert G (2010) Hydrogen adsorption sites and energies in 2D and 3D covalent organic frameworks. Chem Phys Lett 489:86–91
11. Ding SY, Gao J, Wang Q, Zhang Y, Song WG, Su CY, Wang W (2011) Construction of covalent organic framework for catalysis: Pd/COF-LZU1 in Suzuki-Miyaura coupling reaction. J Am Chem Soc 133:19816–19822
12. Tilford RW, Gemmill WR, Zur Loye HC, Lavigne JJ (2006) Facile synthesis of a highly crystalline, covalently linked porous boronate network. Chem Mater 18:5296–5301
13. Musa MAA, Yin CY, Savory RM (2010) Synthesis and textural characterization of covalent organic framework-1: Comparison of pore size distribution models. Mater Chem Phys 12:5–8
14. Sing KSW, Everett DH, Haul RAW, Moscou L, Pierotti RA, Rouquerol A et al (1985) Reporting physisorption data for gas/solid systems with special reference to the determination of surface area and porosity. Pure Appl Chem 57–4:603–619
15. Li YW, Yang RT (2010) Hydrogen storage in metal-organic frameworks and covalent-organic frameworks by spillover. AlChe J 54:269–279

One Pot Synthesis of γ-Valerolactone from D-Glucose Over $H_3PW_{12}O_{40}/ZrO_2$ Catalyst

Dorairaaj Sivasubramaniam and Nor Aishah Saidina Amin

Abstract In this study, two different wt% of $H_3PW_{12}O_{40}/ZrO_2$ catalysts were synthesized, characterized and screened for the synthesis of γ-valerolactone (GVL) from D-glucose. The catalysts involved were 10 % $H_3PW_{12}O_{40}/ZrO_2$ and 15 % $H_3PW_{12}O_{40}/ZrO_2$ prepared through wet impregnation method. Catalyst characterization using FTIR, BET and NH_3-TPD revealed that the catalytic performances were dominantly influenced by the acid sites of the catalyst in the production of Levulinic acid (LA) from D-glucose via acid hydrolysis reaction. High acidity greatly influenced the catalytic performance for the production of GVL. These catalysts were tested based on five sets of reaction time (8, 10, 12, 14 and 16 h) at a fixed temperature of 180 °C with 1:1 catalyst to feedstock ratio. 10 % $H_3PW_{12}O_{40}/ZrO_2$ exhibited high performance at 16 h of reaction time with 22.3 % yield compared to 15 % $H_3PW_{12}O_{40}/ZrO_2$ with only 14.8 % yield of GVL.

Keywords γ-valerolactone · Solid acid catalyst · $H_3PW_{12}O_{40}$ · ZrO_2 · One pot reaction

Introduction

Biomass can serve as an alternative renewable source for the sustainable production of energy and chemicals such as LA and γ-valerolactone [1]. The largest part of biomass is carbohydrates. The carbohydrates are particularly used for the conversion to platform chemicals [2]. Carbohydrates can be converted to

D. Sivasubramaniam · N.A. Saidina Amin (✉)
Chemical Reaction Engineering Group (CREG), Faculty of Chemical Engineering, Universiti Teknologi Malaysia, 81310 UTM Skudai, Johor, Malaysia
e-mail: noraishah@cheme.utm.my

D. Sivasubramaniam
e-mail: dorai_90@yahoo.com

M.A. Hashim (ed.), *ICGSCE 2014*, DOI 10.1007/978-981-287-505-1_36

chemicals including ethanol, n-butanol, sorbitol, hydroxymethylfurfural, and dimethylfurfural [3–5].

GVL is one of the platform chemicals considered as one of the most dominant reproducible chemicals for the production of other bio-based chemicals [6]. It is use as liquid fuel, food additive, solvent and organic intermediate in the synthesis [7]. Generally, there are two steps involved in the production of GVL: (1) dehydration of C_6-sugars to form LA, (2) hydrogenation of LA to form GVL.

GVL has most ideal properties known as sustainable liquid fuel, which is involved in the production of both energy and carbon-based chemicals such as pentanoic acid and 5-nonane. It is a renewable, easy and safe to store and for transportation. It has a low melting point (−31 °C) and a high boiling point (207 °C).

The carbohydrate such as cellulose, hemicelluloses and lignin are polysaccharides utilized in the hydrolysis reaction to form monomer (D-glucose), which is utilized for the synthesis of many bio-based chemicals. The objective of this paper is to synthesize GVL using D-glucose in presence of $H_3PW_{12}O_{40}/ZrO_2$, prepared through wet impregnation method, in one pot reaction. The catalysts were characterized to determine the influence of acid sites and surface area on the catalytic performance [8].

Experimental

Chemicals

All chemicals were supplied from Merck, Germany. The Keggin-type heteropolyacid used in the reaction was phosphotungstic acid ($H_3PW_{12}O_{40}$), Zirconium oxide (ZrO_2) was used for the catalyst preparation. Distilled water was used in the catalyst preparation and as solvent for reactions. Formic acid (FA) was used as homogeneous acid catalyst and hydrogen supply for the reaction mixture. Standard solutions of glucose, γ-valerolactone GVL and methanol were used for product analysis.

Catalyst Preparation

The catalyst was prepared via impregnation method. Intially, zirconium oxide (ZrO_2) was mixed with phosphotungstic acid ($H_3PW_{12}O_{40}$) according to wt% of the catalyst. 10 g of zirconia (ZrO_2) and 20 ml of water followed by 10 or 15 wt% of $H_3PW_{12}O_{40}$ was added into the mixture. The mixture was then stirred at 60 °C until water molecules evaporated to form solid catalyst. Finally, the solid form of catalyst was calcined at 500 °C for 3 h.

Catalyst Characterization

FTIR spectra were used to identify the chemical bonds and functional groups in the sample. The FTIR spectra (KBr pellets) were recorded using Perkin Elmer in the range of 400–4000 cm^{-1}. The acidity properties of the two catalysts were analysed using NH_3 temperature programmed desorption (NH_3-TPD). Brunauer-Emmet-Teller (BET) method was applied to determine the surface area, pores volume and size distribution using Micrometrics ASAP2020 analyzer with isotherm nitrogen adsorption.

One Pot Reaction

The conversion of D-glucose to GVL was carried out in presence of 48 ml water and 2 ml FA as hydrogen carrier for the hydrogenation reaction as reported by Heeres and Chunai [9] and the reaction was performed in a high pressure stirred autoclave reactor at 180 °C. Finally, the product was analyzed using GC-FID to identify the yield of products formed.

Results and Discussion

Catalyst Characterization

Fourier Transform Infrared Spectroscopy (FTIR), Fig. 1 shows the FTIR spectra for parent ZrO_2, 10 % $H_3PW_{12}O_{40}$ and 15 % $H_3PW_{12}O_{40}$. The peak exhibited at 1077 cm^{-1} represents the bond between Zr–O–P in its structure after the impregnation of ZrO_2 with $H_3PW_{12}O_{40}$. Moreover, when the loading of $H_3PW_{12}O_{40}$ increases, weak peak appears at 420 cm^{-1}inferring formation of the impregnated catalyst for the similar bond.

Brunauer-Emmett-Teller (BET)

The BET values in Table 1 indicate that impregnation of $H_3PW_{12}O_{40}$ on ZrO_2 increased the surface area of the solid catalyst but no significant difference between this two loadings. The surface areas were 14.36 and 14.52 m^2/g for 10 and 15 % $H_3PW_{12}O_{40}/ZrO_2$, respectively. Based on the total surface area obtained, it was confirmed that surface does not influence the catalytic performance for the production of GVL.

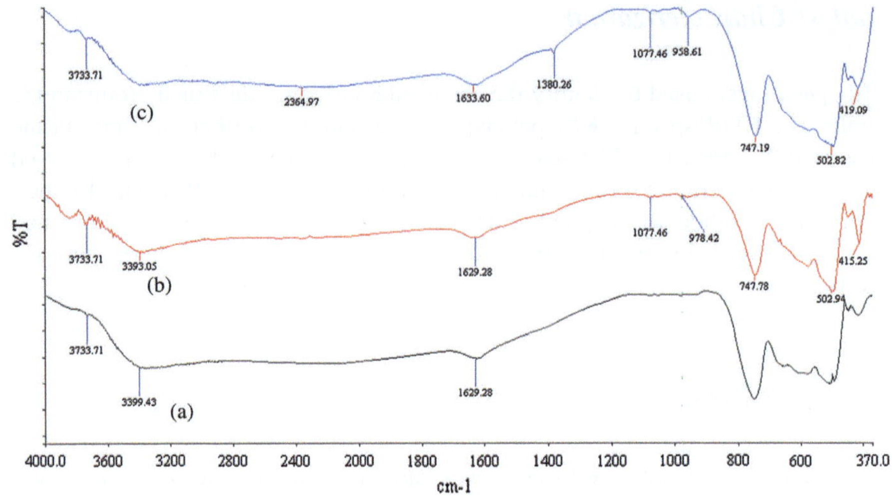

Fig. 1 FTIR spectra for **a** ZrO_2, **b** 10 % $H_3PW_{12}O_{40}/ZrO_2$, **c** 15 % $H_3PW_{12}O_{40}/ZrO_2$

Table 1 BET results for $H_3PW_{12}O_{40}$ impregnate on ZrO_2 at different loadings

Catalyst	Specific surface area (m^2g^{-1})
$H_3PW_{12}O_{40}$	4.55
ZrO_2	14.95
10 % $H_3PW_{12}O_{40}/ZrO_2$	14.36
15 % $H_3PW_{12}O_{40}/ZrO_2$	14.52

Ammonia-Thermal Programmed Desorption (NH_3-TPD), The acidity of the catalyst highly influenced the acid hydrolysis reaction for the production of LA from D-glucose. Table 2 compares the acidity of the parent catalyst with the impregnated ones. The acidity of the impregnated catalyst was higher compared to the parent catalysts due to the introduction of $H_3PW_{12}O_{40}$ on ZrO_2. Generally, $H_3PW_{12}O_{40}$, known as super acid [10], contains strong Bronsted acid sites known as active sites for the catalyst. 10 % $H_3PW_{12}O_{40}/ZrO_2$ catalyst displayed the highest acidity [8].

Dehydration/Hydrogenation of D-glucose to produce γ-valerolactone (GVL), The reaction temperature was fixed at 180 °C and the reaction time was varied from 8 to 16 h [8]. The reaction was conducted based on five sets of reaction time for studying the effect of time on GVL yield in presence of catalyst. In this study we

Table 2 NH_3-TPD profile for the $H_3PW_{12}O_{40}/ZrO_2$ with different loading

Catalyst	Acidity (mmol/g)
$H_3PW_{12}O_{40}$	1.68
ZrO_2	0.27
10 % $H_3PW_{12}O_{40}/ZrO_2$	0.52
15 % $H_3PW_{12}O_{40}/ZrO_2$	0.31

found that the GVL yield increased with time [9] due to high amount of LA were converted to ring closure structure of GVL. When compared this work with the study demonstrated by Son et al. [11], the yield of GVL obtained was low due to different feedstock used. Son et al. utilized D-fructose as the feedstock for the production of GVL and obtained high yield (48 %) but in this study, the feedstock that been utilized was D-glucose for the one pot reaction. The production of GVL direct from D-glucose took a long reaction pathway that was already made shorten into one pot. Hence, the yield of GVL was affected. Figures 2 and 3 exhibit the yield of GVL under various time taken from 8 to 16 h in presence of 10 and 15 % $H_3PW_{12}O_{40}/ZrO_2$.

10 % $H_3PW_{12}O_{40}/ZrO_2$ for GVL production, Based on Fig. 2, the highest yield of GVL was 22.3 % at 16 h of reaction time. This reaction time was suitable to produce GVL with better yield in one pot reaction. Impregnated 10 % $H_3PW_{12}O_{40}/ZrO_2$ was utilized in the production of GVL from D-glucose. 10 % $H_3PW_{12}O_{40}/ZrO_2$ was an appropriate wt% for the production of GVL. Moreover, it was proven that this catalyst was capable of forming a ring closure structure of GVL directly from D-glucose without the need to eliminate LA In this reaction, FA was used as

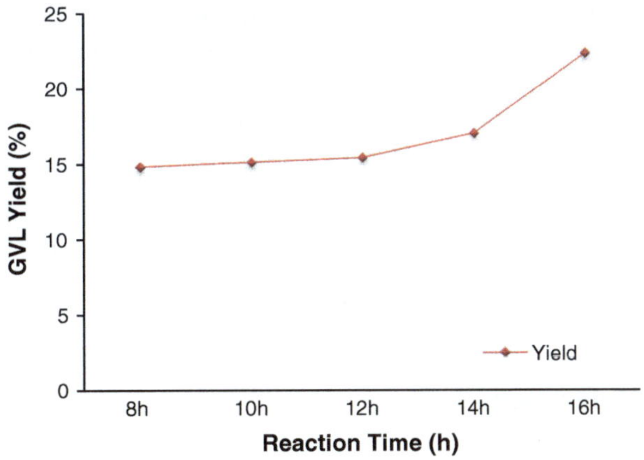

Fig. 2 Effect of the reaction time to GVL yield in the presence of 10 % $H_3PW_{12}O_{40}/ZrO_2$ at 180 °C

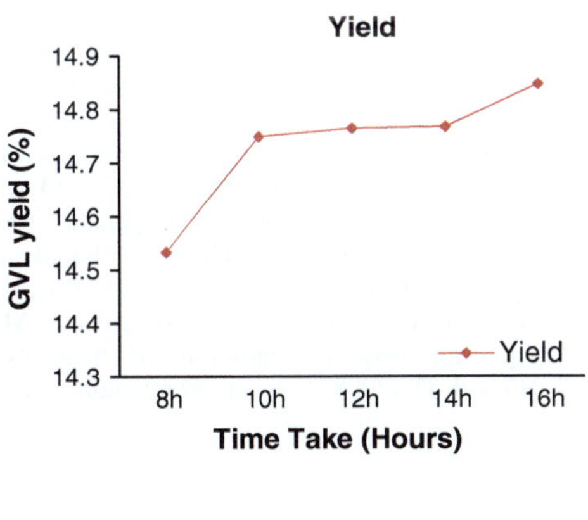

Fig. 3 Effect of the reaction time to GVL yield in the presence of 15 % $H_3PW_{12}O_{40}/ZrO_2$ at 180 °C

hydrogen source in the conversion of LA (intermediate product) to GVL. Moreover, FA also was utilized as a homogeneous acid catalyst for the production of LA from D-glucose via acid hydrolysis.

15 % $H_3PW_{12}O_{40}/ZrO_2$ for GVL production, Fig. 3 depicts the GVL (% yield) from D-glucose in the presence of 15 % $H_3PW_{12}O_{40}/ZrO_2$. The highest yield of GVL was 14.8 % at 16 h. When compared to 10 % $H_3PW_{12}O_{40}/ZrO_2$, 15 % $H_3PW_{12}O_{40}/ZrO_2$ produced lower yield of GVL. This maybe attributed to the acidity of the catalyst, which influenced the production of LA from acid hydrolysis reaction of D-glucose. Based on the NH_3-TPD result, 10 % $H_3PW_{12}O_{40}/ZrO_2$ have higher acidic sites than 15 % $H_3PW_{12}O_{40}/ZrO_2$. Recently, 16.6 wt% of GVL via one pot reaction scheme of giant reed in the presence of Ru/C with niobium phosphate was reported [12].

Conclusion

GVL can be obtained from one pot reaction of D-glucose. FA was utilized as a homogeneous catalyst for the dehydration reaction to LA from D-glucose and hydrogen source for the hydrogenation reaction of LA to produce GVL. It was found that 10 % $H_3PW_{12}O_{40}/ZrO_2$ was suitable for simultaneous dehydration/hydrogenation of D-glucose which produced 22.3 % of GVL in 16 h reaction when compared to 15 % $H_3PW_{12}O_{40}/ZrO_2$.

Acknowledgment The authors would like to express their gratitude for the financial support from Universiti Teknologi Malaysia under Research University Grant (RUG) vote no 07H14 and Ministry of Higher Education (MOHE).

References

1. Huber GW, Iborra S, Corma A (2006) Synthesis of transportation fuels from biomass: chemistry, catalysts, and engineering. Chem Rev 106(9):4044–4098
2. Deng J, Wang Y, Pan T, Xu Q, Guo QX, Fu Y (2013) Conversion of carbohydrate biomass to g-valerolactone by using water-soluble and reusable iridium complexes in acidic aqueous media. Chem Sus chem 6(1163–1167):1163
3. Roman-Leshkov Y, Barrett CJ, Liu ZY, James A. Dumesic JA (2007) Production of dimethylfuran for liquid fuels from biomass-derived carbohydrates. Nature 447:982–985
4. Huber GW, Cortright RD, Dumesic JA (2004) Renewable alkanes by aqueous-phase reforming of biomass-derived oxygenates. Angew Chem 116(12):1575–1577
5. Huber GW et al (2005) Production of liquid alkanes by aqueous-phase processing of biomass-derived carbohydrates. Science 308(5727):1446–1450
6. Horvath IT, Mehdi H, Fábos V, Boda L, Mika LT (2008) c-Valerolactone—a sustainable liquid for energy and carbon-based chemicals. Green Chem 10:238–242
7. Horvath IT (2008) Solvents from nature. Green Chem 10(10):1024–1028

8. Tang X, Hu L, Sun Y, Zhao G, Hao W, Lin L (2013) Conversion of biomass-derived ethyl levulinate into γ-valerolactone via hydrogen transfer from supercritical ethanol over a ZrO_2 catalyst. RSC. Advances 3:10277–10284
9. Heeres H, Handana R, Chunai D, Rasrendra CB, Girisuta B and Heeres HJ (2009) Combined dehydration/(transfer)-hydrogenation of C6-sugars (D-glucose and D-fructose) to c-valerolactone using ruthenium catalysts. Green Chem 11:1247–1255
10. Sharma YC, Singh B, Korstad J (2011) Advancements in solid acid catalysts for ecofriendly and economically viable synthesis of biodiesel. Biofuels, Bioprod Biorefin 5(1):69–92
11. Son PA, Nishimura S, Ebitani K (2014) Production of [gamma]-valerolactone from biomass-derived compounds using formic acid as a hydrogen source over supported metal catalysts in water solvent. RSC Advances 4(21):10525–10530
12. Raspolli Galletti AM et al (2013) From giant reed to levulinic acid and gamma-valerolactone: a high yield catalytic route to valeric biofuels. Appl Energy 102(0):157–162

Carbon Cryogel from Lignin-Furfural as Acid Catalyst in Esterification of Oleic Acid

Muzakkir Mohammad Zainol, Nor Aishah Saidina Amin, Mohd Asmadi and Norzita Ngadi

Abstract Carbon cryogel (CC) has been synthesized from commercial lignin with furfural via acid catalyzed sol-gel polycondensation reaction step. Freeze drying and carbonization were performed on the gels for preparation of the CC. Effect of water and H_2SO_4 concentration on carbon synthesis was studied where lignin to water (L/W) ratio and acid concentration of 1.0 and 8 M, respectively were found suitable for CC synthesis. The selected CC exhibited higher thermal stability which can be used in thermal reaction process. CC has potential as super acid solid catalyst with acidity and surface area of 14.07 mmol/g and 330.35 m^2/g, respectively for esterification of free fatty acid to produce 91.3 wt.% FAME yield.

Keywords Carbon cryogel · Acid catalyst · Lignin-furfural · Esterification · Oleic acid · Biodiesel

Introduction

Carbon gels are different types of carbons from carbon family. There are three types of carbon gels known as carbon aerogel, cryogel and xerogel. These types of gels (or resins) are synthesized through sol gel polycondensation process by using phenol or its derivatives with formaldehyde or its derivatives. Direct replacement of

M. Mohammad Zainol · N.A. Saidina Amin (✉) · M. Asmadi · N. Ngadi
Chemical Reaction Engineering Group (CREG), Faculty of Chemical Engineering, Universiti Teknologi Malaysia, 81300 UTM, Johor Darul Takzim, Malaysia
e-mail: noraishah@cheme.utm.my

M. Mohammad Zainol
e-mail: muzakkirzainol@gmail.com

M. Asmadi
e-mail: asmadi@cheme.utm.my

N. Ngadi
e-mail: norzita@cheme.utm.my

M.A. Hashim (ed.), *ICGSCE 2014*, DOI 10.1007/978-981-287-505-1_37

Introduction

Solid Polymer Electrolytes (SPEs) are among the important classes of macromolecules. These electrolytes are charged macromolecules containing a large number of ionisable or ionic groups [1]. The ionic conductivity of SPEs regularly takes place through two distinct events: the first is due to the charge migration of ions between the coordinate sites of the host polymer and the second is associated with the polymeric chain segmental motion [2]. The charge migration was reported strongly associated by various factors such as crystallinity of the material, simultaneous cation and anion motions and the ion pair formation [3]. In general the conductivity of SPEs coupled to the reduction of the crystalline domain of polymer host with the high ionic dissociation of salts and also reducing in ion pair. To date, studies on the electrical properties of SPEs are a must before it will propose for suitable potential application in electrochemical devices [4]. So far, various type of polymer host has been explored to study their electrical behaviour including synthetic and natural polymers. However, research on natural polymers lately become more vital due to these polymers offer good electrical properties, possess high mechanical strength and able to solvate salt into ion [5]. Consequently, in this work we used cellulose acetate (CA) as a polymer host to prepare the SPEs. The selection of this polymer host is due to the present of C=O functional groups in their structure with the oxygen atom possess a lone pair of electron [6]. In addition, cellulose is abundant material, low cost, good mechanical properties and biodegradable.

Experimental

Solid polymer electrolytes (SPEs) were prepared by solution casting method. Cellulose acetate (CA) with 39.8 wt% acetyl content were used as polymer host, ammonium iodide (NH_4I) salt (Aldrich—purity 99 %) were used as ion sources and acetone (Aldrich) as a solvent. NH_4I was dried for about 2 h at 100 °C prior used. CA and NH_4I with different stichiometric ratio were dissolved in 30 mL acetone for 48 h at room temperature until both the polymer and salt were completely dissolved. The homogeneous solutions were then cast onto glass petri dishes and allowed to dry to form films at room temperature. The films were kept in desiccators with silica gel desiccants for further drying before being characterized. The ionic conductivity of the samples was performed by using HIOKI 3532-50 LRC Hi-Tester in frequency range between 100 Hz and 1 MHz at room and elevated temperature. The samples were sandwiched between two stainless steel electrodes (SUS316) with force 9.8 N and placed in a SH-221 Bench-Top type temperature and humidity chamber at constant humidity 40 % RH. The electronic transference number were taken using an Auto-Lab PGSTAT30 potentiostat/galvanostat analyzer with a constant direct current (DC) potential 10 mV applied voltage across a symmetric SUS316 electrodes with cell configuration of SUS316/SPE/SUS316. The X-ray diffraction (XRD) studies were carried out by means of X'Pert PRO X-ray diffractometer in 2θ range of 5° ≤ 2θ ≤ 70°.

Result and Discussion

According to the sample preparation as described above, a series of SPEs film with NH_4I salt of 5–30 wt% was obtained. It was found that the films are opaque and mechanically stable. The crystalline properties of the films were analyzed by XRD result. Figure 1 shows the XRD pattern of the CA–NH_4I films with different salt concentration, pure NH_4I salt and pure CA. As can be seen in Fig. 1a, the NH_4I salt are in their crystalline form with 2θ peaks at 21.2°, 24.5°, 34.9°, 41.2°, and 56.7° and as-prepared CA film Fig. 1b also has obviously crystalline with amorphous hump peaks centre at 2θ = 8.5° and 17.9° respectively. Upon the addition of NH_4I onto CA the XRD spectrum Fig. 1c–h demonstrates all the crystalline peaks possessed by NH_4I and CA has disrupted. Disappearing of all of these peaks indicates that the salt and polymer is well mixed and form polymer-salt complexes. Since the CA has C=O functional group in their structure, and the oxygen atom possess a lone pair of electrons, the formation polymer-salt complexes may takes place via $C=O \leftarrow NH_4^+$ coordination bond. This complexation wills results the dissociation of ammonium salt into ions and the absent of excess salt can be seen in the CA–NH_4I XRD

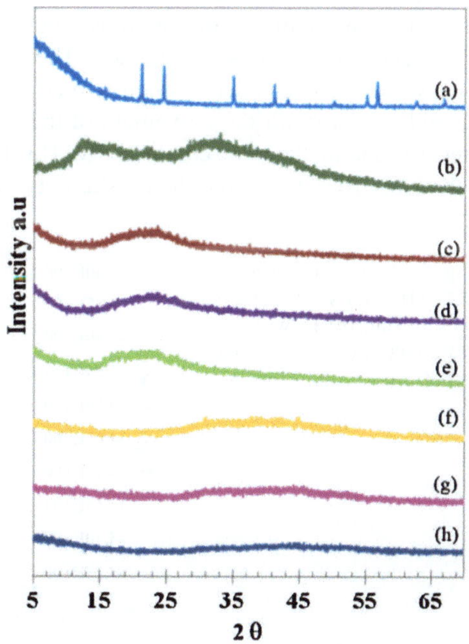

Fig. 1 XRD patterns of: (*a*) pure NH_4I, (*b*) pure CA, and (*c*)–(*h*) different concentration of NH_4I salt (5–30 wt%) on CA–NH_4I system

spectrum [5]. In contrary the individual characteristic peaks of CA in the CA–NH_4I films are not revealed. The amorphous humps also become less prominent and broadened upon the addition of NH_4I. It can be inferred that all complexes samples are less crystalline and the amorphous phase become dominant. The reducing of crystalline nature in CA–NH_4I complexes is expected to exhibit the highest ionic conductivity and the sample with composition of 25 wt% NH_4I salt displayed the most plateaus in the XRD patterns.

The ionic conductivity of SPEs was obtained from impedance measurement. Figure 2 shows the Arrhenius plots of conductivity depending on the NH_4I salt content. As observed, the ionic conductivity of SPEs increases with the increasing salt concentration up to maximum value 1.99×10^{-4} S cm^{-1} at room temperature when 25 wt% NH_4I was added. The increase in conductivity is due to the increase in the number of mobile ions present in the samples. These mobile ions originate from the dissociation of NH_4I salt also were complexed with the polymer host as a result was enhanced the amorphousness of sample. Accordingly, the polymer-salt complexes will favour the ions transport which associated with both the segmental motion of the polymer chain and the long-range path provided by the ions in the amorphous phase hence rising in conductivity [7]. In addition the complexation also will reduce the columbic force between the cation and anion of the salt to form ions aggregation. However, the ionic conductivity was found to decrease at higher salt loading (more than 25 wt%). This phenomenon can be attributed to the re-association of the ions into natural ions aggregates [8, 9]. It is believed that the re-association of ions is strongly connected to the complexation site.

When the dissociated salt is high, the coordination between the NH_4^+ ions and the polymer host is striking and left less coordination site provided by the polymer. Therefore, when more salt was added, higher probability of ions association will take place and form an ions aggregation. The aggregated ion will then consume a large free volume environment, which is fundamentally needed by the polymer chain to flexible. As a result the movement of the polymer chain flexibility will reduce and hence reducing the ionic conductivity. This basis is well agreed with the XRD result, which higher salt loading the crystallinity of the SPE also was increased.

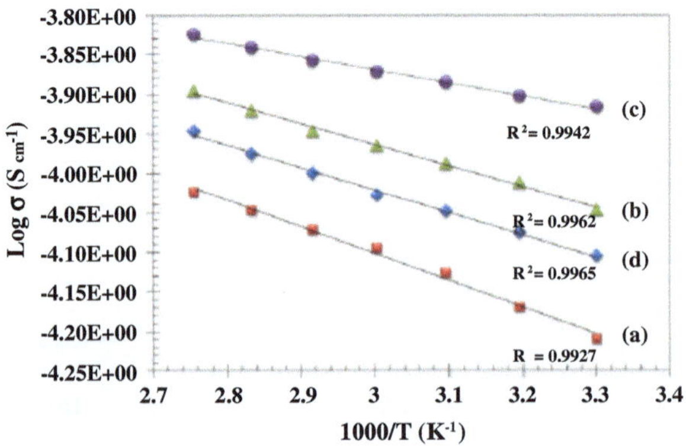

Fig. 2 Arrhenius plot for CA–NH_4I SPEs with different wt% of salt: (*a*) 15 wt%, (*b*) 20 wt%, (*c*) 25 wt% and (*d*) 30 wt%

Usually in the case of SPEs the thermally assisted behaviour is observed for the temperature dependence conductivity. Therefore, according to the Arrhenius law:

$$\sigma = \sigma_o \exp\left(-\frac{E_A}{kT}\right) \tag{1}$$

where σ_o is the exponential factor, E_A is the activation energy, k is the Boltzmann constant, and T is the absolute temperature) the plots of log σ versus $1000/T$ will be manifested the possible mechanism of ionic conduction in the SPEs system. Based on this plot as shown in Fig. 2, it was observed the conductivity increases monotonically as the temperature rises indicating the temperature play a significant role to enhance the ionic mobility. This effect can be interpreted as the increments in the segmental and rotational structural motions of the polymeric chain [8].

Regression values (R2) obtained is close to unity suggesting that all the points lie on a straight line. Thus, the graph profile can be considered as Arrhenius plot and subsequently the nature of ionic transport in the SPEs sample is quite similar to that of in ionic crystals, which the ions able to jump from one vacant site into neighbouring vacant sites and thereby increase conductivity [10, 11]. The E_A of the SPEs was obtained from the slope of the plot and was tabulated in Table 1. The value shows energy required to transport conducting species in such SPEs dominate equally with the component present the system which correlates to the ionic conductivity.

Figure 3 shows the plots of the imaginary electrical modulus (Mi) versus frequency (f) at ambient temperature for various salt concentrations in CA–NH_4I complexes. In polymeric materials, the complex electric modulus has been used to investigate the conductivity relaxation phenomena. From physical point of view the conductivity relaxation process is allied to the applied electric field and be able to discriminate between electrode polarization and ionic relaxation process. As seen in Fig. 3, the Mi plot showed a relaxation a peak present in all samples indicates the SPEs in the present study are ionic conductors [10]. The shifting of these peaks towards higher frequencies manifested the conductivity increased with salt concentration. The frequency range, within which the $(M_i)_{max}$ peaks occur is indicative of the transition of ions from a long range to a short range mobility at increasing frequency and can be defined by the condition $2\pi f_{max}\tau = 1$, where τ is the conductivity relaxation time and f_{max} is the frequency at which $(M_i)_{max}$ occurred.

Table 1 The ionic conductivity and the electric modulus parameters of CA–NH_4I systems

Sample name	Capacitance, C (F) × 10^{-15}	Relaxation time, τ (s) × 10^{-4}	Conductivity, σ (S cm^{-1}) × 10^{-4}	Activation energy, E_A (eV)
A	17.6	5.39	0.98	0.3374
B	1.40	0.50	1.61	0.2664
C	0.70	0.24	1.99	0.1673
D	7.03	2.70	1.33	0.2837

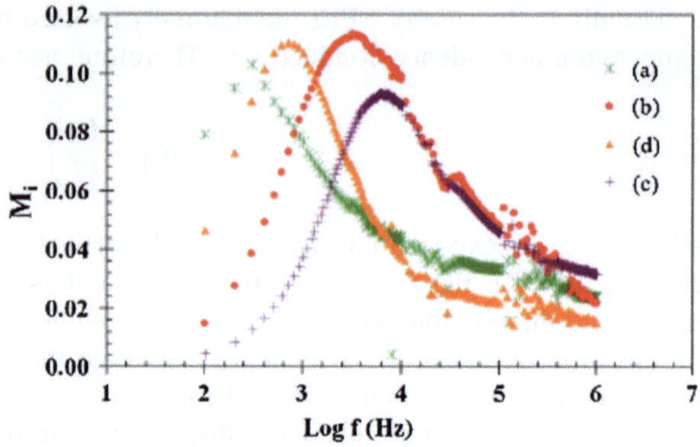

Fig. 3 Imaginary part of electric modulus (M_i) versus frequency (f) for CA–NH_4I complexes with: (*a*) 15 wt%, (*b*) 20 wt%, (*c*) 25 wt% and (*d*) 30 wt% of NH_4I content

The equivalent frequency at $(M_i)_{\max}$ is known as the relaxation frequency. The capacitance value of the relaxation peak can be calculated from the equation:

$$M_{i(\max)} = \frac{\varepsilon_o}{2C} \tag{2}$$

These values are tabulated in Table 1. The capacitance value was found to be in the range 10^{-14}–10^{-15} F indicates that the dispersion peaks observed are attributable to the bulk effect of the material. This suggests that the relaxation peaks is dominated by ions and not at the polarization of electrode-electrolytes interface.

To justify the conductivity process mainly contributed by the ions, the transference number was calculated based on Wagner's DC polarization technique [12]. This method was used to analyse the mobile species in the electrolyte. The polarization current was monitored as a function of time on the application of DC potential across the cell. The current decay immediately and asymptotically approaches steady state. The total ionic transference number was calculated from the polarization current versus time plots as shown in Fig. 4 by using the standard equations are as follows:

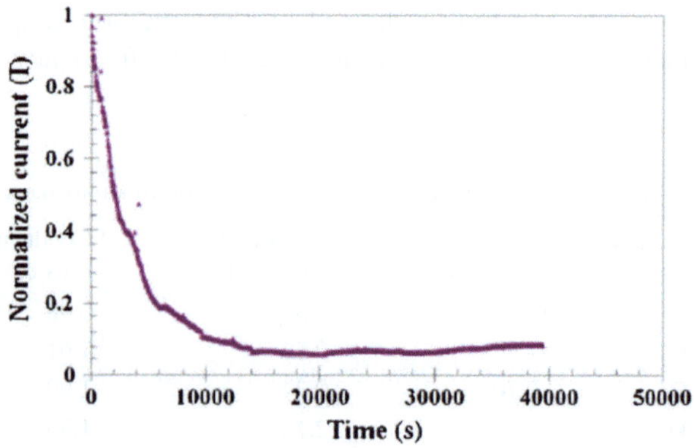

Fig. 4 Chronoamperometry of polarization current as a function of time for the sample CA (75 wt%):NH_4I (25 wt%) under constant voltage of 10 mV at room temperature

$$t_{ion} = 1 - I_f/I_i \quad (3)$$

$$t_{ele} = 1 - t_{ion} \quad (4)$$

where I_i is the initial current and If is the final residual current. The ionic transference number of SPE containing 25 wt% salts was found to be 0.916 suggesting that the charge transport in this electrolyte film confirmed mostly due to the ions. Thus, it can be conclude that the increase in the conductivity is attributed to the increase in the number of ions, and only a negligible contribution comes from the electron.

Conclusion

Solid Polymer electrolytes (SPEs) based on different weight percentage ammonium iodide (NH_4I) doped onto cellulose acetate (CA) were successful prepared and characterised. The SPE with the composition CA (75 wt%):NH_4I (25 wt%) exhibits the highest ionic conductivity of 1.99×10^{-4} S cm^{-1} at room temperature. The electrolyte system has an ionic transference number ion of 0.916 and suggests that charge transport in this polymer electrolytes system is due to ions. XRD spectral analysis shows CA was complexed with NH_4I and formed polymer salt complexes with reduce crystallinity of pristine materials.

Acknowledgment The authors would like to thank to the Ministry of Education Malaysia (KPM), Research Management Institute (RMI), Institute of Science (IOS) and Faculty of Applied Sciences (FSG), Universiti Teknologi MARA for funding this research.

References

1. Gel YS, Jin SH (2004) A self-doped ionic conjugated polymer: poly (2-ethynylpyridinium-n-benzoylsulfonate) by the activated polymerization of 2-ethynylpyridine with ring-opening of 2-sulfobenzoic acid cyclic anhydride. Bull Korean Chem Soc 256:777–778
2. Pradhan DK, Choudhary RNP, Samantaray BK (2009) Studies of dielectric and electrical properties of plastized polymer nanocomposite electrolyte. Mater Chem Phys 115:557–561
3. Pradhan DK, Choudhary RNP, Samantaray BK (2008) Studies of dielectric relaxation and AC conductivity behaviour of plasticized polymer nanocomposite electrolytes. Int J Electrochem Sci 3:597–608
4. Armand M (1994) The history of polymer electrolyte. Solid State Ionic 69:309–319
5. Khiar ASA, Puteh R, Arof AK (2006) Conductivity studies of a chitosan-based polymer electrolyte. Physica B 373:23–27
6. Sreekanth T, Jaipal Reddy M, Subramanyam S, Subba Rao UV (1999) Ion conducting polymer electrolyte film based on (PEO + KNO_3) system and its application as an electrochemical cell. Mater Sci Eng B 64:107–112

7. Chen-Yang YW, Chen HC, Lin FJ, Chen CC (2002) Polyacrylonitrile electrolytes a novel high-conductivity composite polymer electrolyte based on PAN, $LiClO_4$ and α-Al_2O_3. Solid State Ionic 150:327–335
8. Mellende BE, Albinsson I (1996) Ion association in alkali triflate salt poly(propylene oxide) polymer electrolytes. In: Chowdari BVR, Dissanayake MAKL, Careem MA (eds) (1996) Solid state ionics: new developments. World Scientific, Singapore, p 97
9. Rajendran S, Mahendran O, Mahalingam T (2002) Thermal and ionic conductivity studies of plasticized PMMA/PVdF blend polymer electrolytes. Eur Polym J 38:49–55
10. Khiar ASA, Arof AK (2010) Conductivity studies of starch-based polymer electrolytes. Ionics 16:123–129
11. Ramesh S, Arof SAK (2001) Ionic conductivity studies of plasticized poly (vinyl chloride) polymer electrolytes. Mater Sci Eng B 85:11–15
12. Wagner H, Richert R (1997) Thermally stimulated modulus relaxation in polymer: method and interpretation. Polymer 38:5801–5806

Synthesis, Characterization and Corrosion Inhibition Studies of 2-Methylbenzoylthiourea Derivatives

Noor Khadijah Mustafa Kamal, Adibatul Husna Fadzil and Karimah Kassim

Abstract The effect of new corrosion inhibitors, namely *N*-(2-methoxyphenyl)-*N'*-(2-methylbenzoyl) thiourea (A1). *N*-(3-methoxyphenyl)-*N'*-(2-methylbenzoyl) thiourea (A2), *N*-(4-methoxyphenyl)-*N'*-(2-methylbenzoyl)thiourea (A3) and *N*-(2-methylbenzoyl)-*N'*-phenylthiourea (A4) on mild steel in 1.0 M H_2SO_4 with variation concentrations of inhibitors: 1×10^{-5}, 1×10^{-4} and 1×10^{-3} M was tested by using weight loss method and linear polarization resistance (LPR) technique. These compounds were characterized using elemental analyser, FTIR, ^1H and ^{13}C NMR spectroscopy. Result obtained for IR showed the important bands of thiourea compound such as ν(N–H), ν(C=O), ν(C–N) and ν(C=S) which were observed around 3200, 1670, 1250 and 700 cm^{-1}. The ^1H chemical shifts for CONH and CSNH had been observed around 11 and 12 ppm respectively. For the ^{13}C resonance, the signal of C=O and C=S had been observed at 180 and 206 ppm. The presence of the inhibitors significantly reduced the corrosion rate of the mild steel. The most promising inhibitors action is at 1×10^{-3} M and was achieved by compound A1 where the inhibition efficiency is 82.4–86.2 %. The effectiveness of the compounds was affected by the presence and absence of methoxy as the substituents in the compounds.

Keywords Methylbenzoylthiourea · Corrosion inhibitor · Spectroscopic studies · Corrosion inhibition studies · Inhibition efficiency

N.K. Mustafa Kamal (✉) · K. Kassim
Faculty of Applied Sciences, Universiti Teknologi MARA, 40450 Shah Alam, Selangor, Malaysia
e-mail: noorkhadijah@ymail.com

K. Kassim
e-mail: karimah@salam.uitm.edu.my

A.H. Fadzil
Faculty of Applied Sciences, Universiti Teknologi MARA (Perak), Kampus Tapah, Tapah Road, 35400 Tapah, Perak, Malaysia
e-mail: adibatul@perak.uitm.edu.my

K. Kassim
Institute of Science, Universiti Teknologi MARA, 40450 Shah Alam, Selangor, Malaysia

M.A. Hashim (ed.), *ICGSCE 2014*, DOI 10.1007/978-981-287-505-1_40

Introduction

Mild steel is highly susceptible to corrosionsince it is widely used as structural material that involved in industrial processes such as acid pickling, acid cleaning, acid descaling and oil-well acidizing [1–4]. In order to reduce the corrosion activity, organic inhibitors containing nitrogen, oxygen, sulphur and/or phosphorus atom in their structure are the most efficient inhibitor [5, 6]. The inhibitors which have both nitrogen and sulphur atoms simultaneously like thiourea provide a better inhibition performance [6]. Apart from that, the presences of aromatic rings in the compounds make it more effective inhibitors in acid solutions. Heterocyclic ring structure which contains nitrogen and oxygen atoms can enhance greater adsorption on metal surface [7]. Therefore, heterocyclic structure of thiourea is the best organic compound that can be used as the corrosion inhibitor. In this study, a series of thiourea derivatives were synthesized and applied as corrosion inhibitors in order to investigate the inhibition efficiency of the compounds (Fig. 1).

Experimental

Physical Measurements

This experiment was carried out under an ambient atmosphere and no special requirements were taken to exclude air or moisture during the experimentation. All chemicals were purchased from Sigma Aldrich or MERCK and used as received

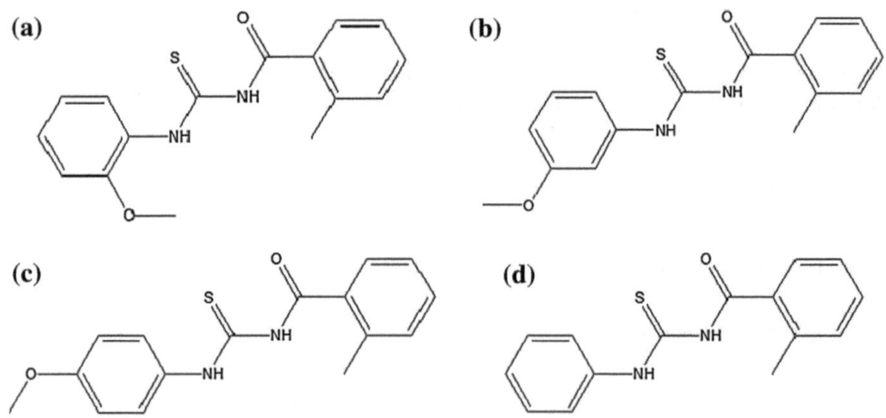

Fig. 1 The chemical structure of **a** *N*-(2-methoxyphenyl)-*N'*-(2-methylbenzoyl)thiourea, A1; **b** *N*-(3-methoxyphenyl)-*N'*-(2-methylbenzoyl)thiourea, A2; **c** *N*-(4-methoxyphenyl)-*N'*-(2-methylbenzoyl)thiourea, A3 and **d** *N*-(2-methylbenzoyl) -*N'*-phenylthiourea, A4

without further purification. The elemental analyses of compound were performed by CHNS Analyzer Flash EA 1112 series. Infrared spectra were recorded using FTIR Perkin Elmer 100 Spectrophotometer in the spectral range of 4000–400 cm^{-1} by using KBr pellet. ^{1}H and ^{13}C NMR spectra were obtained from Bruker Advance III 300 Spectrometer at room temperature.

Synthesis of N-*(2-Methoxyphenyl)*-N'*-(2-Methylbenzoyl) Thiourea, A1*

Freshly prepared 2-methylbenzoyl chloride (5 mmol) in 15 ml acetone was added to equal molar of ammonium thiocyanate. The solution was stirred about 20 min. At the same time, a solution with *o*-anisidine and aniline (5 mmol) in acetone were prepared and added to the previous solution. The reaction solution was heated under reflux for 2 h. The solution was poured into a beaker containing some ice cubes. The resulting precipitate was collected by filtration, washed with ethanol and re-crystallized by using acetonitrile. This compound was obtained as dark brown solids with percentage yield of 56.1 %. IR (KBr pellet, cm^{-1}): ν(N–H) 3184.23, ν(C=O) 1678.54, ν(C–N) 1249.76, ν(C=S) 739.52, ν(OCH_3) 754.18. ^{1}H NMR δ 2.42 (s, 3H, CH_3); 6.90–7.55 (m, Ar–H); 11.70 (d, H, CONH); 12.90 (s, H, CSNH); 3.89 (s, 3H, OCH_3). ^{13}C NMR δ 19.89 (CH_3); 120.25–136.35 (aromatic ring); 206.97 (C=O); 178.19 (C=S); 56.54 (OCH_3). *Anal.* Calc. For $C_{16}H_{16}N_2SO$: C, 64.00; H, 5.33; N, 9.33; S, 10.67. Found: C, 61.27; H, 3.55; N, 9.01; S, 7.81.

Synthesis of N-*(3-Methoxyphenyl)*-N'*-(2-Methylbenzoyl) Thiourea, A2*

In the similar manner as described above, compound A2 was obtained by the reaction of 2-methylbenzoyl isothiocyante with *m*-anisidine, as light yellow solids (69.5 %). IR (KBr pellet, cm^{-1}): ν(N–H) 3184.23, ν(C=O) 1674.42, ν(C–N) 1279.21, ν(C=S) 739.52, ν(OCH_3) 899.51. ^{1}H NMR δ 2.49 (s, 3H, CH_3); 7.20–7.60 (m, Ar–H); 11.70 (d, H, CONH); 12.55 (s, H, CSNH); 3.78 (s, 3H, OCH_3). ^{13}C NMR δ 19.93 (CH_3); 122.47–136.49 (aromatic ring); 206.99 (C=O); 179.25 (C=S); 54.23 (OCH_3). *Anal.* Calc. For $C_{16}H_{16}N_2SO$: C, 64.00; H, 5.33; N. 9.33; S, 10.67. Found: C, 63.01; H, 5.31; N, 8.96; S, 7.62.

Synthesis of N-*(4-Methoxyphenyl)-*N'*-(2-Methylbenzoyl) Thiourea, A3*

This compound was prepared as described above using 2-methylbenzoyl isothiocyante with *p*-anisidine, as dark purple solids (67.6 %). IR (KBr pellet, cm^{-1}): ν(N–H) 3160.02, ν(C=O) 1680.42, ν(C–N) 1238.62, ν(C=S) 737.75, ν(OCH_3) 831.09. ^{1}H NMR δ 2.50 (s, 3H, CH_3); 7.25–7.70 (m, Ar–H); 11.65 (d, H, CONH); 12.40 (s, H, CSNH); 3.78 (s, 3H, OCH_3). ^{13}C NMR δ 19.92 (CH_3); 114.26–136.44 (aromatic ring); 206.93 (C=O); 180.01 (C=S); 55.76 (OCH_3). *Anal.* Calc. For $C_{16}H_{16}N_2SO$: C, 64.00; H, 5.33; N, 9.33; S, 10.67. Found: C, 63.65; H, 4.70; N, 9.12; S, 8.11.

Synthesis of N-*(2-Methylbenzoyl) -*N'*-Phenylthiourea, A4*

The compound A4 was obtained by the reaction of 2-methylbenzoyl isothiocyante with aniline in similar manner as described above, as white solids (69.1 %). IR (KBr pellet, cm^{-1}): ν(N–H) 3222.11, ν(C=O) 1673.80, ν(C–N) 1253.53, ν(C=S) 729.27. ^{1}H NMR δ 2.45 (s, 3H, CH_3); 7.20–7.75 (m, Ar–H); 11.70 (d, H, CONH); 12.60 (s, H, CSNH). ^{13}C NMR δ 19.93 (CH_3); 124.70–136.49 (aromatic ring); 206.97 (C=O); 179.43 (C=S). *Anal.* Calc. For $C_{15}H_{13}N_2SO$: C, 64.91; H, 4.83; N, 10.41; S, 10.67. Found: C, 64.58; H, 3.05; N, 10.09; S, 7.81.

Corrosion Inhibition Studies

Weight loss method, The mild steel coupons were used as specimen with surface area of 4 cm^2. The coupons were abraded with emery paper and washed with acetone followed by distilled water and finally dried. The coupon were immersed for a week in 0.1 M sulphuric acid, H_2SO_4 solution with and without inhibitors. The concentrations of the inhibitors used for the inhibition studies are 1×10^{-5} M, 1×10^{-4} M to 1×10^{-3} M. After the immersion, the weight loss of each coupon was recorded. Each samples was triplicate. The IE% was calculated by using the following formula:

$$\text{IE\%} = \frac{(\text{WL without inhibitor} - \text{WL with inhibitor})}{\text{WL without inhibitor}} \times 100\,\%$$

where WL is weight loss of mild steel.

Linear polarization technique, This method was carried out in a three-electrode cells using AUTOLAB instrument equipped with NOVA software. A saturated calomel electrode (SCE) was used as a reference electrode and a graphite electrode as a counter. The working electrode was prepared by embedding a rod with mild steel coupon in epoxy resin, with an exposed surface area of 0.065 cm^2. The surfaces were polished with emery paper and washed with distilled water for electrochemical studies. Prior to analysis, the working electrode was immersed in the tested solution for 30 minutes to achieve oxidation stability.

Result and Discussion

The FTIR spectra for all four compounds had shown the expected IR frequencies for the important functional groups at a range of 3200 cm^{-1} for N–H, 1670 cm^{-1} for C=O, 1250 cm^{-1} for C–N and 700 cm^{-1} for C=S. These assignments for thiourea derivatives were confirmed by [8, 9]. The functional group of methoxy ($-OCH_3$) were found at different frequencies between compound A1, A2 and A3. A strong peak was observed at 754.18 cm^{-1} in A1 indicates the presence of methoxy group in *ortho-* positions, a medium intensity peak in A2 at 899.51 cm^{-1} for *meta-* positions and a strong peak was observed in A3 at 831.09 cm^{-1} for *para-* positions [10]. In A4, there is no peak that shows the existence of methoxy since it is a free phenyl ring.

The ^{1}H NMR data reveal the existence of methoxy proton between 3.78 and 3.89 ppm. One resonance was observed at 2.42–2.50 ppm as singlet shows that proton of methyl substituents attached to phenyl ring existed. The aromatic proton was found in the range of 6.90–7.75 ppm as distinctive multiplet due to *o-*,*m-* and *p-* substituent methoxy groups at the phenyl ring and the overlapping proton in the aromatic groups. The proton related to CONH and CSNH were observed at 11 and 12 ppm, respectively. The signals are slightly different because of intramolecular hydrogen bonding in NH and the deshielding of aromating ring [9]. In ^{13}C NMR spectra, carbon spectra for CONH and CSNH were found around 180 and 210 ppm, respectively.

For the corrosion inhibition studies, it was found that all four compounds were able to inhibit corrosion activity in H_2SO_4 solution. From Table 1 and Fig. 2, it can be observed that as the concentration of the compounds increased, the inhibition efficiency (IE%) also increased. The highest IE% is 82.4 % which was achieved at concentration 1×10^{-3} by compound A1. From the LPR technique, electrochemical parameters such as corrosion current density (i_{corr}), corrosion potential (E_{corr}), cathodic (b_c) and anodic (b_a) Tafel slopes and inhibition efficiency (IE%) has been listed in Table 2. In the presence of the inhibitor, the i_{corr} of mild steel were smaller than the i_{corr} without the inhibitor. This means that there is an adsorption of the inhibitor on mild steel surface [11]. It is also found that compound A1 has the highest inhibition efficiency which is 86.2 % as seen in Table 2. The results from both weight loss and LPR are in good agreement. Thus, this shows that compound A1 is a potential corrosion inhibitor. The efficiency of the compounds to act as inhibitors was affected by the presence of methoxy substituent. The different positions of the methoxy give different reactivity of the compounds towards the

Table 1 Parameters of mild steel corrosion activities in 1.0 M H_2SO_4 solution with and without inhibitors

Compound	Concentration (M)	Average weight loss (g)	Inhibition efficiency (IE %)	Corrosion rate (g/cm^2 day)
Blank	–	0.0279	–	9.96×10^{-4}
A1	1×10^{-5}	0.0227	18.6	8.11×10^{-4}
	1×10^{-4}	0.0191	31.5	6.82×10^{-4}
	1×10^{-3}	0.0049	82.4	2.92×10^{-5}
A2	1×10^{-5}	0.0201	27.9	7.18×10^{-4}
	1×10^{-4}	0.0167	40.1	5.96×10^{-4}
	1×10^{-3}	0.0086	69.2	3.07×10^{-4}
A3	1×10^{-5}	0.0238	14.7	8.05×10^{-4}
	1×10^{-4}	0.0222	20.4	7.93×10^{-4}
	1×10^{-3}	0.0069	75.3	2.46×10^{-4}
A4	1×10^{-5}	0.0247	11.5	8.82×10^{-4}
	1×10^{-4}	0.0202	27.6	7.21×10^{-4}
	1×10^{-3}	0.0083	70.3	2.96×10^{-4}

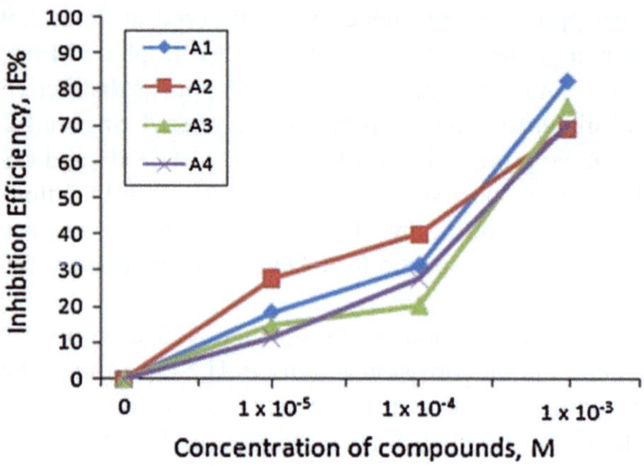

Fig. 2 Graph of inhibitory efficiency percent (IE%) versus concentration of inhibitors by using weight loss method

Table 2 Corrosion parameters for mild steel in 1.0 M sulphuric acid with and without inhibitor

Compound	E_{corr} (mV)	i_{corr} (mA/cm^2)	b_a (mV/dec)	b_c (mV/dec)	IE%
Blank	−425.75	2.61	132.20	38.92	–
A1	−346.62	0.36	30.80	8.93	86.2
A2	−278.92	1.03	222.51	43.77	60.5
A3	−373.60	0.54	58.53	27.81	79.3
A4	−375.94	0.56	63.97	29.34	78.5

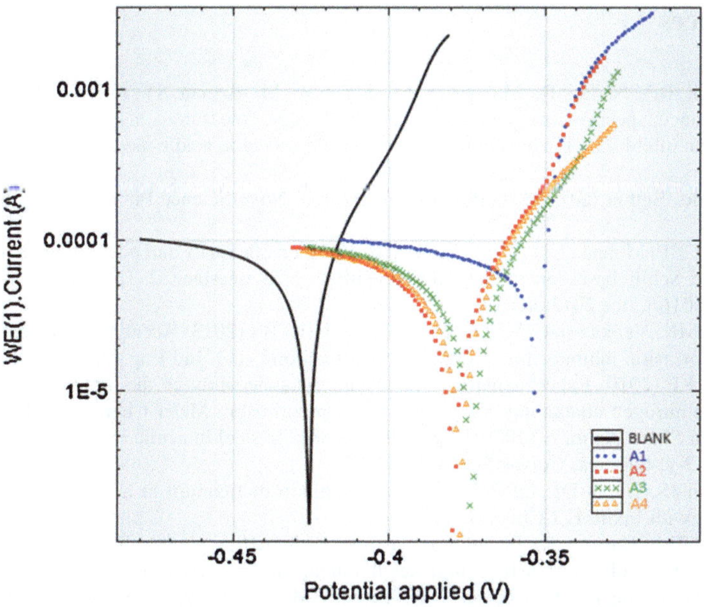

Fig. 3 Polarization curves of 1 × 10^{-3} M A1, A2, A3 and A4 in 1.0 M H_2SO_4

inhibition activity. As methoxy is electron donating group, it is more favours to *o*- and *p*- positions because it able to donate electrons to the benzene group and achieve the compounds' stability to form a protective layer on the metal surface. The order of the effectiveness of the compounds as corrosion inhibitors are A1 > A3 > A4 > A2 (Fig. 3).

Conclusion

Four 2-methylbenzoyl thiourea derivatives namely *N*-(2-methoxyphenyl)-*N'*-(2-methylbenzoyl)thiourea (A1), *N*-(3-methoxyphenyl)-*N'*-(2-methylbenzoyl)thiourea (A2), *N*-(4-methoxyphenyl)-*N'*-(2-methylbenzoyl)thiourea (A3) and *N*-(2-methylbenzoyl) -*N'*-phenyl thiourea (A4) were successfully synthesized, characterized and have the potential to inhibit corrosion. Compound A1 was found to be the potential corrosion inhibitor with the highest inhibition efficiency which is 82.4–86.2 %.

Acknowledgment The authors are grateful to the Ministry of Higher Education of Malaysia for the research grant no. 600-RMI/RAGS 5/3 (3/2012)and the Faculty of Applied Sciences, Institute of Science and Universiti Teknologi MARA for providing research facilities.

References

1. Markhali BP, Naderi R, Mahdarian M, Sayebani M, Arman SY (2013) Electrochemical impendance spectroscopy and electrochemical noise measurements as tools to evaluate corrosion inhibition of azole compounds on stainless steel in acidic media. Corros Sci 75:269–279
2. L. Emilio, Bethni (ed) (2007) Progress in research. Nova Science Publisher Inc., New York, p 160
3. Kumar CBP, Mohana KN (2013) Corrosion inhibition efficiency and adsorption characteristics of some Schiff bases on mild steel/hydrochloric acid interface. J Taiwan Inst Chem Eng. doi:10.1016/j.jtice.2013.08.017
4. Pavitra MK, Venkatesha TV, Kumar MKP, Shylesha BS (2013) Accalyphaterta leaf extract as green corrosion inhibitor for mild steel in hydrochloric acid. Ind Eng Chem Res 52:728–772
5. Khaled KF (2010) Experimental and molecular dynamic study on the inhibition performance of some nitrogen containing compounds for iron corrosion. Mater Chem Phys 124:760–767
6. Hosseini SMA, Azimi A (2009) The inhibition of mild steel in acidic medium by 1-methyl-3-pyridin-2-yl-thiourea. Corros Sci 51:728–732
7. Chauhan JS, Gupta DK (2009) Corrosion inhibition of titanium in acidic media containing fluoride with bixin. E J Chem 6:975–978
8. Halim NIM, Kassim K, Fadzil AH, Yamin BM (2012). Synthesis, characterisation and antibacterial studies of Cu(II) complexes thiourea. Malays J Anal Sci 16(1):56–61
9. Yusof MSM, Jusoh RH, Khairul WM, Yamin BM (2010). Synthesis and characterization a series of N-(3,4-dichlorophenyl)-N'-(2,3 and 4-methylbenzoyl) thiourea derivatives. J Mol Struct 975:280–284
10. Lampman GM, Pavia DL, Kriz GS, Vyvyan JR (2010) Spectroscopy international edition, 4th edn. Cengage Learning, Canada
11. Fouda AEE, Hussein A (2012) Role of some phenylthiourea derivatives as corrosion inhibitors for carbon steel in HCl solution. J Korean Chem Soc 56(2):264–273

Performance Studies of Anode Microbial Fuel Cells Using *Geobacter Sulfurreducens* as a Biocatalyst

Noor Fazliani Shoparwe, Muhammad Hekarl Uzir and Suhairi Abdul Sata

Abstract Microbial fuel cells (MFC) is a bio-electrochemical device that converts organic substrate into electricity by electrogenic bacteria. In the MFC, anode compartment plays an important role in achieving high power density and thereby improving the cell performance. In this study, *Geobacter sulfurreducens* was used as a biocatalyst to catalyze the generation of electricity by using acetate as an oxidizing agent and sodium fumarate as a reducing agent. The polarization curve was determined after MFC system was reached at open circuit voltage (OCV) around 0.80 V and gives maximum power production of about 0.22 mW. The cyclic voltammetry (CV) method was performed in order to evaluate the oxidation and reduction processes in a batch of MFC before and after attachment of *Geobacter sulfurreducens* cells and it was proved by scanning electron microscope (SEM) images. Overall, this study provides a convenient way for evaluating the performance of MFC.

Keywords *Geobacter sulfurreducens* · Microbial fuel cells · Batch · Single chamber · Cyclic voltammetry

N.F. Shoparwe · M.H. Uzir · S. Abdul Sata (✉)
School of Chemical Engineering, Engineering Campus, Universiti Sains Malaysia, 14300 Nibong Tebal, Seberang Perai Selatan, Pulau Pinang, Malaysia
e-mail: chhairi@usm.my

N.F. Shoparwe
e-mail: fazliani_usm@yahoo.com

M.H. Uzir
e-mail: chhekarl@usm.my

M.A. Hashim (ed.), *ICGSCE 2014*, DOI 10.1007/978-981-287-505-1_41

Introduction

Microbial Fuel Cell (MFC) is a bioreactor that converts chemical energy from chemical bonds of organic compounds into electricity based on catalytic reactions of microorganism under anaerobic conditions. *Geobacter sulfurreducens* are one of the most effective microorganisms used acetate as a substrate and electrodes as the sole electron acceptor in MFC to generate electricity. The half and overall reactions of an MFC using acetate as the substrate can be presented as shown below [1]:

Anodic Oxidation reaction:

$$CH_3COO^- + H^+ + 2H_2O \rightarrow 2CO_2 + 8H^+ + 8e^- \quad (1)$$

Cathodic Reduction reaction:

$$2O_2 + 8e^- + 8H^+ \rightarrow 4H_2O \quad (2)$$

Overall reaction:

$$CH_3COO^- + H^+ + 2O_2 \rightarrow 2CO_2 + 2H_2O \quad (3)$$

Research relating to MFC is rapidly increasing and involves researchers from several different fields, due to its bioelectrochemical nature. Many researches have been conducted are looking at ways in which to evaluate and increased MFC performance. An improved understanding of the catalytic activity of the bioanode as a function of solution conditions would contribute to enhanced MFC operation and performance. However, the electrochemistry of the bioanode is complex due to changing microbial communities and interactions within these communities in the anode biofilm. One of the techniques that have been focused by researchers nowadays is using electrochemical techniques. Cyclic voltammetry (CV) is gains in popularity as a rapid method of evaluating the performance of electron transfer mechanisms that are used in the fuel cell. CV is an electroanalytical technique in which a varying voltage potential is applied to an electrode in an electrochemical system while the resulting current is measured [2]. CV has been applied to MFC in an increasing number of studies and has been widely used for characterization of electrochemically active compound in MFC [1, 3–8]. The strategy to applied CV for the study of anodic bioelectroactalytic electron transfer in MFC has been described well by several authors [7, 9–11]. In this study, overall performance studies of anode batch microbial fuel cells were evaluated using *Geobacter sulfurreducens* as a biocatalyst and graphite felt as an anode electrode in a single-chamber MFC. The performance of MFC was determined based on the interaction between the cell biofilm of *Geobacter sulfurreducens* and anode electrode surface via polarization curve and CV method.

Experimental

Geobacter sulfurreducens strain PCA (ATCC 51573), used in this study was obtained from the American Type Culture Collection, ATCC (Rockville, MD). The strain was subcultured in the laboratory at and incubated under anaerobic condition at 32 °C. The growth medium was prepared according to the protocol supplied by the ATCC consisted the following (per liter): 1.5 g of NH_4Cl, 0.6 g of NaH_2PO_4, 0.1 g of KCl, 2.5 g of $NaHCO_3$, 10.0 ml Wolfe's vitamin mix solution and 10.0 ml Modified Wolfe's mix mineral solution, respectively. The medium pH was initially adjusted to 6.8 and was autoclaved at 121 °C for 15 min. The anode compartment of MFCs was flushed with N_2/CO_2 (80–20 %) stream before inoculation at least 1 h and a mild flow was then maintained during the experiment. The electrochemical reactor was inoculated (10 % v/v) with cell suspensions that had an optical density at 620 nm around 0.2–0.3 during the late of exponential phase. The single-chamber MFC was designed and fabricated using glass (Pyrex). Schematic diagram of the design used in this research are depicted in Fig. 1. Single chamber design consist of graphite felt as anodic electrode, where graphite as a cathode electrode was deposited to one side to ensure the directly exposed to air. The working volume of each chamber was about 500 ml. A proton exchange membrane (PEM) (Nafion 117, DuPont Co., USA) was used to separate the anode and cathode compartment and allowing transfer of H^+ ions through the membrane was held in place by rubber gaskets.

In all experiments, the MFC was connected to the multi-potentiostat (VMP3 Biologic SA). Three-electrode systems (working electrodes (WE) and counter electrodes (CE) and reference electrode (RE)) were implemented in this study. For the polarization experiment, the value of current and voltage were measured at variable external resistance, R_{ext} between (50–150 Ω). These values were measured

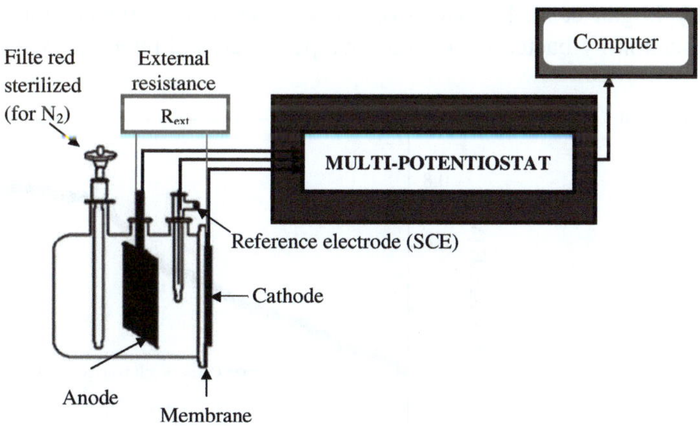

Fig. 1 Schematic diagram of the glass single-chamber MFC connected to multi-potentiostat and computer

after open circuit voltage (OCV) reached a steady state optimum voltage. CV experiment was performed under OCV conditions and the voltage was changed from −1.0 to 1.0 V against standard calomel electrode (SCE) in forward and reverse scans 10 mVs^{-1}. Samples were withdrawn at regular time intervals and analyzed for cell and acetate concentrations. Cell growth was monitored using spectrophotometer (Thermo Spectonic Genesys 20) at an optical density of 660 nm (OD_{660}). A correlation was established relating OD_{660} to dry weight per liter of culture broth (gl^{-1}). One unit of OD_{660} corresponds to 0.532 ± 0.0035 gl^{-1} of dry cell weight (DCW). Acetate concentration was performed using High Performance Liquid Chromatography (HPLC). HPLC analyses were performed with a Chromatography (HPLC) instrumentation system equipped with Supelcolis CL-SAXI column (250 × 4.8 mm). The mobile phase was acetonitrile-water (3:97 % (v/v)) at pH 3.0 adjusted with H_3PO_4. The surface morphology of anode electrode before and after attachment of *Geobacter sulfurreducens* cells were observed using scanning electron microscope (SEM) (Oxford INCA/ENERGY.350). After the end experiment, the anode electrode was collected and fixed overnight with paraformaldehyde and glutaraldehyde in a buffer solution at pH 5.4 and 4 °C. The sample was coated with gold and observed with a SEM.

Result and Discussion

In this study, the electricity was produced by the *Geobacter sulfurreducens* growth in the anaerobic anode compartment by using acetate as a substrate. Initially, in order to determine the optimum voltage inside the single-chamber of MFC system, the open circuit voltage (OCV) (without any load) was determined (Fig. 2). It was found that the maximum value of OCV was achieved around 0.80 V after 4 days of fermentation time. This voltage was slightly lower than the theoretical value obtained by Logan et al. [1]. However, this value was significant might due to energy utilization by bacteria, electrode overpotential, and internal resistance [12].

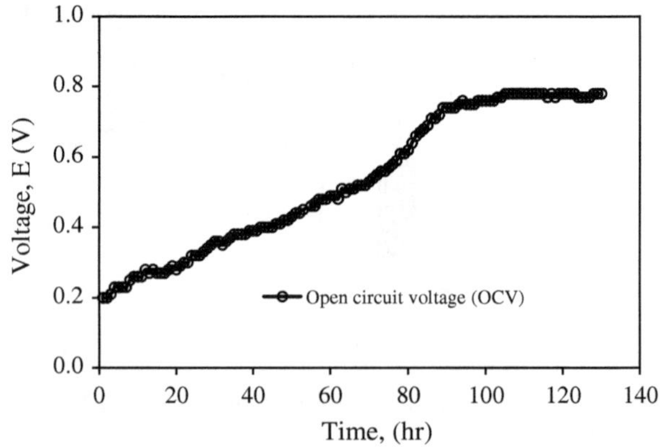

Fig. 2 Open circuit voltage (OCV)

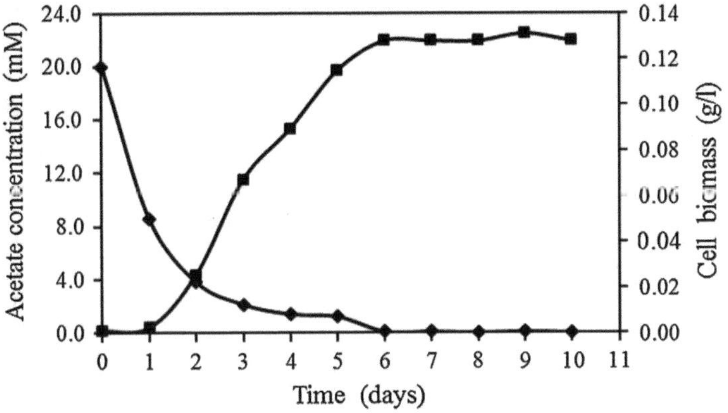

Fig. 3 Time-course profile of growth *Geobacter sulfurreducens* and acetate utilization in a batch anode MFC

A time-course profiles of anaerobic cultivation of *Geobacter sulfurreducens* using 20 mM of initial acetate concentration in anode compartment of MFCs is shown in Fig. 3. A classical sigmoidal growth trend for the *Geobacter sulfurreducens* cells, in which, an initial lag phase (~24 h), which varied in duration were observed. This was followed by the exponential growth phase, where cell growth increased exponentially and a stationary phase accordingly. Polarization curve as a function of current, potential and power production was obtained and shown in Fig. 4.

The polarization curve is the most common method of representing the performance of fuel cells. The polarization curve can be divided into three main region; activation, ohmic and mass transport losses [13]. The power performance curve was calculated from the polarization curve, which represents the relationships between power generation and given current. The polarization curve as a function of current, potential and power measured at variable external resistance R_{ext} between

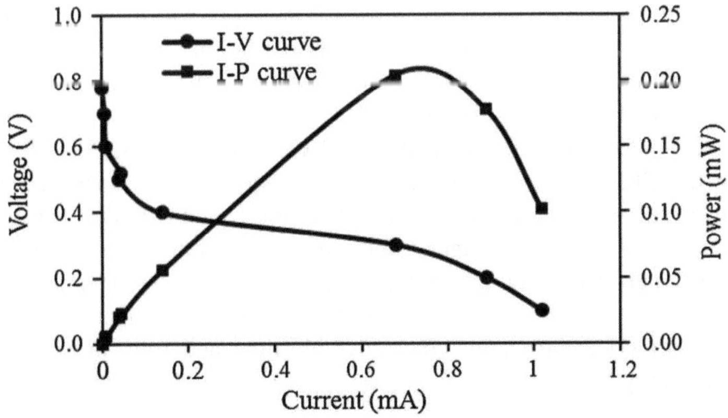

Fig. 4 Polarization curve of batch anode MFC

Fig. 5 Cyclic votammogram at scanning rate 10 mVs^{-1} (before inoculation and after 6 days of anaerobic fermentation)

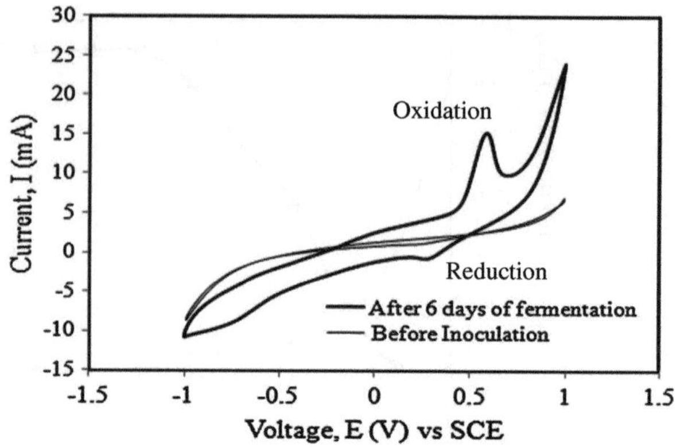

(50–150 KΩ). The dominance of activation loss was observed from the initial step decrease in the voltage from OCV value at 0.80 to 0.39 V. The subsequent slope of the voltage decreased from 0.18 to 0.9 mA and was almost linear, which might indicate the dominance of ohmic losses. The relatively lower drop in voltage from 0.18 to 0.05 V and maximum current obtained at lower resistance reveals a lower potential drop and lower mass transfer losses at the electrode. The effective electron discharge observed at lower resistances is due to the further potential drop and slow stabilization of the voltage at lower resistance [14]. The curve depicts a maximum power of 0.22 mW at the resistance 350 Ω. After reaching the optimum power production, power generation showed a decreasing trend with an increases in R_{ext}. This is consistent with the literature [15–19], an indicated typical MFC behavior.

CV is extensively used to examine the performance of MFC throughout the oxidation and reduction properties. It also have been used to evaluate the bio-electrocatalytic activity of cell attachment on the anode surface of electrode toward substrate oxidation. In this study, CV experiment was carried out after the current reached its maximal value in a batch MFC process (Fig. 5). As can be shown CV in Fig. 5, none of the redox peaks were obtained in the sterile culture medium before the inoculation. However, after six days of anaerobic fermentation, the cyclic voltammogram obtained showed very sharp oxidation and reduction peaks at potential of 0.61 and 0.32 V (*vs* SCE) which indicated the electrochemical activity of the attached cell on the anode electrode.

The results before and after attachment of cell *Geobacter sulfurreducens* were proved by the SEM (Fig. 6a, b). The SEM images indicated surface and morphology information about the used electrode in the anode compartment before and after attachment of cell on the anode surface. The obtained images demonstrated microorganisms grown on the graphite felt surface. Based on Logan and co-worker studies [1, 20, 21], when microorganisms interact with conductive surface (electrode), they formed electroactive microbial biofilms and transfer the electron to generate electricity.

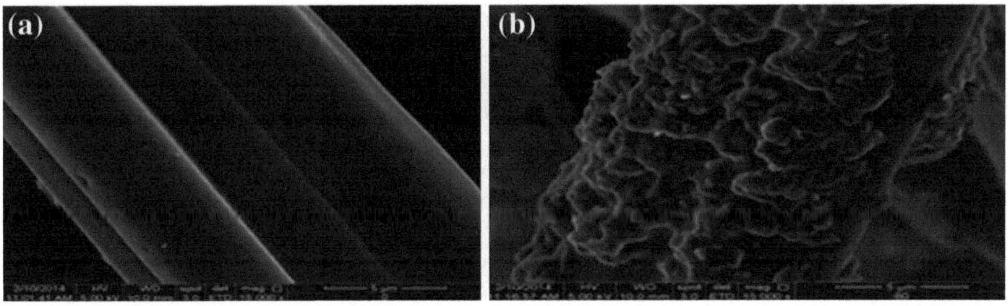

Fig. 6 SEM results **a** before attachment and **b** after attachment at 15,000 x magnification

Conclusion

In this study performance of anode MFC was evaluated in a single-chamber design using *Geobacter sulfurreducens* as a biocatalyst. The polarization curve analysis was performed in order to characterize the voltage losses involved in the system. The optimum power production was reached about 0.22 mW at an external resistance of 350 Ω. The oxidation and reduction processes occurring at the anodes were confirmed using the CV method and SEM analysis. This research provided a potential approach to evaluate the overall performance of MFC.

Acknowledgment This research was supported by funding through the Fundamental Research Grant Scheme (FRGS) (Acc Number: 6071232) by Ministry of Higher Education, Malaysia and Postgraduate Research Grant Scheme (PRGS) (Acc Number: 8046005) by Universiti Sains Malaysia (USM).

References

1. Logan BE, Hamelers B, Rozendal R, Schröder U, Keller J, Freguia S, Aelterman P, Verstraete W, Rabaey K (2006) Microbial fuel cells: methodology and technology. Environ Sci Technol 40:5181–5192
2. Nicholson RS (1965) Theory and application of cyclic voltammetry for measurement of electrode reaction kinetics. Anal Chem 37:1351–1355
3. Zhou M, Chi M, Luo J, He H, Jin T (2011) An overview of electrode materials in microbial fuel cells. J Power Sour 196:4427–4435
4. Liu X-W, Sun X-F, Huang Y-X, Sheng G-P, Zhou K, Zeng RJ, Dong F, Wang S-G, Xu A-W, Tong Z-H (2010) Nano-structured manganese oxide as a cathodic catalyst for enhanced oxygen reduction in a microbial fuel cell fed with a synthetic wastewater. Water Res 44:5298–5305
5. Kim JR, Jung SH, Regan JM, Logan BE (2007) Electricity generation and microbial community analysis of alcohol powered microbial fuel cells. Bioresour Technol 98:2568–2577
6. Rabaey K, Ossieur W, Verhaege M, Verstraete W (2005) Continuous microbial fuel cells convert carbohydrates to electricity. Water Sci Technol 52:515–523
7. Marsili E, Rollefson JB, Baron DB, Hozalski RM, Bond DR (2008) Microbial biofilm voltammetry: direct electrochemical characterization of catalytic electrode-attached biofilms. Appl Environ Microbiol 74:7329–7337

8. Pham CA, Jung SJ, Phung NT, Lee J, Chang IS, Kim BH, Yi H, Chun J (2003) A novel electrochemically active and Fe (III)-reducing bacterium phylogenetically related to Aeromonas hydrophila, isolated from a microbial fuel cell. FEMS Microbiol Lett 223:129–134
9. Fricke K, Harnisch F, Schröder U (2008) On the use of cyclic voltammetry for the study of anodic electron transfer in microbial fuel cells. Energy Environ Sci 1:144–147
10. Marsili E, Sun J, Bond DR (2010) Voltammetry and growth physiology of *Geobacter sulfurreducens* biofilms as a function of growth stage and imposed electrode potential. Electroanalysis 22:865–874
11. Harnisch F, Freguia S (2012) A basic tutorial on cyclic voltammetry for the investigation of electroactive microbial biofilms. Chem Asian J 7:466–475
12. Oh S, Kim J, Joo J, Logan B (2009) Effects of applied voltages and dissolved oxygen on sustained power generation by microbial fuel cells. Water Sci Technol 60:1311
13. Logan BE (2008) Microbial fuel cells. Wiley, New York
14. Nimje VR, Chen C-C, Chen H-R, Chen C-Y, Tseng M-J, Cheng K-C, Shih R-C, Chang Y-F (2012) A single-chamber microbial fuel cell without an air cathode. Int J Mol Sci 13:3933–3948
15. Bond DR, Lovley DR (2003) Electricity production by *Geobacter sulfurreducens* attached to electrodes. Appl Environ Microbiol 69:1548–1555
16. Cho HK, Trinh NT, An JY, Kim BW (2008) Optimizing the power density generated from a microbial fuel cell using *Geobacter sulfurreducens* with acetate. J Biotechnol 136 (Supplement):S485
17. Nevin KP, Richter H, Covalla S, Johnson J, Woodard T, Orloff A, Jia H, Zhang M, Lovley D (2008) Power output and columbic efficiencies from biofilms of *Geobacter sulfurreducens* comparable to mixed community microbial fuel cells. Environ Microbiol 10:2505–2514
18. Kim M-S, Lee Y-J (2010) Optimization of culture conditions and electricity generation using *Geobacter sulfurreducens* in a dual-chambered microbial fuel-cell. Int J Hydrogen Energy 35:13028–13034
19. Trinh NT, Park JH, Kim B-W (2009) Increased generation of electricity in a microbial fuel cell using Geobacter sulfurreducens. Korean J Chem Eng 26:748–753
20. Logan BE (2009) Exoelectrogenic bacteria that power microbial fuel cells. Nat Rev Microbiol 7:375–381
21. Logan BE, Regan JM (2006) Electricity-producing bacterial communities in microbial fuel cells. Trends Microbiol 14:512–518

Effects of Chain Length on the Thermotropic and Lyotropic Phase Behaviours of Maltosides by Small-angle X-ray Diffraction Study

Hairul A.A. Hamid, Rauzah Hashim, John M. Seddon and Nicholas J. Brooks

Abstract Industrially useful surfactants such as alkyl polyglucosides (APGs) and synthetic branched-chain glycosides are interesting biosurfactants and offer an interesting insight into life-science applications. Hence, research presented here a study of the thermotropic and lyotropic phase of 2-hexyldecyl β-D-maltoside (β-Mal-$C_{10}C_6$) and the corresponding 2-decyltetradecyl β-D-maltoside (β-Mal-$C_{14}C_{10}$) over a range of temperatures (0–80 °C). Small-angle X-ray diffraction (SAXS) and optical polarizing microscopy (OPM) have been used to characterize the mesophase behaviour. This study has shown that in dry conditions β-Mal-$C_{10}C_6$ and β-Mal-$C_{14}C_{10}$ adopt a lamellar crystalline phase (L_c), on the other hand, in excess water these compounds form a fluid lamellar phase (L_α) upon heating and cooling. The study also reported the effects of temperature (0–80 °C) on the d-spacing value using SAXS, and we estimated the surface area per molecule and water content of the fluid lamellar phase (L_α) in the fully hydrated state for both β-Mal-$C_{10}C_6$ and β-Mal-$C_{14}C_{10}$. The former gave the surface area per molecule and water content of 57 Å^2 and 25 wt % respectively, however, the latter forms 69 Å^2 and 20 wt %.

Keywords Thermotropic phase · Lyotropic phase · Lamellar phase

H.A.A. Hamid (✉)
School of Chemistry and Environment, Faculty of Applied Sciences, Universiti Teknologi MARA, 40450 Shah Alam, Malaysia
e-mail: h.amani@salam.uitm.edu.my

H.A.A. Hamid · R. Hashim
Department of Chemistry, Faculty of Science, University of Malaya, 50603 Kuala Lumpur, Malaysia
e-mail: rauzah@um.edu.my

J.M. Seddon · N.J. Brooks
Department of Chemistry, Imperial College London, London SW7 2AZ, UK
e-mail: j.seddon@imperial.ac.uk

N.J. Brooks
e-mail: nicholas.brooks@imperial.ac.uk

M.A. Hashim (ed.), *ICGSCE 2014*, DOI 10.1007/978-981-287-505-1_42

Introduction

New generation sugar-based surfactants (glycolipids) such as alkyl glycosides are made from renewable resources. Biodegradability, highly value-added and material versatilities are, hence, obtained [1, 2]. These surfactants, therefore, exhibit a varied potential application and are gaining importance progressively especially in cellular membrane's fields [3, 4]. Membranes mainly composed of a variety of lipids and most lipid membranes are built from phospholipids [5]. An interesting membrane component which is glycolipids has recently received much attentions. Its pervasive presence in nature has made tantalizing hints that the glycolipids play a pivotal role the same as phospholipids. Both component, are amphitropic liquid crystals which contain hydrophilic and hydrophobic parts. As such, these characteristics have enabled them to self-organize in a wide variety of phases showing lipid polymorphisms in both thermotropic (dry state) and lyotropic (presence of solvent) conditions [6–9].

In addition to provide more understanding of membrane functions study, the present work reports the amphiphilic liquid crystals Guerbet branched-chain maltosides with two different types of hydrophobic chains known to produce phases in the absence (thermotropic) and presence of excess water (lyotropic) as shown in Fig. 1. Combination of two techniques which were small-angle X-ray diffraction and optical polarizing microscopy had been used during investigations. Moreover, the study also presented the d-spacing value at different temperatures (0–80 °C) for both compounds in dry and excess water states. The area per molecule and water content were then estimated for both compounds.

Materials and Methods

Synthesis of β-D-Maltosides

Chemically synthesized *β*-maltosides were prepared [10] and a higher purity (98 %) of these compounds was estimated by ^{1}H-NMR and TLC techniques. The HPLC-grade water was used in all excess water sample preparations.

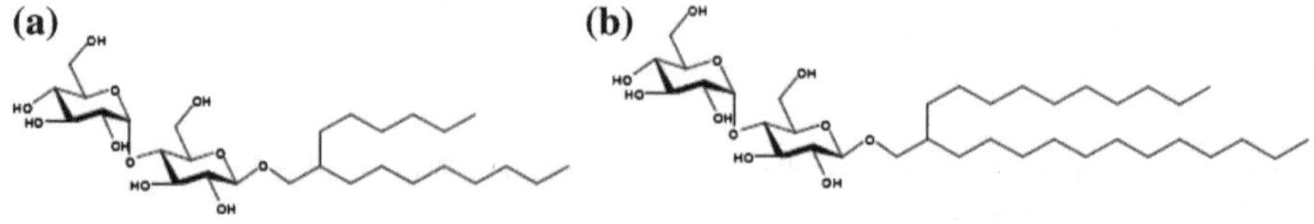

Fig. 1 Chemical structures of the two surfactants: **a** *β*-Mal-$C_{10}C_6$, **b** *β*-Mal-$C_{14}C_{10}$

X-Ray Diffraction

Philips generator was used for X-ray measurements. It was operated at 40 kV and 30 mA that capable to generate X-rays. Line X-ray beam was focused due to the isolation of Cu radiation by quartz crystals monochromator. Copper block, which consisted of X-ray capillary for holding the prepared-sample [11], was heated and cooled at temperatures in the range of 0–80 °C. The heating or cooling rate was at 10 °C per hour. The diffraction pattern was recorded on a Kodak Biomass MS autoradiography film. The film was inserted into a film holder which was able to move at the rate of 0.06 mm per minute. It was developed using standard procedure and analysed using Axcess software [12].

Calculation of Fully Hydrated Lα Phase

The area per molecule [13] and volume fraction of water of L_α phase from the measured d-spacings in dry and excess water conditions, and density of the maltosides [14] can be estimated by using the following formula: Calculation of estimated volume fraction of water is:

$$\frac{\upsilon_W}{\upsilon_{total}} = \frac{d_W}{d_0} = \frac{\Delta d}{d_{excess}} \tag{1}$$

where υ_W and υ_{total} are the partial specific volumes of water and total (lipid + water), d_w and Δd are the water layer thickness, and d_0 and d_{excess} are d-spacings in dry and excess water conditions. Volume percent of water is approximately the same as weight percent of water if the density of the maltosides is equal to the density of water.

Optical Polarizing Microscopy

Thermotropic and lyotropic liquid crystal phases were determined by optical polarizing microscopy (OPM). Nikon Eclipse E600 polarizing microscope combined with a Nikon DXM1200F digital camera was used to take sample images. As Nikon Act-1 (version 2.70) recorded the actual images, Linkam heating stage controlled by Linkam Controller software was utilised to heat and cool the sample.

Dry Sample Preparation

A dry sample was carefully transferred to a glass microscope slide and covered with a glass cover slip. The sample was placed on the hot stage and further investigated by controlling the temperature.

Hydrated Sample Preparation

Sample was placed on the glass microscope slide and a drop of water was added. The mixture was then covered with a glass cover slip and sealed with a silicone sealant. Heating and cooling were employed on the sample under investigation.

Results and Discussions

Small-Angle X-Ray Diffraction

The X-ray of β-maltosides consist of chain length C_{16} (β-Mal-$C_{10}C_6$) and C_{24} (β-Mal-$C_{14}C_{10}$) were identified. As can be seen from Fig. 2a, c, only the first-order peak of the lamellar phase is visible for both compounds in the dry state upon heating and cooling at temperatures ranging from 0 to 80 °C. This phase is a lamellar liquid crystalline phase (L_c) with d-spacings of 30 and 36 Å, respectively. The occurrence of the same phase over the temperature range is confirmed by an optical polarizing microscopy results (Fig. 3a, c). It is well known that chain branching stabilizes non-lamellar curved phases [15]; in contrast, we observe that the longer chain lipids (β-Mal-$C_{10}C_6$ and β-Mal-$C_{14}C_{10}$) form only lamellar L_c phase.

Figure 2b, d shows the X-ray diffraction results of β-Mal-$C_{10}C_6$ and β-Mal-$C_{14}C_{10}$ compounds in excess water with the d-spacings of 40 and 45 Å,

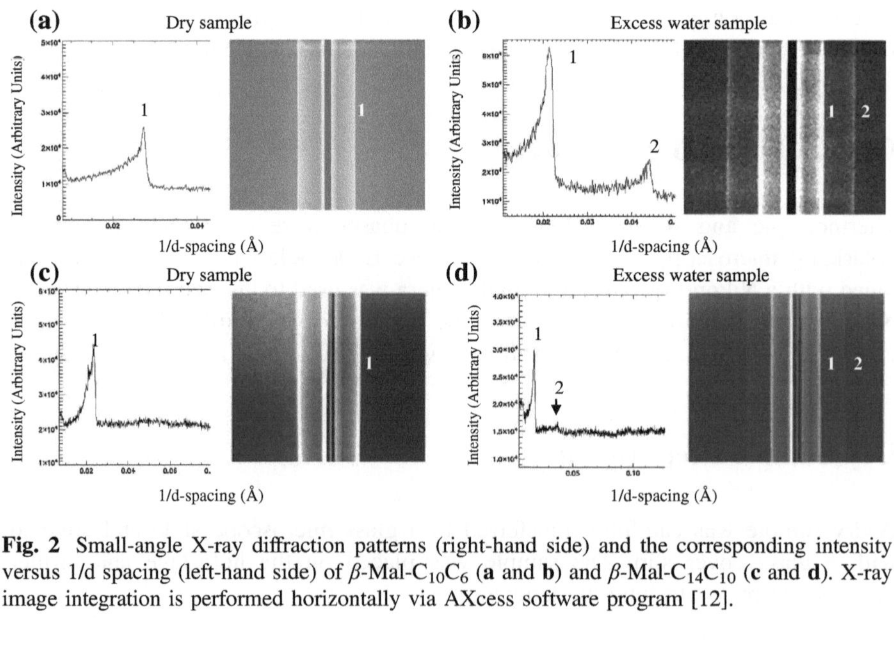

Fig. 2 Small-angle X-ray diffraction patterns (right-hand side) and the corresponding intensity versus 1/d spacing (left-hand side) of β-Mal-$C_{10}C_6$ (**a** and **b**) and β-Mal-$C_{14}C_{10}$ (**c** and **d**). X-ray image integration is performed horizontally via AXcess software program [12].

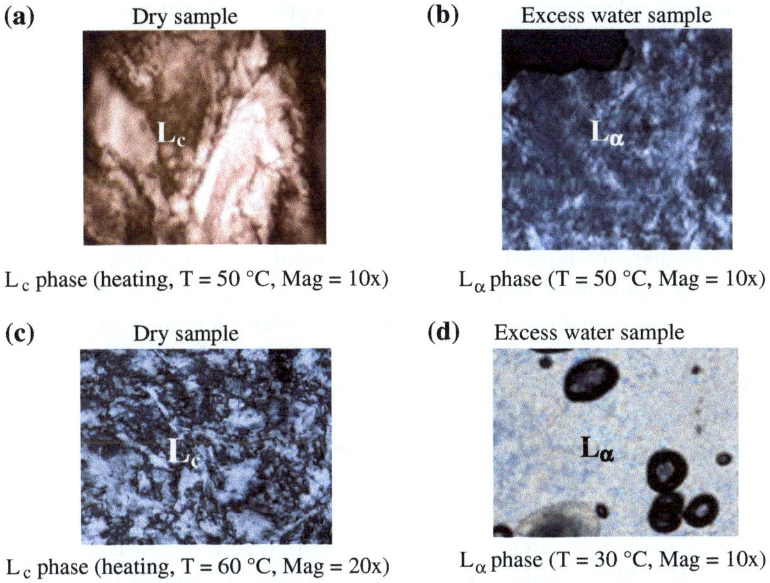

Fig. 3 Optical textures of β-Mal-$C_{10}C_6$ (**a** and **b**) and β-Mal-$C_{14}C_{10}$ (**c** and **d**)

respectively. The equally spaced peaks (ratio 1 and 2) fluid lamellar phases (L_α) are detected in both compounds. The same type of phase remains in existence throughout the variation of temperatures which is confirmed by the optical polarizing microscopy as indicated in Fig. 3b, d.

Optical Polarizing Microscopy

Birefringent optical textures were investigated for both compounds with the temperatures ranging from 0 to 80 °C. Fig. 3 illustrates sample of the dry and excess water conditions. On heating and cooling, both β-Mal-$C_{10}C_6$ and β-Mal-$C_{14}C_{10}$ exhibit a lamellar crystalline phase (L_c) in the dry conditions and a fluid lamellar phase (L_α) in the samples containing excess water.

Hydration of the Fluid Lamellar L_α Phase

In the calculation of the estimated area per molecule and weight percent of water of the fully hydrated maltoside bilayers, the X-ray d-spacings of the dry and excess water conditions are plotted as functions of temperatures upon heating and cooling between 0 and 80 °C (data not shown) in order to determine the water layer thickness (Δd) for each maltoside.

Table 1 The structural parameters of the fluid lamellar L_α phase of β-maltosides at full hydrations varying at temperatures 0–80 °C (heating and cooling)

Lipid	Phase	Chain length	Water layer thickness (Δd) [Å]	Area per molecule [Å^2]	Weight % of water
β-Mal-$C_{10}C_6$	L_α	C_{16}	9	57	25
β-Mal-$C_{14}C_{10}$	L_α	C_{24}	10	69	20

In excess water, for both maltosides, upon heating and cooling between 0 and 80 °C the only stable state formed is a fluid lamellar L_α phase as demonstrated in Table 1. This can be explained by the increased chain length region surface area which lowers the sugar domain packing stress [14]. Eventhough, the type of phase occurred is the same for both β-Mal-$C_{10}C_6$ and β-Mal-$C_{14}C_{10}$, the concominent increases in area per molecule are observed from 57 (former) to 69 Å^2 (latter). These are considerably lower than the obtainable value 72.2 Å of fully hydrated 1,2-dioleoyl-*sn*-glycero-3-phosphatidylcholine (DOPC). Indeed, the larger molecular weight of DOPC (786) indicates the greater molecular volume (1303 Å^3) [16] and thus, requires greater an area per molecule. This result is consistent with those β-Mal-$C_{10}C_6$ (M: 567 g/mol) and β-Mal-$C_{14}C_{10}$ (M: 679 g/mol) which exhibit an increment in the area of the molecule when the chain length of the hydrocarbons and the molecular volumes are increased [14].

One of the important factors that affects the water spacing is the fluctuation pressures of the interbilayer interactions exist in the fully hydrated L_α phase [16]. Moreover, McIntosh and Simon [17] found that the flexible L_α lipid with bigger fluctuation pressure resulted in a larger water spacing in comparison with rigid gel or subgel lipid. As shown in Table 1, the water layer thickness L_α phase of β-Mal-$C_{10}C_6$ and β-Mal-$C_{14}C_{10}$ are 9 and 10 Å, respectively. This small increment in Δd arises probably due to the higher interbilayer fluctuation pressures as mention earlier and a change in free energy leads to the layer to bend [18].

Twenty five weight percent of water of β-Mal-$C_{10}C_6$ and 20 wt % of water of β-Mal-$C_{14}C_{10}$ indicate the solvation of both maltosides. It is interesting to note that, the latter shows a reduction in weight percent of water as the chain length increases. This suggests that the interaction factors needed to be considered are the freely water entering the fluid space between bilayers, the intercalated water into the head groups bilayers [19]; and the interbilayer interactions namely van der Walls interaction [20], fluctuation pressure [16], hydration force [21] and the steric thickness [22].

Conclusions

Knowledge of the fundamental parameters, for instance, the surface area per molecule and amount of water per lipid molecule [11] are essential to biomembrane of lipids. Hence, in this work, the thermotropic and lyotropic phase behaviours of β-Mal-$C_{10}C_6$

and β-Mal-C $_{14}C_{10}$ upon heating and cooling (0–80 °C) have been determined . In dry conditions, at all temperatures, both maltosides exhibit L_c phase with d-spacing values 30 and 36 Å, respectively. On the contrary, in excess water the former (d-spacing 40 Å) and the latter (d-spacing 45 Å) produce a fluid lamellar phase. Values of area per molecule and water content which are 57 Å, 25 wt % (β-Mal-$C_{10}C_6$) and 69 Å, 20 wt % (β-Mal-$C_{14}C_{10}$) have been estimated in the fully hydrated L_α phase.

Acknowledgement This work was supported by UiTM-scholarship, UM.C/HIR/MOHE/SC11 and EPSRC (UK)-Platform grant EP/G00465X. We thank them for all financial and moral support given.

References

1. Andreozzi P, Gente G, Mesa CL (2009) Solution behavior of alkyl glycosides and related compounds. In: Ruiz CC (ed) Sugar-based surfactants fundamentals and applications. CRC Press Taylor & Francis, New York, pp 21–59
2. Vill V, Hashim R (2002) Carbohydrate liquid crystals: structure–property relationship of thermotropic and lyotropic glycolipids. Curr Opin Colloid Interface Sci 7:395–409
3. Hill K, LeHen-Ferrenbach C (2009) Sugar-based surfactants for consumer products and technical applications. In: Ruiz CC (ed) Sugar-based surfactants fundamentals and applications. CRC Press Taylor & Francis, New York, pp 1–20
4. Lopez O, Cocera M, Coderch L, Barbosa-Barros L, Parra JL, Maza ADL (2009) Applications of alkyl glucosides in the solubilization of liposomes and cell membranes. In: Ruiz CC (ed) Sugar-based surfactants fundamentals and applications. CRC Press Taylor & Francis, New York, pp 585–626
5. Karp G (2010) Cell biology, six edn. Wiley, New Jersey
6. Mannock DA, McElhaney RN (2004) Thermotropic and lyotropic phase properties of glycolipid diastereomers: role of headgroup and interfacial interactions in determining phase behaviour. Curr Opin Colloid Interface Sci 8:426–447
7. Dumoulin F, Lafont D, Boullanger P, Mackenzie G, Mehl GH, Goodby JW (2002) Self-organizing properties of natural and related synthetic glycolipids. J Am Chem Soc 124:13737–13748
8. Minden HMV, Branburg K, Seydel U, Koch MHJ, Garamus V, Willumeit R, Vill V (2000) Thermotropic and lyotropic properties of long chain alkyl glycopyranosides. Part II. Disaccharide head group. Chem Phys Lipids 106:157–179
9. Brooks NJ, Hamid HAA, Hashim R, Heidelberg T, Seddon J, Conn CE, Husseini SMH, Zahid NI, Hussen RD (2011) Thermotropic and lyotropic liquid crystalline phases of Guerbet branched-chain β–D-glucosides. Liq Cryst 38:1725–1734
10. Hashim R, Abdalla Hashim HH, Rodzi NM, Hussen RSD, Heidelberg T (2006) Branched chain glycosides: Enhanced diversity for phase behavior of easily accessible synthetic glycolipids. Thin Solid Films 509:27–35
11. Hamid HAA, Hashim R, Seddon JM, Brooks NJ (2014) Lyotropic phase behaviour and structural parameters of monosaccharide and disaccharide Guerbet branched-chain β-D-glycosides. Adv Mater Res 895:111–115
12. Seddon JM, Squires AM, Conn CE, Ces O, Heron A, Mulet X, Shearman GC, Templer RH (2006) Pressure-jump X-ray studies of liquid crystal transitions in lipids. Philos Trans Roy Soc A 364:2635–2655

13. Seddon JM, Templer RH, Warrender NA, Huang Z, Cevc G, Marsh D (1997) Phosphatidylcholine-fatty acid membranes: effects of headgroup hydration on the phase behaviour and structural parameters of the gel and inverse hexagonal (H_{II}) phases. Biochim Biophys Acta 1327:131–147
14. Nguan HS, Heidelberg T, Tiddy G, Hashim R (2010) Quantitative analysis of the packing of alkyl glycosides: a comparison of linear and branched alky chains. Liq Cryst 37:1205–1213
15. Nilsson F, Soderman O, Johansson I (1998) Four different C_8G_1 alkylglucosides: anomeric effects and the influence of straight vs branched hydrocarbon chains. J Colloid Interface Sci 203:131–139
16. Tristam-Nagle S, Petrache HI, Nagle JF (1998) Structure and interactions of fully hydrated dioleoylphosphatidylcholine bilayers. Biophys J 75:917–925
17. McIntosh TJ, Simon S (1993) Contributions of hydration and steric (entropic) pressures to the interactions between phosphatidylcholine: experiments with the subgel phase. Biochemistry 32:8374–8384
18. Seddon JM, Templer RH (1993) Cubic phases of self-assembled amphiphilic aggregates. Philos Trans Roy Soc Lond A 344:377–401
19. McIntosh TJ, Simon S (1986) Area per molecule and distribution of water in fully hydrated dilauroylphosphatidylethanolamine bilayers. Biochemistry 25:4948–4952
20. Costigan SC, Booth PJ, Templer RH (2000) Estimation of lipid bilayer geometry in fluid lamellar phases. Biochim Biophys Acta 1468:41–54
21. McIntosh TJ (1996) Hydration properties of lamellar and non-lamellar phases of phosphatidylcholine and phosphatidylethanolamine. Chem Phys Lipids 81:117–131
22. Kucerka N, Tristam-Nagle S, Nagle JF (2005) Structure of fully hydrated fluid phase lipid bilayers with monounsaturated chains. J Membr Biol 208:193–202

Thermal Properties Comparison Between Alumina Filled and Organic Nano-crystal Filled UPR/EPS Composite

Syed Anas Bin Syed Mustafa, Rahmah Mohamed and Husni Bin Rustam

Abstract The preparation and characterization of Unsaturated Polyester Resin (UPR) filled recycled Expanded Polystyrene (EPS) composite systems were systematically investigated. Additives such as Alumina and Organic Nano-crystal (ONC) were added to the composite for infrared reflection enhancement. The effect of different weight percentage of additive on the thermo-physical properties, e.g. thermal conductivity and physical density were determined. The fabricated composite undergoes thermal analysis by using a hand-held thermal probe which yield thermal conductivity, thermal resistivity and heat capacity data for analysis. The results obtained from comparing the use of ONC additive against alumina addition in the composite system revealed and concluded that ONC could impart better heat insulation to the studied composite using a lower concentration. It can be inferred that this is might be due to alumina being a metal oxide, which are more heat conductive compared to ONC. Further studies on the material can be done to determine the actual mechanism of how ONC and Alumina enhances thermal insulation by reducing thermal conductivity and increasing heat capacity of the composite.

Keywords EPS · UPR · Alumina · Organic nano-crystal · Thermal properties

---

S.A.B. Syed Mustafa (✉) · H.B. Rustam
Faculty of Applied Science, Universiti Teknologi MARA (UiTM), 40450 Shah Alam, Selangor, Malaysia
e-mail: ssyed_anas@hotmail.com

H.B. Rustam
e-mail: husnirustam43@gmail.com

R. Mohamed
GREEN Polymer Research Group, Faculty of Applied Science (FSG), Universiti Teknologi MARA (UiTM), 40450 Shah Alam, Selangor, Malaysia
e-mail: greenkayangan@gmail.com

M.A. Hashim (ed.), *ICGSCE 2014*, DOI 10.1007/978-981-287-505-1_43

Introduction

Unsaturated Polyester Resin (UPR) matrix composites remains a common material in use under a broad technology fields such as land and naval construction applications. Wan Nur Bahirah [1] postulated that this is due to UPR had proven itself an economical thermo-set material with excellent process ability, good cross-linking tendency as well as mechanical properties once fully cured.

Expanded Polystyrene (EPS) is a common polymeric material used in impact cushioning purposes. According to Vaidya et al. [2], EPS sheets has been used as core sandwich material for composite core door shutters, an alternative aimed to replace the use of wooden door shutters in building.

Studies had been done on the use of EPS incorporated into UPR matrices with relevant diluents to enhance its impact performance. The need to toughen UPR is very important for its efficient use in construction applications as indicated by Gryshchuk et al. [3].

Fahlman [4] described a nano-crystal as any single-crystalline nano-sized materials with at least one of its dimension be less than 100 nm. Song et al. [5] had indicated that metal oxides such as alumina are currently the commonly utilized nano-crystals in imparting certain spectral and thermal properties in polymeric materials, and are even constituted as an important ingredient of certain protective thin films. However, organic nano-crystals (ONC) had the potential to replace their metal oxide counterparts as noted by Aloui et al. [6], and hence prompting their use was employed in this study.

This paper was written to compare the effects of ONC and metallic alumina integration had on the thermo-physical properties of the polymeric composite systems. By comparing the resultant changes appearing in each addition, we may determine which additives results in the best thermo-physical enhancement.

Experimental

Materials and Methodology

The selected thermo-set matrix material used for fabricating these composite systems is Reversol UPR system, consisting of vinyl ester oligomers with 41–44 % styrene content, bulk density of 1.12 g/cc, and the viscosity of 450–600 cps. Methyl Ethyl Ketone Peroxide (MEKP) and Cobalt solution used as initiator and promoter respectively, were supplied together with the UPR resin by Revertex Sdn Bhd. These ingredients are similar to materials used by Rashidan et al. [7]. The EPS or styrofoam filler was obtained from waste material; alumina and the silane coupling agent was supplied by Merck KGaA.

This paper was produced from a study conducted via qualitative analysis of the sample manufactured. The fabricated sample were tested via thermal probe to determine their thermal conductivity, thermal diffusivity and volumetric heat capacity; whereas the density of the sample were determined by a densitometer.

Composite Fabrication and Testing

Recycled EPS, fixed at 10 % weight percent (10 wt%) of UPR resin, was blended by utilizing a high speed agitation mixer until complete dissolution was attained. This is to enable the gaseous and solid contaminates present are easily eliminated from the mixture via gravity settling in 24 h.

Additives were added before the samples were prepared through utilizing the hand lay-up processes onto an aluminium mould. The silica aerogel content was fixed at 0.2 wt% while the ONC content was varied up to 0.2 wt%. The alumina impregnated samples followed similar procedures where silica aerogel content was fixed at 3 wt% while alumina content was varied up to 5 wt%.

The process was followed by adding 3 wt% of MEKP initiator and suitable amount of cobalt as accelerant to cure the EPS/UPE composite. Gelation time was set to be about 5 min. Once fully gelled, complete curing was done under compression for 1 h at 70 °C. These steps are similar to the process used by Rashidan et al. [7].

The thermal property of samples was then determined by using a KD2 Pro Handheld Thermal Probe which adheres to ASTM D5334-08 and IEEE Standard 422-03 [8, 9]. The density of the samples was obtained using a digital densitometer.

Result and Discussion

The thermal conductivity showed a minimum in both systems at 0.122 and 0.147 W m^{-1} K^{-1} respectively, which are results from UPR-EPS samples. Aerogel introduction in the ONC system generates a significant thermal conductivity spike compared to the alumina system. ONC addition increases the thermal conductivity, k above the baseline value in the ranges between 0.131 and 0.227 W m^{-1} K^{-1}; alumina addition also produces similar trends which ranges between 0.227 and 0.313 W m^{-1} K^{-1} (Tables 1 and 2).

The heat capacity (C_p) values obtained from thermal probe technique showed some increase for the silica aerogel sample; the ONC system experienced some significant increase before stabilizing, while the alumina system reports gradual increment before fluctuating. The thermal diffusivity reported from the probe also showed similar trend, where ONC system was relatively stable compared to the alumina system. The bulk density of samples from both systems was uniform despite differences in additives used (Fig. 1).

Cutnell and Johnson [10] explains how the thermal properties in this study are related. Thermal conductivity relates to the amount of thermal energy could pass

Table 1 KD2 pro handheld probe data for ONC filled

Samples	Conductivity (W/m K)	Diffusivity (mm^2/s)	Volumetric heat capacity (MJ/m^3 K)	Density (g/cm^3)
UP	0.122	0.092	1.112	1.230
UPA 0.2	0.254	0.122	1.810	1.210
UPAO 0.05	0.183	0.105	1.853	1.220
UPAO 0.1	0.131	0.098	1.546	1.228
UPAO 0.2	0.227	0.102	1.657	1.240

Table 2 KD2 pro handheld probe data for alumina filled

Samples	Conductivity (W/m K)	Diffusivity (mm^2/s)	Volumetric heat capacity (MJ/m^3 K)	Density (kg/m^3)
UE	0. 147	0.093	1.231	1.221
UEA 3.0	0. 165	0.103	1.434	1.206
UEAL 0.5	0.227	0.146	1.547	1.237
UEAL 1.0	0.300	0.286	1.050	1.234
UEAL 3.0	0.234	0.160	1.467	1.252
UEAL 5.0	0.313	0.283	1.105	1.266

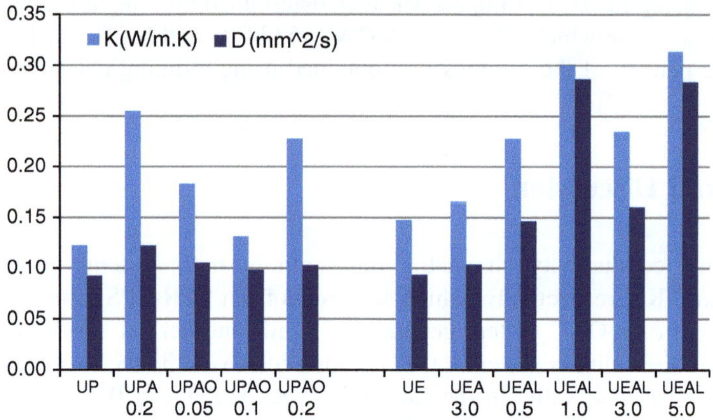

Fig. 1 Comparative thermal conductivity and resistivity for systems tested

through a material for a given length and temperature difference; a higher value indicates how much easier and/or larger amount of heat may be transferred. Volumetric heat capacity relates to how much a material could store heat for a given volume and temperature change; the higher value indicates a larger heat retention capability.

Diffusivity however is a product of thermal conductivity against heat capacity, and it indicates the thermal inertia of a material or how fast a material flows heat compared to storing it. High diffusivity causes a material to move heat across better than storing it. The benefit of this attribute depends heavily on the material's application. Applications requiring quick removal of heat favors materials with high diffusivity while heat shielding applications favors the opposite. Metals are usually associated with high level of diffusivity.

As illustrated by the two charts, few patterns had emerged. Addition of Aerogel to the base UPR-EPS mix (UP and UE respectively) increases all three thermal properties, quite significantly at the lower concentration of 0.2 wt% compared to 3.0 wt%. This might indicate that either the higher concentration have better dispersal and spread, or that the lower concentration is more volatile to the composite system.

Addition of ONC seems to lowers the conductivity to a certain level while maintaining a stable diffusivity level. alumina addition however gradually increases both conductivity and diffusivity in lock-step. Both system as a result had their respective heat capacity level reacted to produce the correlation evident in Fig. 2. This effect might be caused by the nature of the additives used; alumina is a metal oxide which are more heat conductive compared to ONC which are organic in nature, which then was imparted to the systems.

It could be inferred that the ONC system are suitable for use in heat shielding applications since their stable diffusivity lessen temperature increase of layers below the material. Hence, the capability of ONC to reduce heat transfer might also be inferred to be better than alumina while maintaining stability of the system.

Conclusion that could be derived that ONC had shown to impart better thermal insulation in the form of lower thermal conductivity and higher heat capacity, all at a lower concentration compared to alumina. ONC also had an advantage of retaining the diffusivity of the composite at low, stable level.

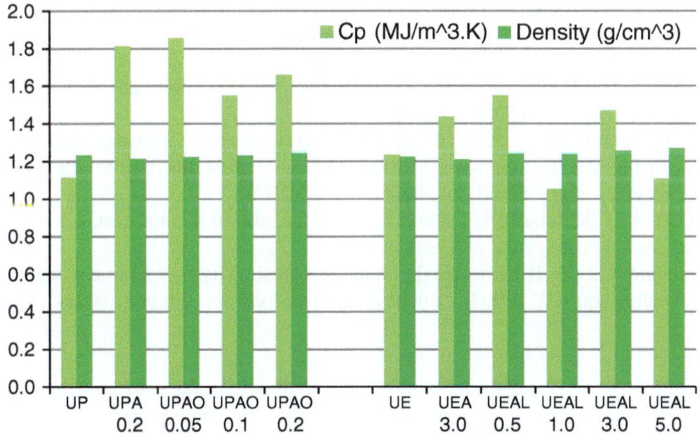

Fig. 2 Comparative heat capacity and bulk density for systems tested

Conclusion

It could be concluded that the addition of alumina and ONC have mixed effects on the thermal characteristics of the composites tested. ONC had shown to be able to impart better thermal insulation capability to the composite at a lower concentration compared to alumina integration. More research are required to fully understand its true mechanism, but the aim of the paper is to compare the effect of ONC and alumina addition on the thermal performance of the compounded composite.

Acknowledgment I would like to acknowledge the technicians, and staffs for their assistance during the composite fabrication and testing phases; and to my colleagues and lecturers for discussions related to this work. Special acknowledgement for Assoc. Prof. Rahmah Mohamed for her guidance and direction in producing this publication. I also wish to thank Universiti Teknologi MARA (UiTM) for their research funding under the Research Intensive Faculty (RIF) fund [File:600-RMI/DANA 5/3/RIF (64/2012)].

References

1. Wan Nur Bahirah W (2011). Effect of curing and flame retardant system on properties of polyester-filled EPS composite (Master's thesis). March
2. Vaidya UK, Hosur MV, Earl D, Jeelani S (2000) Impact response of integrated hollow core sandwich composite panels. Compos Part A: App Sci Manuf 31(8):761–772
3. Gryshchuk O, Jost N, Karger-Kocsis J (2002) Toughening of vinylester–urethane hybrid resins by functional liquid nitrile rubbers and hyperbranched polymers. J Polymer 43:4763–4768
4. Fahlman BD (2007) Materials chemistry, vol 1. Springer Mount Pleasant, Mt. Pleasant, pp 282–283
5. Song JE, Kim YH, Kang YS (2006) Preparation of indium tin oxide nanoparticles and their application to near IR-reflective film. Curr Appl Phys 6(4):791–795
6. Aloui F, Ahajji A, Irmouli Y, George B, Charrier B, Merlin A (2007) Inorganic UV absorbers for the photo-stabilisation of wood clear-coating systems: comparison with organic UV absorbers. Appl Surf Sci 253(8):3737–3745
7. Rashidan K, Rahmah M, Nor ZIZ, Mohd HM (2009) Effect of antioxidant and flame retardant additives on thermal degradation and flammability of SBR filled EPS/UPR composite. In: Malaysia polymer international conference (MPIC 2009)
8. ASTM D5334-08, Standard test method for determination of thermal conductivity of soils and rock by thermal needle probe procedure
9. IEEE Standard 422-03, IEEE thermal conductivity/resistivity measurement standards
10. Cutnell JD, Johnson KW (2004) Cutnell & Johnson physics, 6th edn. Wiley, NEW York

Part VI
Safety, Policies and Regulations

Hierarchy of Controls Analysis for Equipment Failures Prevention

Nor Afina Eidura Hussin, Anwar Johari, Kamarizan Kidam and Haslenda Hashim

Abstract Equipment failures play significant roles in most accidents that occur and recur in the chemical process industry. Many risk reduction strategies were applied to prevent these accidents but the accidents keep on happening. In this study, 50 equipment-related accidents are analyzed based on their comprehensive accident investigation reports, retrieved from several accident databases to determine the preferred accident prevention strategies of the industry. Based on the analysis, 91 % out of 590 recommended corrective actions are procedural strategies. Other layers of hierarchy of controls are only 9 % i.e. active-engineered (3.4 %), inherently safer (3.1 %), and passive-engineered (2.5 %). An analysis using Analysis of Variance (ANOVA) shows a significant relationship exists between the groups of hierarchy of controls. By emphasizing on the most reliable group of hierarchy of controls for equipment failures prevention such as inherently safer, passive-engineered, and active-engineered accident prevention strategies, the preference for procedural strategies in the industry may be reduced.

Keywords Accident prevention · Equipment failures · Hierarchy of control · Risk reduction

N.A.E. Hussin · A. Johari · K. Kidam
Faculty of Chemical Engineering, Institute of Hydrogen Economy, Universiti Teknologi Malaysia, 81310 Johor Bahru, Johor, Malaysia
e-mail: afina.eidura@yahoo.com

A. Johari (✉) · K. Kidam · H. Hashim
Department of Chemical Engineering, Faculty of Chemical Engineering, Universiti Teknologi Malaysia, 81310 Johor Bahru, Johor, Malaysia
e-mail: anwar@cheme.utm.my

K. Kidam
e-mail: kamarizan@cheme.utm.my

H. Hashim
e-mail: haslenda@cheme.utm.my

M.A. Hashim (ed.), *ICGSCE 2014*, DOI 10.1007/978-981-287-505-1_44

Introduction

The history of accident prevention in the chemical process industry (CPI) shows that different approaches of risk reduction strategies have been implemented. Previously, accident prevention emphasizes either on design, technical, or procedural strategies. In the 1800s, a CPI plant with little instrumentation and means of protection only emphasized on procedural aspects for accident prevention. One of the procedural risk reduction examples is the one-legged stool strategy in nitroglycerin production. In this case, the operators had to sit on one-legged stools while watching over the production of highly exothermic nitroglycerin in large-stirred pots. If the heat was not removed by cooling and stirring, the reaction became uncontrollable and might lead to an explosive decomposition of the nitroglycerin. Hence, the operators had to watch the temperature closely. If they fell asleep, they fell off and injured themselves or at worst, could lead to fatality [1].

In 1960s, a great change in CPI occurred with process operating conditions (e.g. temperature, pressure etc.) became more severe; the energy stored in the process increased; problems in areas such as material construction and process control became more taxing; and the plants grew in size with a factor of about 10, were often single stream. Relatively sophisticated instrumentation provision was developed to run a process under extreme conditions and close to the limits of safety thus causing higher accident rates [2]. The focus of accident prevention shifted to technical and design-oriented. Later, from 1980s onwards, the trend in accident prevention was mostly utilizing the outer layers of protection by adding add-on engineered either passive or active; and procedural control strategies. However, the risk reduction approaches were only effective to a certain extent. Nowadays, the focus of loss prevention is human and organizational-related which emphasis on the safety management system and safety culture to overcome fluctuating accident rate issues. The accident rates however remain persistently high. Thus, this paper analyzes the application of risk reduction strategies in solving equipment failure problems that commonly lead to the occurrence and recurrence of accidents in the industry.

Hierarchy of Controls

Risk reduction strategies can be classified into four different layers i.e. layers of protection (LOP). The layers are inherently safer, passive-engineered, active-engineered, and procedural. In managing risk, the most reliable LOP is the inner most layer which is the inherently safer, followed by passive-engineered, active-engineered, and procedural strategies, respectively. Thus, the priority in risk management strategy is inherently safer > passive > active > procedural. By changing the design and operation at the earlier stages, the capital and operating costs required are much cheaper than the latter stages. Only, procedural control strategies require low relative costs compared to other stages but the reliability of the strategies is the lowest and the modification is difficult to mark since the complexity increases throughout the process lifecycle [3].

Inherently Safer

Inherently safer is the premier strategy for hazard avoidance and control at its source through design changes. By inherent safety principles, elimination is used to avoid hazard by design; intensification, substitution or attenuation is used to reduce the severity of the hazard; and simplification of process or plant is used to reduce the likelihood of the hazard occurring. Using moderation principle, the existing processing condition is changed to an inherently safer by manipulating temperatures, pressures, concentrations and physical states of the chemicals. Substitution is done by selecting safer and compatible chemicals. Use of safer chemical reduces the severity of accidents.

A process plant can be simplified using credible equipment and establishment of a reliable or self-regulating system. The system reduces the need for redundant systems and complicated controls. Thus, a simpler and more robust design is the key for reliable and safer chemical plant operations. Meanwhile, error tolerance is effectively being used as inherently safer strategy to solve problem related to wrong material selection, corrosion, chemical reactivity, incompatibility, and sub-standard equipment application. Majority of the corresponding corrective actions amend the existing technical and design deficiencies that led to process failure. The idea is to redesign the default part of the plant (or equipment) to a robust and accident-resistible design.

Passive-Engineered

Add-on layers are mainly installed as passive or active engineered safety protection systems. Passive strategy employs systems that remain static and do not perform any fundamental operations. This passive-engineered risk control further reduces the likelihood and consequences of accident by using passive safety protection such as dikes, containment and fire wall. The passive-engineered modifications are mostly related to layout, mechanical/physical aspects, design specification changes, additional equipment, equipment modification, and friendlier design.

The common errors in plant layout are the incorrect arrangement and safety distance between main processing equipment which eventually increase the severity of damage due to domino effect. The detailed layout changes for safety distance normally involve redesigning and repositioning of piping system, and reshaping of specific equipment or parts. Other safety issues related to plant layout are equipment accessibility and visual obstacles are improved using proper organization of plant layout. The mechanical and passive engineering control changes material of construction for a better robust, and corrosion-and-high-pressure-resistant process equipment. The design specification is applied to equipment with changes in process condition, fire and explosion ratings. Additionally, plant and equipment modifications are carried out in order to improve human-machine interface leading to user-friendlier process which reduces the occurrence and recurrence of human-related errors.

Active-Engineered

Active add-on engineered strategies use active systems that depend on timely hazard detection and initiation (i.e. utilizes safety devices that respond to process changes) to further reduces the accidents using relief valves, controllers, detectors and alarms. For controlling risk, active-engineered control requires additional devices to sense and indicates process variables, valves, etc. either by adding or removing the instrumentation and automation of the equipment. Among the common active-engineered control strategies are modifying control setting specification, supplementing additional instrumentation, and introducing new protection systems.

To specify control setting, majority of actions are conducted to accurately set the safety limits for flow rate, temperature, pressure, density and speed of the automation system. The correct number of control instrumentations is important for early detection of process deviation. A number of temperature and pressure-related accidents have been reported due to lack of sensors or detectors e.g. chemical reactors require adequate number of detectors with correct positioning. In addition, the process equipment needs relief and mitigation systems to avoid accidents such as Seveso and Bhopal. The accidents had severe consequences as a result of under-designed protection and mitigation systems. Thus, suitable protection and mitigation systems based on the worst-case scenario with adequate design capacity are essential to manage process hazards and reduce risks.

Procedural

Procedural or human and organizational-oriented risk control usually focuses on safe operation including training, supervision, procedure, work instructions, inspection and maintenance. This operator and maintenance procedures should be the last resort, especially for control and mitigation where the chance of errors or failure is high. Among the highest procedural corrective actions were proper documentation, improved management system, continuous monitoring/supervision, and training. The gathered data indicates that poor organizational support of work system and ineffective management systems have potential to cause accidents.

Confusion might arise from poor or unclear documentation and increases chances of operators to perform incorrect work sequences and taking shortcuts. To deal with these attitudes, effective safety management system and safety culture education are essential in promoting safety awareness among the CPI players.

Research Approach

In this study, 50 comprehensive accident investigation reports from 1998 to 2012 were extracted from U.S. Chemical Safety and Hazard Investigation Board (CSB) and U.S. National Transportation Safety Board (NTSB) databases [4, 5]. The data mean was 2005. In the paper, the emphasis is on the recommended corrective actions of the reports. Statistical analysis on the hierarchy of controls is conducted. Then, the preferred risk reduction criteria are discussed and lessons learned are established for effective accident prevention of the CPI.

Analysis of Results

The corrective action section of the accident reports was analyzed to determine the applied risk reduction strategies of the CPI. The analysis has showed that the industry normally takes several corrective actions for multiple causation accidents. However, procedural strategies were mostly recommended as the corrective actions. In this study, 590 corrective actions had been suggested by the boards. Out of these 590 corrective actions, 91 % were procedural-based. From all the recommended corrective actions, active-engineered was 3.4 %, followed by inherently safer (3.1 %). Passive-engineered strategies were the least options used with only 2.5 % of the total recommended corrective actions. Further details on the applied hierarchy of controls are summarized in Table 1.

Process hazards and their risks can be managed effectively through layer of protection. The statistics indicated that procedural strategies were most preferred by the industry. The layer is less reliable than inherent safety, passive-engineered, and active-engineered controls [5]. Based on the analysis, 3.1 % of the corrective actions were inherently safer strategies. Inherently safer approach uses material and process conditions that are less hazardous to eliminate and mitigate hazard. Four types of inherently safer used were minimization (44.4 %), moderation (27.8 %), simplification (16.7 %), and substitution (11.1 %). Minimization was used to limit energy generation capabilities by using smaller amount of hazardous substances. Among others recommended inherently safer strategies were replacing a hazardous substance with a less hazardous one such as the use of air or pigging with air instead of natural gas blow for cleaning fuel gas piping (substitution) and sodium

Table 1 Classifications for the applied hierarchy of controls

Percentage							
Inherently safer 3.1 % out of 590		Passive-engineered 2.5 % out of 590		Active-engineered 3.4 % out of 590		Procedural 91 % out of 590	
Minimization	44.4	Protective system	60.0	Instrumentation	55.0	Communication	22.9
Moderation	27.8	Design changes	13.3	Mitigation system	30.0	Regulations and guidance	15.3
Simplification	16.7	Layout	6.7	Protective system	15.0	Training and education	12.7
Substitution	11.1	Sizing	6.7			Inspection	11.5
		Equipment modification	6.7			Management system	10.1
		Additional equipment	6.7			Emergency preparedness	5.4
						Work mechanism	4.8
						Documentation	3.2
						Enforcement	2.6
						Cooperation	2.4
						Expertise and consultation	2.0
						Management of change	1.9
						Maintenance	1.5
						Monitoring and supervision	1.1
						Research	1.1
						Contractor performance	0.9
						Cleaning and housekeeping	0.6

hypochlorite as a biocide in cooling water treatment instead of chlorine (substitution); and the use of appropriate materials for wastewater treatment (moderation), and revision of operating conditions to reduce risk of over-chlorination (moderation) for safer process condition.

The add-on engineering controls are used to further reduce the likelihood and consequences of accident. Passive-engineered controls do not require any device to sense or actively respond to process changes but active-engineered controls require devices to monitor and mitigate a hazard. In the study, 2.5 % passive-engineered controls were used and 3.4 % were active-engineered controls. For passive-engineered, protective system (60 %), design changes (13.3 %), layout (6.7 %), sizing (6.7 %), equipment modification (6.7 %), and additional equipment (6.7 %) were recommended. Passive-engineered controls further reduced hazard and risk by using firewalls and blast-resistant construction; adding protective fireproofing for fire rack support steel near process unit containing highly pressurized flammables; and ensuring that penetrations of partitions, floors, walls, and ceiling were sealed dust-tight. Meanwhile, instrumentations were mostly used as active-engineered controls (i.e. 55 %). Other active-engineered controls were mitigation system (30 %), and enhanced protective system (15 %). Active-engineered controls established adequate layers of protection system by using additional interlock and shutdown, air monitoring devices, automated audible alarms, level indicators, automatic controls, post-ignition deflagration detection, and damage control devices.

Procedural safety systems are administrative controls which include standard operating procedures, safety rules and procedures, operator training, management systems, and emergency response procedures. Human and organizational-oriented strategies for safe operation were mostly recommended by the study. The procedural strategies were 91 % of all the corrective actions suggested. Communication was the highest strategies recommended by the boards i.e. 22.9 %, followed by development or amendment of safety regulations and guidance (15.3 %). Training and education (12.7 %), inspection (11.5 %), and management system (10.1 %) were also parts of these implemented corrective actions. Other less prioritized procedural strategies were emergency preparedness (5.4 %), work mechanism (4.8 %), documentation (3.2 %), enforcement/implementation (2.6 %), cooperation (2.4 %), expertise and consultation (2 %), management of change (1.9 %), maintenance (1.5 %), monitoring/supervision (1.1 %), research (1.1.%), contractor safety performance (0.9 %), and cleaning and housekeeping (0.6 %).

Discussion

In the study, communication issues were commonly addressed to ensure timely transmission of critical safety information to responding personnel. Therefore, safety alerts and health bulletin were published to warn owners and operators on potential hazards and risks of the industry. The safety alerts advised them on their responsibilities for accident prevention such as recommending the use of inherently safer design features, describing sufficient security measures, and recommending the use of hazards signs to identify the fire and explosion hazards.

The importance of implementing hierarchy of controls includes to comprehensively control all potential ignition sources and continuously monitor hazards at appropriate locations and elevations; to train and certify emergency response personnel; to publish the technical guidance addressing the safe operating procedures; to avoid occurrence and recurrence of accidents by analyzing the key findings, cause, recommendations of the reports to shareholders, membership, and workforce; and to establish a timely notification community procedures in the event of chemical release that save life and public properties.

Establishment of safety regulations and guidance are important in preventing accidents. The involvement of several governmental agencies such as the Occupational Safety and Health Administration (OSHA), the Environmental Protection Agency (EPA), National Fire Protection Association (NFPA), and other regulatory bodies in the reports has contributed to these issues. Cooperation between the agencies is needed to develop comprehensive regulations and standards, and avoid any misunderstandings. Training and education are recommended to the industry players to increase safety awareness and skills of their personnel. By training and education, accidents caused by human errors could be reduced. Regular inspections are applied to ensure that the facilities and work mechanism are complied with safety regulations and guidance. Management systems are updated to cope with several issues such as changing of work mechanism, modified equipment or facilities, new staffing, and other organizational changes.

In general, the emphasis on procedural accident prevention strategies does not provide adequate hazard elimination and risk reduction. The human reliability is not high enough and often leads to improper problem-solving, inappropriate actions, and ill-timed responses. The CSB and the NTSB suggested various procedural corrective actions with the involvement of governmental agencies, industry players, research institutes, and other non-governmental agencies. The industry should reconsider the implementation of the inner most layers of hierarchy of controls to prevent accidents thus resulting in safety and cost benefits of the CPI. Moreover, one-way ANOVA using Excel on the data shows that $F_{statistic} \gg F_{critical}$ that indicates a significant relationship exists between the groups of hierarchy of controls ($137.52 \gg 2.653$). The probability is small that the difference or relationship happened by chance. In the analysis, p-value is 0, less than the critical alpha level ($p < 0.05$).

Conclusion

In the CPI, the preference for outer layer of hierarchy on controls results in unbalanced risk management approach. Only 9 % of the 590 recommended corrective actions were applying the inner layer controls compared to 91 % of the outer layer countermeasures. As the reliabilities of the hierarchy of controls are decreasing from inherently safer to procedural strategies i.e. from inner to outer layer, the accident rate of the CPI is still high due to the less reliable corrective actions been implemented. To reduce the accident rate, the CPI should focus on the inherently safer and passive-engineered i.e. the inner layer accident prevention, thus avoiding human unreliability and creating safer work environment.

Acknowledgment Special thanks to Zamalah Scholarship of Universiti Teknologi Malaysia (UTM), Malaysia for funding this research.

References

1. Kletz TA, Amyotte P (2010) Process plant—a handbook for inherent safer design. CRC Press, Boca Raton
2. Lees FP (1996) Loss prevention in the process industries, 2nd ed. Butterworth Heinemann, London
3. CCPS (1998) Guidelines for design solutions for process equipment failures. AIChE, New York
4. Information on http://www.csb.gov/investigations/completed-investigation
5. Information on http://www.ntsb.gov/investigations/reports

Analysis the Effect of Explosion Efficiency in the TNT Equivalent Blast Explosion Model

Zulkifli Abdul Rashid, Azil Bahari Alias, Ku Halim Ku Hamid, M. Shahnor Bani and Mohanad El Harbawi

Abstract In this study, the analysis will focus firstly on the analysis of the effect of an increment of explosion yield, η value between 2 and 9 % and its correlation toward the parameter peak side-on overpressure p^{o}, the duration of the positive phase t_d compared to diagnostic features of explosion damage results translated according to criteria developed by V.J. Clancey. Secondly, the aim is to validate the recommended methods of yield factor by Less [1] with estimated results (η between 2 and 9 %) by using TNT model in comparison with the physical observation damage of VCE incident impact at Flixborough in 1974.

Keywords Investigation effect · Explosion efficiency · TNT equivalency

Introduction

Explosions are the most serious hazards which may occur in petrochemical industries. Historical evidence has shown that these explosion incidents in petrochemical and refining industries caused a high level of damage scale. For example, Flixborough explosion in1974, uncontrolled leakage of about 30 tons of cyclohexane causing death to 28 men and serious damage to onsite infrastructure. Another serious incident also happened at PEMEX LPG Ixhuatepec, Mexico City, which over 500 people died and over 4000 people injured and also resulted in

Z.A. Rashid (✉) · A.B. Alias · K.H. Ku Hamid
Faculty of Chemical Engineering, Universiti Teknologi MARA, 40450, Shah Alam, Malaysia
e-mail: zulmas06@yahoo.com.my

M.S. Bani
Petronas Carigali Sdn Bhd, Miri, Sarawak, Malaysia
e-mail: shahnor.bani@petronas.com.my

M.E. Harbawi
Faculty of Chemical Engineering, King Saud Riyadh, Riyadh, Saudi Arabia
e-mail: mohanad_75@yahoo.com

M.A. Hashim (ed.), *ICGSCE 2014*, DOI 10.1007/978-981-287-505-1_45

severe damage to the immediate surroundings with window break-age up to 4 km from the source [1–4]. Other examples are incident at BP Texas, 2005, Bukom refinery, Singapore 2011, Chevron refinery, Richmond United States, 6 August 2012, Amuay refinery and Venezuala. Based on previous incidents analysis, even though precaution has been taken to reduce accident, when a new petrochemical plant was built or an existing plant was undergoing process maintenance or modification process route, these types of devastating accidents were still happened. In this paper, the consequences impact from the magnitude of explosion will be analyzed especially on the effect of explosion efficiency in the TNT equivalent model.

Method of Analysis

An explosion is a process involving the production of blast wave resulting from a rapid release of energy. The result from this blast wave, a pressure disturbance is generated into the surrounding medium and air becomes heated due to its compressibility. This phenomena leads to an increase in the velocity of sound and causing the front disturbance becomes steepen as it travels through the air. As a results the pressure and the density of air is increased until a maximum pressure wave is developed at reflected incident distance. For cases of unconfined vapor cloud explosion, the energy release in the blast wave generally is only small fraction of the energy theoretically available from the combustion of all the material in the cloud. The ratio of the actual energy in the explosion to that theoretically available is termed the 'explosion efficiency'. Typically the value of explosion efficiencies is in the range of 1–10 %. For an explosion such as the bursting of a pressure vessel due to high gas pressure at ground level the explosion energy absorbed by the ground is generally smaller than for a condensed phase explosion. Thus the value of the yield ratio lies nearer to 2. However, it is not always clear what assumption should be made when concerning the yield ratio, and therefore it may be necessary to exercise experience judgment. To investigate the effect of explosion efficiency, TNT equivalent model is applied. TNT equivalency explosion model is the most popular model for estimating far field damage because it is simple and the correlation strength among the parameters in the model was improvised with a detailed experimental process from trials done using explosives. In TNT equivalency model, value from the fraction of total energy of the explosion used in the shock wave will be calculated before it is converted into the equivalent mass of TNT, W_{TNT} (kg):-

$$\mathrm{W_{TNT}} = \frac{\mathrm{W_{gas}}\eta \Delta \mathrm{H_{c(gas)}}}{\mathrm{E_{(TNT)}}} \tag{1}$$

Where W_{TNT} is the equivalent mass of TNT (kg) that would produce the same effects as the explosion, η represents the explosion yield (dimensionless); it is generally accepted that, taking as a basis for calculation the total quantity of vapor in the cloud. W_{gas} is the total mass of flammable gas in the cloud, $\Delta H_{c(gas)}$ is the lower heat of combustion of the flammable material (kJ/kg), and $E_{(TNT)}$is the energy of explosion of TNT (approximately 4680 kJ/kg). However, the TNT equivalent models have tendency to differ in few conditions, as discussed by several researchers such as Brasie and Simpson 1968, Harris and Wickens 1989 [5, 6] at conditions are (1) the mass of vapour participating in the explosion, (2) the yield factor of the explosion and (3) the value used for the energy of explosion of TNT. For instance, the value of energy of explosion of TNT quoted by Brasie and Simpson [5], varied between 1800 BTU/lb (4190 kJ/kg) and 2000 BTU/lb (4650 kJ/kg). However, Harris and Wickens [6] quoted the energy level of TNT explosion limited to 4.6 MJ/kg. This is also observed in method used in TNT model equivalent by Eichler and Napadensky (1977), Prugh (1987) [7, 8]. Finally, CCPS [9] in 1994, reviewed and stated that the values used for the explosion energy of TNT should lie in the range of 4140–4680 kJ/kg. The value of equivalent TNT mass, was subsequently used to calculate the equivalent effects of explosions occurring at the same normalized distance, as expressed in Eq. 2:-

$$Z_e = \frac{R}{(W_{TNT})^{1/3}} \tag{2}$$

where Z_e is the scaled distance (m/kg$^{1/3}$) and R is the real distance (m). To get a better estimation of potential structural and building damage and human casualties from air shock waves, there are parameters which are principally use to explain the explosion scenario. The principal parameters of the blast wave from a TNT explosion are the peak side-on overpressure p^o, the impulse of the positive phase i_s, the duration of the positive phase t_d and the arrival time t_a. Overpressure P^o, (kPa) was calculated using the equations as in Eqs. (3) and (4), where W_{TNT} = TNT equivalency (kg), P^s = scaled side on overpressure, P_a = atmospheric pressure and Z_e (scaled distance) = actual distance (m)/$W^{1/3}$, The appropriate are included in these two equations.

$$P^s = \frac{P^o}{P_a} \tag{3}$$

Where P^o is calculated by using an Eq. 4

$$P^0(\text{kPa}) = P_a \cdot \frac{1616\left[1 + \left(\frac{Z_e}{4.5}\right)^2\right]}{\sqrt{1 + \left(\frac{Z_e}{0.048}\right)^2}\sqrt{1 + \left(\frac{Z_e}{0.32}\right)^2}\sqrt{1 + \left(\frac{Z_e}{1.35}\right)^2}} \tag{4}$$

Table 1 Estimation of common structures damage based on overpressure result (these values are only approximations)* [10]

Pressure (*kPa*)	Damage
0.21(E_1)	Occasional breaking of large glass windows already under strain
0.69(E_2)	Breakage of small windows under strain
1.03(E_3)	Typical pressure for glass breakage
2.07(E_4)	"Safe distance" (probability 0.95 of no serious damage below this value); projectile limit; some damage to house ceilings; 10 % window glass broken
2.76(E_5)	Limited minor structural damage
3.4–6.9(E_6)	Large and small windows usually shatter; occasional damage to window frames
4.8(E_7)	Minor damage to house structures
6.9(E_8)	Partial demolition of houses, made inhabitable, corrugated asbestos shatters; corrugated steel or aluminium panels, fastenings fail, followed by buckling; wood panels (standard housing), fastenings fail, panels blow in
6.9–13.8(E_9)	Partial collapse of walls and roofs of houses, concrete or cinder block walls, not reinforced, shatter
13.8(E_{10})	50 % destruction of brickwork of houses
13.8–20.7 (E_{11})	Frameless, self-framing steel panel buildings demolished; rupture of oil storage tanks
17.2(E_{12})	Cladding of light industrial buildings ruptures
20.7–27.6 (E_{13})	Wooden utility poles snap; tall hydraulic presses (40,000 Ib) in buildings slightly damaged
34.5–48.2 (E_{14})	Nearly complete destruction of houses
68.9(E_{15})	Probable total destruction of buildings; heavy machine tools (7000 Ib), moved and badly damaged, very heavy machine tools (12,000 Ib) survive

Positive phase duration **t,** (ms) is calculated using Eq. 5

$$t_d(ms) = W^{1/3}.\frac{980\left[1+\left(\frac{Z_e}{0.54}\right)^{10}\right]}{\left[1+\left(\frac{Z_e}{0.2}\right)^{3}\right]\left[1+\left(\frac{Z_e}{0.74}\right)^{6}\right]\left[1+\left(\frac{Z_e}{6.9}\right)^{2}\right]^{1/2}} \tag{5}$$

Values peak overpressure will be interpreted according to the criteria developed by V.J. Clancey [10] as shown in Table 1.

For the above analysis, flammable propane was used as the study material with the data heat of combustion for propane at 46013 kJ/kg.

Results and Discussion

In this analysis, the quantity of propane involved in the explosion incident was 13,000 kg and the value of explosion efficiency used was within 1–10 % [1, 9, 11]. Even though majority of explosion in chemical process plant occur at ground level with 2 % of explosion efficiency (η) value, as recommended by Brasie and Simpson [5], the probability to use η, of 2 % directly in TNT equivalent model for VCE incident consequences can be argued. According to Brasie and Simpson [5] the value of 2 % (η) or less in the range 1.52 %, is applied when the energy in the blast wave is expended and the detonation impact may result crater formation at the ground level of explosion source. For that reason, the recommended guidelines to determine yield factor values used in the TNT equivalent model as proposed by Less [1] still need to be improved based on the suitability of method selection with the incident scenario. Meanwhile according to Phillips H. [12] the significant investigation of an explosion impact incident for the equivalent mass of TNT is obtained by site examination on the damage effects. Therefore the past field investigation report for vapor cloud explosion incident cases such as Flixborough in 1974 [1, 9, 13], Port Hudson Missouri in 1970 [13], Ludwigshafen, at BASF plant Germany in 1948 [13] and several others VCE incident were referred [1, 9, 13]. It is observed that the severity of damage on the structural building around the explosion is obviously varies [1, 10, 13–15]. The correlation trend impact on the structural damage differs depending on the type of explosion as observed in the method developed by Merrifield and Mackenzie [16], in which the researchers have established a relationship between damage and distance, also known as destruction profile. Mendonca et al. [17] also used the same method introduced by Merrifield [16] to estimate the TNT equivalence of a 15-ton single base powder explosion and predict the possible damage in buildings and structures over distance. He is also observed that the effect of an explosion was not only caused by the blast damage, but is also by the thermal effect, flying fragment, ground shock and from crater formation. However, not all of the above effects occur in every explosion. Many of the data on the effects of explosions gathered from studies of military and industrial explosives [1]. From these studies, the pressure waves created by explosive materials were measured at a different distance and parallel to the recorded building location. To test field analysis, parameter correlation between the size of explosive, distance, flame speed and the effect of obstacle is optimized into simplified value impact represent as peak overpressure [18]. Therefore in this analysis, the effect of explosion yield on the peak overpressure (kPa) results is obtained based on the visual simulation of structural building damage as developed by V. J. Clancey [10]. The results analysis is better, since previously the impact for all VCE incidences and the first indicator for all VCE incidences were derived from the physical structure damage observation. Even though the recommended guideline by Less [1] is available to choose an appropriate explosion yield applied to different scenario, this guideline still have deficiency. The usage of explosion yield still becomes a major discussion in TNT equivalent model calculation. For instance in CCPS [9], Crowl and Louvar had discussed in their book [19], for reactive source such as flammable cloud of propane, the recommended explosion yield was 0.05. However, the recommended η value, still does not yield an accurate indicator. It is well known that, not every flammable mass in the cloud involved in the explosion. This indicate that the higher the η value, the potential fraction mass involved in the explosion will be also higher. However this is also depending on how effective and reactive the flammable material mixing with air to produced a complete combustion. The confinement and turbulent will caused the flame velocity to speed up the unburnt fuel and air homogenous mixing process and these will increase the cloud detonation. Therefore in this study, the analysis will focus firstly on the analysis of the effect of increment of η value between 2 and 9 % and its correlation toward the parameter peak side-on overpressure p^o, the duration of the positive phase t_d compared to visual damage developed by Clancey [10] as shown in Table 2. Secondly, the aim is to validate the recommended guidelines by Less [1] with estimated results (η between 2 and 9 %) by using TNT model in comparison with the physical observation damage of VCE incident impact di Flixborough, [1, 13, 14]. Table 2 shows the results from the above analysis.

Based on Table 2, it is shown that the values of peak overpressure receptor at points 50, 100, 150, 200 to 500 m (from the source of explosion) increases exponentially when η (%) increases. Meanwhile, time duration phase for the value of P^o, (kPa) at the same points do no show obvious changes, for example at the value of η (between 2 and 9 %) at impact distance of 50 m is between 0.048 and 0.032 s, compared to the positive duration phase measured in other model such as Lees 2005. When the denominator value is $[1 + (z/0.02)^3][1 + (z + 0.74)^6][1 + (z/6.9)^2]^{1/2}$, this will give the t_d value between 27 and 29 s with 2 s difference. Even though, this value is still short for any emergency plan to be undertaken by the expected victim to find a safe shelter, the possibility of escape is better compared to the model proposed by Naval Weapon Center of United States. This probability might be happened since more time is needed for vapor formation when the quantity of material involved is large [η (%) increase], and also is due to the duration to achieve the explosion. Meanwhile, at the formation of explosion and its destructive effect, the magnitude of effect is shown to be gradually increase from E4 till E15 due to various factors such as fraction to flash off to form vapour cloud detonation is increase with the increment of η value between 2 and 9 %, effect of temperature due to exothermic reaction is probably occur with the increment of η value and has a tendency for flammable cloud to initiate the burn faster due to the decreasing of lower limit of flammability (LEL) value of the cloud. Since the case incident occur in open air so the chances of mass to be unsaturated condition is less therefore when [η (%) increase] the flammable range to mix with air will be also increase, thus the possibility of energy transfer to initiate reaction in fresh air by compression in high velocity of shock wave is increase over a very short time. The energy released in the combustion process will be longer and maintains under the magnitude of shock front [1, 13] if the duration of overpressure is increased.

Table 2 Peak overpressure versus scaled distance for blast wave pressure from an explosion of 13,000 kg of propane

η (%)	R (m)	P°, (kPa)	T_d (s)$^{Less/USEPA}$	Diagnose
2	50	53.51	0.054/26.95	E14
	100	15.44	0.143/41.83	E9
	150	8.82	0.200/48.66	E8
	200	6.22	0.236/51.97	E7
	300	3.96	0.274/54.80	E6
	400	2.92	0.293/55.90	E5
	500	2.32	0.305/56.44	E4
3	50	71.27	0.048/27.7	E15
	100	19.08	0.141/44.7	E10
	150	10.52	0.208/53.4	E8
	200	7.30	0.251/57.9	E8
	300	4.59	0.300/61.9	E6
	400	3.37	0.325/63.5	E5
	500	2.66	0.341/64.3	E4
4	50	87.94	0.044/28.1	E15
	100	22.36	0.139/46.7	E11
	150	11.99	0.212/56.8	E8
	200	8.21	0.262/62.2	E8
	300	5.10	0.319/67.3	E7
	400	3.73	0.350/69.4	E6
	500	2.94	0.369/70.4	E5
5	50	103.85	0.041/28.4	E15
	100	25.41	0.136/48.1	E11
	150	13.32	0.214/59.4	E8
	200	9.02	0.269/85.7	E8
	300	5.54	0.334/71.7	E7
	400	4.03	0.369/74.2	E6
	500	3.18	0.391/75.5	E5
7	50	134.04	0.035/28.8	E15
	100	31.11	0.130/50.0	E12
	150	15.73	0.214/63.1	E9
	200	10.44	0.278/70.9	E8
	300	6.31	0.356/78.7	E7
	400	4.56	0.399/82.1	E6
	500	3.58	0.426/83.8	E6
9	50	162.60	0.032/29.0	E15
	100	36.43	0.124/51.3	E13
	150	17.91	0.213/65.9	E10
	200	11.70	0.282/74.9	E8
	300	6.96	0.371/84.2	E8
	400	5.00	0.421/88.4	E7
	500	3.92	0.453/90.5	E6

To get the most appropriate effect of explosion yield in TNT model, since the recommended methods (which is a compilation of analysis result by few researchers) as proposed by Less [1] still have some uncertainty, therefore the analysis VCE which occurred in Flixborough in 1974 as published by Hoiset et al. [20] and Sadee et al. [14] including site incident observation [1, 14] were referred.

Figure 1 shows the comparison of equivalent analysis results of TNT on the effect of vapor cloud explosion which occurred in Flixborough by both researchers [14, 19]. Based on the analysis done, both researchers used value $\eta = 2$ to 3 % in the TNT model by considering the explosion occurred near the field scenario using Brasie and Simpson methods [5] and ACMHH as recommended in the selection of explosion yield table by Less [1]. Meanwhile the value of physical and chemical properties of cyclohexane used for both analysis are the same, except for the different quantity of cyclohexane involved in VCE, whereby Hoiset et al. [20] assumed that 70 % of 40 tons of cyclohexane released was contributed to the formation of the flammable cloud (based on local observations). For Sadee et al. [14], the quantity of cyclohexane released was based on the flash fraction of fuel (0.17 at 423 K) on the basis of actual thermodynamics data as recorded from daily activity of plant operation before incident which is estimated 34,000 kg og cyclohexane. The outcome of results comparison for overpressure by Hoiset et al. [20], Sadee et al. [14] and site incident observation [1, 14] is shown in Fig. 1. Figure 1 shows that value of overpressure from TNT model method estimated by Sadee et al. [14] which was observed to have a tendency to provide the actual condition as occurred in Flixborough in 1974. However the value is still far from the actual picture of damage on the structure building in the plant vicinity and its

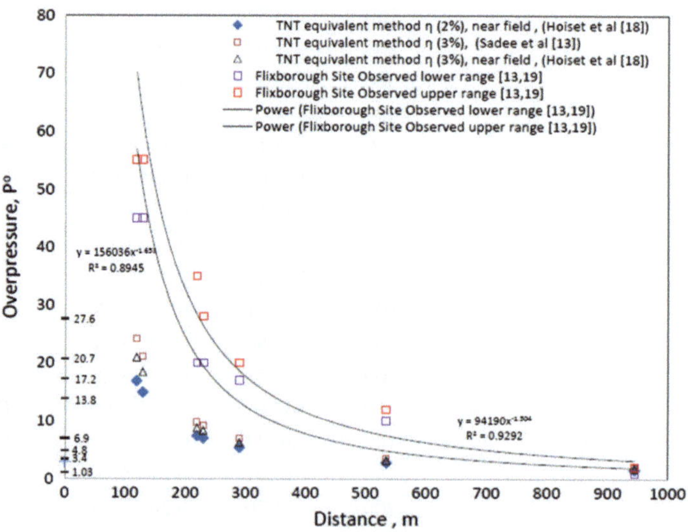

Fig. 1 The comparison of equivalent analysis results of TNT model on the effect of vapor cloud explosion which occurred in Flixborough by Hoiset et al. [19] and Sadee et al. [14] and site observation [1, 14]

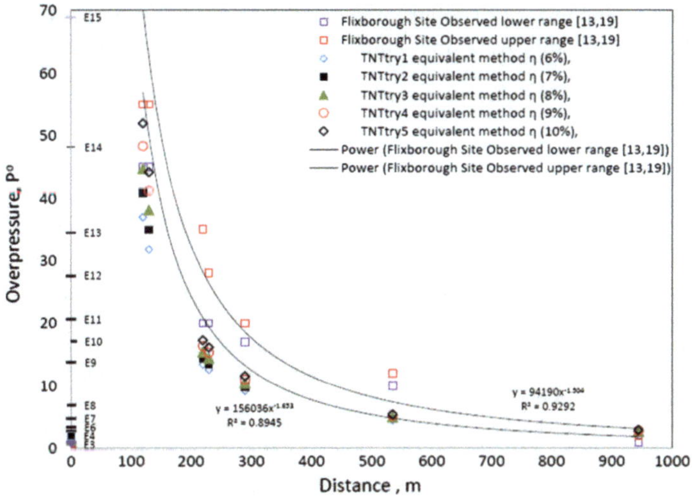

Fig. 2 The comparison of equivalent analysis results of TNT model on the effect of an increment of η value between 2 and 10 % and site observation [1, 14] which occurred in VCE Flixborough with correlation to structure damage by V.J Clancey [10]

surrounding area; within 120 m (offices), 130 m (plant canteen), 220 m, 230 m (machine shop), 290 m (storage tank farm), 535 m (houses) and 945 m (houses). For Fig. 2, the value of explosion efficiency used in TNT model is measured at yield factor (η (%) between 6 % and limit to 10 % [1, 13, 19],) at 535 m and 945 m within the actual range of structural building damage as occurred in Flixborough in 1974. As a results, at η (10 %) factor the overpressure value is closer to the actual scenario of structural damage as referred by site observation between distance 100–300 m. From the results in Fig. 2, also shows that the possibility of mass formation mix to air can happened very fast and lead to a large flammable cloud, therefore the high temperature will accelerate the effect of evaporation rate and finally the flame speed velocity will reach the unburnt material very fast and the pressure effect received by the structural building will be very strong. The site observation structure damage can be seen in agreement with the damage indicator as developed by VJ Clancey [10].

Conclusion

For high yield explosion, the area of destruction scales becomes lesser in direct proportion to the energy release. As a result, the average blast wave arrival time is increasing and allowing more time for escape between the flash, the blast and the sound arrival. The most suitable zone for any land use activities, which is allowed to operate for major hazard installation is proposed at 600 m and above. It is also observed that beyond this zone, the damage impact may contribute to a very minor house damage.

References

1. Lees FP (2005) Loss prevention in the process industries. Butterworth Edited by Sam Mannan
2. Zulkifli RA, Mohanad HE, Rashid Sharif A (2010) Assessment on the consequences of liquefied petroleum gas release accident in the road transportation. J Appl Sci 10(12):1157–1165
3. Zulkifli RA, Mohanad HE, Shariff AR (2009) Development and design of smart advisory system in the accident of transportation of hazardous material via quantitative risk approach. J Occup Safety Health 6(12):48–67
4. El-Harbawi M, Mustapha S, Choong TSY, Abdul Rashid Z, Abdul Rashid S, Sherif AA (2010) SCIA: GIS-based software for assessing the impacts from chemical industrial accidents. ASCE J Pract Periodical Hazard Toxic Radioactive Waste Mgmt 14:104
5. Brasie WC, Simpson DW (1968) Guidelines for estimating damage explosion. Loss Prev 2:91
6. Harris RJ, Wickens MJ (1989) Understanding vapour cloud explosions an experimental study. Comm 1408. Instn Gas Engrs, London
7. Eichler TV, Napadensky HS (1977) Accidental vapor phase explosions on transportation routes near nuclear plants. Rep. J6405 ANL-K77-3776-1. Argonne Nat Lab, Argonne, IL
8. Prugh RW (1991) Quantify BLEVE hazards. Chem Engng Prog 87(2):66
9. Center for Chemical Process Safety (CCPS) (2000) Guidelines for chemical process quantitative risk analysis. 2nd Edition, American Institute of Chemical Engineers, New York
10. Clancey VJ (1972) Diagnostic features of explosion damage. In: Paper presented at the sixth international meeting of forensic sciences, Edinburgh
11. Health and Safety Executive Report (1979) Advisory committee on major hazards 'second report, HMSO publication
12. Phillips H (1981) Unconfined vapour cloud explosions: a new look at Gugan's book. Chem Engr Lond 369:286
13. Center for Chemical Process Safety (CCPS) (2000) Guidelines for evaluating the characteristics of vapor cloud explosions, flash fires and BLEVEs. 1st Edition, American Institute of Chemical Engineers, New York
14. Sadee C, Samuels DE, O'Brien TP (1974) The Characteristics of the explosion of cyclohexane at the NYPRO (UK) Flixborough plant on 1st June 1974. J Occup Accid 203
15. Parker RJ (1975) The Flixborough disaster. Report of the Court Inquiry London
16. Merrifield R, MacKenzie JF (1997) Methodology for estimating the explosion yield of incidents involving conventional or improvised explosives. In: Proceedings of the eighth international symposium on interaction of the effects of munitions with structures
17. Mendonc¸a-Filho LG, Bastos-Netto D, Guirardello R (2008) Estimating the TNT equivalence of a 15-ton single base powder explosion through damaged building profiles analyses. J Hazard Mater 158:599–604
18. Gould KE, Kempo (1971) High explosive field test: explosion phenomena and environmental impacts. Defence Nuclear Agency, USA
19. Crowl DA, Louvar JF (2002) Chemical process safety: fundamental with application, 2nd edn. Prentice Hall, New Jersey
20. Hoiset S, Hjertager BH, Solberg T, Malo KA (2000) Flixborough revisited—an explosion simulated approach. J Hazard Mater 77:1–9

Disparities in Generated Noise Between Predictions and Measurements from Construction Sites

Zaiton Haron, Nadirah Darus and Khairulzan Yahya

Abstract In order to avoid excessive noise exposure and to improve construction environmental performance, noise should be predicted at the planning stage and appropriate mitigation measures should be implemented. There are various noise prediction methods that have been practiced by the respected parties. The simplest and most recommended method of noise prediction is the procedure of BS 5228-1:2009. However, previous studies have claimed that this method was inaccurate because of several factors. Therefore, this study attempts to assess the difference between the prediction using this method and the noise obtained from the measurement. The study was conducted by measuring noise emissions from earthwork, piling work and structural work at selected measurement stations, investigating the individual noise emission of construction equipment and predicting construction noise using the measured data. Several related variables were also measured to identify their effects on outdoor sound propagation. The results of *t*-test showed that all the stations have the significant disparities between prediction and measurement. The highest over prediction of 5 dBA was due to the use of several moving machines during operation processes and high noise emission level of machines. Consequently, this affects the distance between the sound source and the measurement station (geometry factor).

Keywords Construction noise · Noise prediction · Noise measurement · Construction sites

Z. Haron (✉) · N. Darus · K. Yahya
Faculty of Civil Engineering, Universiti Teknologi Malaysia, 81310 Skudai, Johor Bahru, Malaysia
e-mail: zaitonharon@utm.my

N. Darus
e-mail: nddarus@yahoo.com

K. Yahya
e-mail: khairulzan@utm.my

M.A. Hashim (ed.), *ICGSCE 2014*, DOI 10.1007/978-981-287-505-1_46

Introduction

In most developing countries including Malaysia, construction industry is one of the major contributors to economic growth. Construction is a major contributor to the environmental impact and pollution [1]. There are several initiatives have been taken to cater the problem but the effect remains the same. The environmental impacts are including effects on human health [2] due to the noise of the construction process, among others. This is because many plants with loud sound are extensively used without considering the effect on social problems [3].

Early prediction of construction noise at planning stage and implementation of any appropriate noise mitigation measures may help to avoid excessive noise exposure to the public and to improve the environmental performance and construction activity process. Therefore, sound of construction is one of the key elements that are subjected to the environmental impact assessment [4]. The prediction of construction noise will be included in the EIA report by an EIA consultant. However, the methods of construction noise predictions vary between different consultants since there is no established method to be adopted.

The most recommended prediction method is the BS 5228-1:2009 procedures. However, previous studies have claimed that this method is not accurate and a stochastic model was proposed to simulate noise arising from construction activities [5]. It was due to the nature of the sound fluctuations coming from the construction process. Previous studies are very limited and not conclusive. Thus, this paper studies the noise emissions from construction activities, investigates the individual noise emission level of construction equipment, predicts noise from construction activities, assesses the disparity between the noise level prediction and the measured noise level, and subsequently determines the cause of the disparity between the results.

Noise is a set of undesired sound [6–9], and from the construction viewpoint, noise is characterised by the noise emission level from machines, acoustic noise emission characteristics, number of machines concurrently in use and distance between the receiver and machines, and existence of obstruction or reflection between the receiver and the machines. These include the changes in acoustic power during full or idle working condition and movement of the machine when working in a workspace. Furthermore, each stage of construction has different spectrum levels [10], so the noise characteristics must be considered in estimating the impact to humans [4]. The impacts cannot be detected spontaneously but will escalate in the long-term period. Effects of noise on animals should not be taken lightly because it may disturb the ecosystem where they live.

Currently, there are four methods of construction noise prediction in the BS 5228-1:2009 procedures [11]. The first two methods for the stationary machine are activity L_{Aeq} and sound power, and the other two methods for mobile machines on-site are on-site (limited area) and hauling in the streets. Basically, noise levels (L_{Aeq}) at the receiver is predicted by combining three basic elements such as acoustic power of the machine, emission model, and propagation model. The construction noise predictions as described in Appendix F of BS 5228-1:2009, which take into account the sound power machine, the operating facility, the distance between the source and the receiver, the presence of the inspection, the reflected sound, and the attenuation due to the earth's surface. The result of this method is accurate, provided that all the factors mentioned earlier, which are regarded as an input to the model, are accurate. Temperature and wind speed are not included in the BS 5228-1:2009 procedures, unless the distance is greater than 300 m. The wind speed is less than 5 km/h and also had a great impact on the accuracy of the measurement noise.

Without any obstruction between receiver and operation activities, and also no reflection factor, the noise level of the prediction arising from the two plants (excavator and dump truck) of a small site (50 × 50 m) had an insignificant difference or an over prediction of approximately 1 dBA from the measurement [12]. However, based on similar comparison on a larger site and found that, using various plants (five to six plants), there were significant differences in the noise level predictions compared with measurements (an average of 4 dBA, higher than the measured value) because of the effects of plant movement [13]. Over prediction will be felt by contractors who enter a project tender. Although predictions are used just to get approval from the local authorities, the responsibility for compliance will be transferred to the contractor. The over prediction of L_{Aeq} will result in an excessively high bid price and lessen the chances of gaining the contract [14].

Methodology

Figure 1 shows the three selected construction stages of earthworks, substructure (piling), and superstructure located in Kempas and Skudai, Johor. The study was carried out in three phases as in Fig. 2. Phase 1 involves the measurement noise levels from construction activities (L_{Aeq}), and noise level of individual machines involved in construction activities. Because of safety reasons, the average noise emission level was obtained during the machine working at its full load. The instrumentations used are the sound level meter (type 1 and type 2), anemometer, and distometer.

Measurements were carried out accordingly at the selected stations for 1 h with 15 min interval. Stations are free from reflection that can influence the sound transmission. The sound level meter was calibrated before and after measurements, and it was set at 1.2 m above the ground. Noise level (dBA), height of the machine

Fig. 1 Selected sites **a** Earthworks **b** Substructure (piling works) and **c** Superstructure works

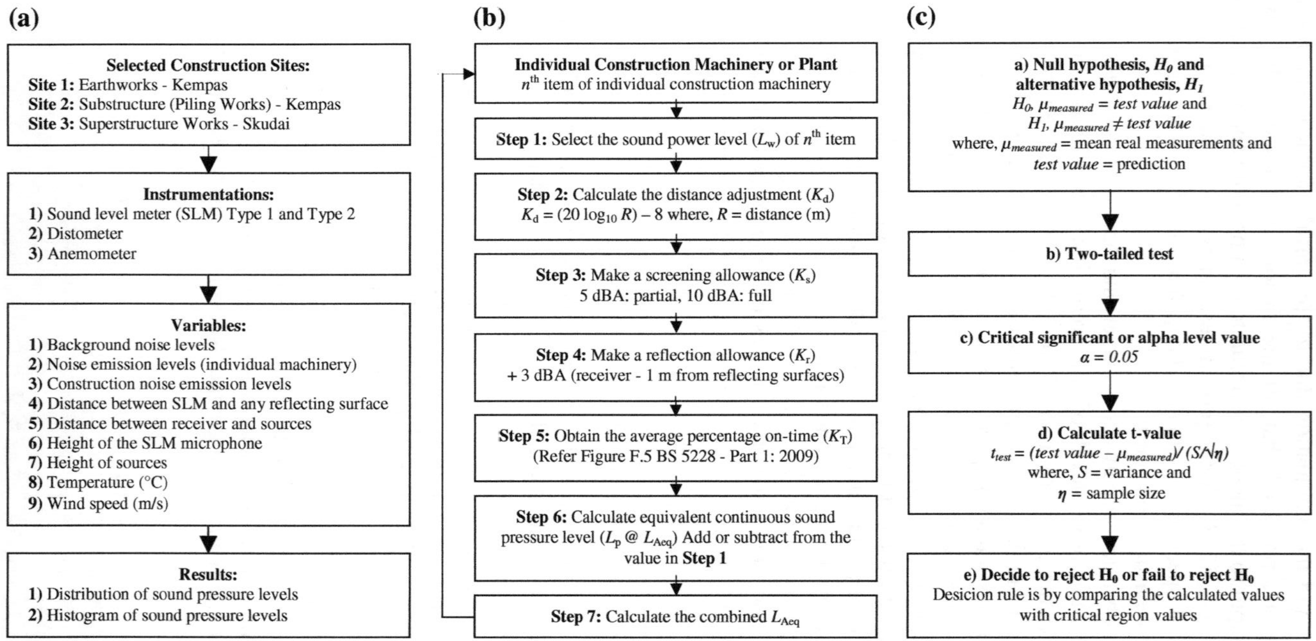

Fig. 2 Methodology **a** Phase 1 **b** Phase 2 and **c** Phase 3

(m), noise sources (m), temperature (°C), and wind speed (m/s) were measured. Measurements were performed at a minimum wind speed and temperature variation to reduce the differences between predictions and measurements.

Meanwhile, Phase 2 involves the noise predictions by using the BS 5228-1:2009 procedures that were carried out for all stations. For this study, the data from Phase 1, that is, the noise emission by individual plant and the average distance between the receiver and the machine were used. For Phase 3, the disparities of the results of noise prediction and the mean values of measurement for all stations were assessed. A *t*-test was conducted to test the significant difference in the mean values of real on-site measurement results and noise prediction. H_0 was rejected if the p value was less than 0.05 or if the t value from calculation was higher than the t value from the t table (critical region).

Result and Discussion

Table 1 shows the values of all variables measured for all measurement stations. These data were obtained to identify the variables that affect sound propagation outdoors. It can be seen that the change in wind speed and temperature is relatively small during the measurement.

Meanwhile, Table 2 summarises the distribution and histogram of the sound pressure level. The highest and lowest noise emissions were generated from the substructure (piling) and superstructure works, respectively. It was caused by several individual construction plants involved for each construction activity. Meanwhile, the level of noise generated from earthworks was within the range of noise level generated from substructures (piling) and structural works. However, the noise level at station E2 (earthwork) was less than the noise from the installation of the main structure because of the fewer number of individual machines operated during noise measurement.

Table 1 Computation of variables measured on-site

Stages of construction	Stations	Background noise levels (dBA)	Noise sources (m)	Temperature (°C)	Wind speed (m/s)
Earthworks	E1	56.20	1.2–2.4	29.8	1.5
	E2	48.90	1.2–2.4	29.6	2.5
	E3	61.50	1.2	30.8	1.8
Substructure (piling works)	P1	66.30	1.2–2.4	27.7	1.5
	P2	63.80	1.2–2.4	32.8	2.0
	P3	61.20	1.2–2.4	31.8	1.8
Superstructure works	S1	53.80	1.2–2.4	32.4	0.5
	S2	55.50	1.2–2.4	33.6	0.7
	S3	60.40	1.0–1.2	29.0	0.5

Table 2 Computation of sound pressure levels for 1 h (5-min time interval)

Stages of construction	Stations	Distributions of sound pressure levels (dBA)		Histogram of sound pressure levels (dBA)			
		Highest	Lowest	Range	Highest frequency of occurrence	Highest	Lowest
Earthworks	E1	$74.90_{\text{4th 5min}}$	$64.70_{\text{9th 5min}}$	65.00–75.00	70.00–71.00	3.143	70.82
	E2	$69.50_{\text{3rd 5min}}$	$50.20_{\text{11th 5min}}$	50.00–70.00	55.00–60.00	6.225	59.61
	E3	$76.30_{\text{6th 5min}}$	$62.00_{\text{10th 5min}}$	<60.00–>80.00	65.00–67.50	4.809	67.94
Substructure (piling works)	P1	$77.60_{\text{5th 5min}}$	$71.80_{\text{2nd 5min}}$	71.00–78.00	75.00–76.00	1.612	74.94
	P2	$77.60_{\text{6th 5min}}$	$69.60_{\text{12th 5min}}$	68.00–80.00	74.00–80.00	3.401	75.65
	P3	$79.70_{\text{11th 5min}}$	$69.00_{\text{8th 5min}}$	68.00–80.00	78.00–80.00	3.323	75.96
Superstructure works	S1	$67.60_{\text{7th 5min}}$	$59.30_{\text{11th 5min}}$	58.00–68.00	64.00–68.00	3.098	63.72
	S2	$70.20_{\text{3rd 5min}}$	$56.70_{\text{10th 5min}}$	<55.00–>75.00	60.00–61.00	3.899	62.44
	S3	$68.80_{\text{2nd 5min}}$	$57.50_{\text{10th 5min}}$	<58.00–>70.00	67.50–70.00	3.619	65.56

The results of the noise emissions from individual construction equipment were used to calculate their sound power level (L_w) as input data to the noise prediction model. Eight machineries involved in earthwork: two crawler excavators, a bulldozer, a dump truck, and four compactors. Meanwhile, ten machines: three piling machine, three crawler cranes, and four crawler excavators were used in the substructure (piling). There were only five machines used for main structural works: two mobile cranes, a backhoe, a concrete mixer truck, and a bar bending and a cutting machine. The highest and the lowest L_w values were generated from the piling machine of the substructure (piling) (2–115 dBA) and the concrete mixer truck (1–92 dBA) from the structural works, respectively. Meanwhile, Table 4 shows the noise predictions for all measurement stations of three construction stages. On the basis of Table 3, station E3 produced the highest noise emissions for work related to earthwork because all machines were operated simultaneously. Meanwhile, stations P3 and S3 generated the highest level of noise emissions to the substructure (piling) and major structural works, respectively.

Table 4 shows the results for all measurement stations construction. On the basis of Table 4, the critical t value ($t_{critical}$) for all measurement stations was 1.812. However, the t value observed for all measurement stations exceeded the $t_{critical}$. Therefore, all H_0 were rejected. Thus, the t-tests showed that there were significant disparities between the noise level of the actual on-site measurements and the noise prediction results for stations E1, E2, and E3.

All stations at the three stages of construction measurement have higher noise prediction results compared with the results of the actual on-site measurements. The disparities between the prediction and the measurement t values of stations E1, E2, and E3 (earthworks) were 3.00, 6.70, and 5.92 dBA, respectively (an over prediction). This result is higher than the previous study [13] because the number of moving machines in this study was more than that of the study. Meanwhile, for stations P1, P2, and P3 in the substructure (piling) stage, the over prediction values were 1.14, 2.40, and 2.45 dBA, respectively. For stations S1, S2, and S3 in the superstructure works stage, the over prediction values were 5.73, 3.15, and 4.24 dBA, respectively. For t-test, the observed values of t_{noise} for all stations were greater than the $t_{critical}$ value of 1.812. Therefore, the hypothesis that measured and predicted noises were equal was rejected. There were significant disparities between real on-site measurement and noise prediction.

It can be seen that the static machines with high noise emission level produced a smaller disparity noise level at the receiver, whereas the construction operation involved several moving plants that produced the largest disparities. The large disparities may be due to a gross simplification in the data input related to the distance between the station and the moving plants. Because the plants have to move around the construction site in real life and because of the level of noise generated from on-site construction equipment, individuals have relied on the variation of the acoustic power of machines during heavy load. In this study, the average acoustic power of machines was used in the prediction.

Table 3 Computation of L_{Aeq} obtained using the BS 5228-1:2009 procedures

Stages of construction	Station	Sub site	Construction plants	L_w (dBA)	Distance (m)	Corrections				Individual L_{Aeq}	L_{Aeq} sub sites	Combined L_{Aeq}
						K_d	K_s	K_r	K_T			
Earthworks	E1	A	CE 1	108	36.5	−39.25	0	0	0	69.75	70.74	73.82
			CE 2	106	20.3	−34.15	0	0	−8.0	63.85		
		B	DT 1	107	56.7	−43.07	0	0	−2.0	63.93	68.05	
			DT 2	107	45.1	−41.08	0	0	0	65.92		
		C	BD 1	101	60.5	−43.64	0	0	−2.0	55.36	67.66	
			DT 3	109	65.8	−44.36	0	0	0	64.64		
			DT 4	108	62.2	−43.88	0	0	0	64.12		
	E2	A	CE 2	106	55.0	−42.81	0	0	−2.5	60.69	60.69	66.31
		B	DT 1	107	50.0	−41.98	0	0	−2.5	62.52	63.78	
			CP 1	98	40.7	−40.19	0	0	0	57.81		
		C	CE 1	108	88.5	−46.94	0	0	−2.5	58.56	58.56	
	E3	A	CE 1	108	45.2	−41.10	0	0	−0.2	66.70	73.04	73.86
			CE 2	106	20.9	−34.11	0	0	0	71.89		
		B	BD 1	101	20.6	−34.28	0	0	−0.5	66.22	66.22	
Substructure (piling works)	P1	A	PM 1	111	42.7	−40.61	0	0	−0.2	70.19	74.23	76.08
			CE 1	104	16.5	−32.35	0	0	0	71.65		
			CC 1	101	37.4	−39.46	0	0	0	61.54		
		B	PM 2	115	55.5	−42.89	0	0	−2.5	69.61	71.49	
			CE 2	105	52.5	−42.40	0	0	0	62.60		
			CE 3	105	40.0	−40.04	0	0	−1.0	63.96		
			CC 2	100	50.0	−41.98	0	0	0	58.02		
	P2	A	PM 1	111	50.0	−41.98	0	0	0	69.02	69.30	78.05

(continued)

Table 3 (continued)

Stages of construction	Station	Sub site	Construction plants	L_w (dBA)	Distance (m)	Corrections K_d	K_s	K_r	K_T	Individual L_{Aeq}	L_{Aeq} sub sites	Combined L_{Aeq}
			CC 1	101	55.0	−42.81	0	0		57.19		
		B	PM 2	115	30.4	−37.66	0	0	−0.5	76.84	76.84	
		C	PM 3	114	65.0	−44.26	0	0	−1.5	68.24	68.[illegible]6	
			CC 3	99	60.0	−43.56	0	0	0	55.44		
	P3	B	PM 2	115	90.0	−47.08	0	0	0	67.92	67.92	78.41
		C	PM 3	114	30.6	−37.71	0	0	−0.5	75.79	78.00	
			CE 3	105	23.2	−35.31	0	0	0	69.69		
			CE 4	106	20.3	−34.15	0	0	−1.0	70.85		
			CC 3	99	18.6	−33.39	0	0	0	65.61		
Superstructure works	S1	A	MC 1	99	30.4	−37.66	0	0	0	61.34	69.45	69.45
			MC 2	109	37.5	−39.48	0	0	−2.5	67.02		
			CM 1	92	40.6	−40.71	0	0	−0.2	51.63		
			BH 1	98	20.5	−34.24	0	0	−0.2	63.56		
	S2	A	MC 2	109	60.5	−43.64	0	0	0	65.36	65.44	65.59
			CM 1	92	49.7	−41.93	0	0	−2.0	48.07		
		B	BH 1	98	90.7	−47.15	0	0	0	50.85	50.85	
	S3	A	MC 2	109	38.1	−39.62	0	0	0	69.38	69.62	69.80
			BH 1	98	44.8	−41.03	0	0	0	56.97		
		B	BB 1	93	28.3	−37.04	0	0	0	55.96	55.96	

Table 4 Test of significant difference in mean equivalent noise levels

Stages of construction	Stations	Test value (dBA)	Means, M ($\mu_{measured}$)	No. of samples(η)	Standard deviation	p	t	Test hypotheses H_0: $\mu_{measured}$ = *test value* H_1: $\mu_{measured} \neq$ *test value*
Earthworks	E1	73.82	70.82	11	3.143	0.010	3.168	H_0 was rejected*
	E2	66.31	59.61	11	6.225	0.005	3.570	H_0 was rejected*
	E3	73.86	67.94	11	4.809	0.002	4.086	H_0 was rejected*
Substructure (piling works)	P1	76.08	74.94	11	1.612	0.040	2.353	H_0 was rejected*
	P2	78.05	75.65	11	3.401	0.042	2.336	H_0 was rejected*
	P3	78.41	75.96	11	3.323	0.035	2.441	H_0 was rejected*
Superstructure works	S1	69.45	63.72	11	3.098	0.000	6.137	H_0 was rejected*
	S2	65.59	62.44	11	3.890	0.023	2.683	H_0 was rejected*
	S3	69.80	65.56	11	3.619	0.003	3.882	H_0 was rejected*

* At a 95% significance level.

Conclusion

This study has evaluated the difference in noises predicted using BS 5228-1:2009 procedures and measured at several stations of three selected stages: earthworks, substructure (piling), and structural works. The *t*-test showed significant differences for all stations. In addition to the high noise emission level of machines, the highest difference (5 dBA) has been shown as the result of the activities that involve several moving machines. Consequently, this affects the distance between the sound source and measurement station. Therefore, in this study, the difference is caused by the movement of the plant which affects the distance between the sound source and measurement station (geometry factor), number of moving machines, and high levels of machine noise emissions.

Acknowledgment The authors would like to express their appreciation for the support of the sponsors with Project No. Q.J130000.2522.03H43 and UTM Zamalah Scholarship.

References

1. Fuertes A, Casals M, Gangolells M, Forcada N, Macarulla M, Roca X (2013) An environmental impact causal model for improving environmental performance of construction processes. J Clean Prod 52:425–437
2. Li X, Zhu Y, Zhang Z (2010) An LCA-based environmental impact assessment model for construction process. Built Environ 45:766–775
3. Manatakis E, Skarlatos D (2002) A statistical model for evaluation and prediction of the noise exposure in a construction equipment area. Appl Acoust 63:759–773
4. Department of Environment (2007) The planning guidelines for environmental noise limits and control. Department of Environment Malaysia, Malaysia
5. Carpenter F (1997) Construction noise prediction at the planning stage of new developments. Build Acoust 3(4):239–249
6. Edworthy J (1997) Noise and its effect on people: an overview. Int J Environ Stud 51:335–344
7. Muzet A (2007) Environmental noise, sleep and health. Sleep Med Rev 11:135–142
8. Hamoda MF (2008) Modelling of construction noise for environmental impact assessment. J Constr Developing Countries 13(1):79–89
9. Fernandez MD, Quintana S, Chavarria N, Ballesteros JA (2009) Noise exposure of workers of the construction sector. Appl Acoust 70:753–760
10. Ballesteros MJ, Fernandez MD, Chavarria N, Ballesteros JA, Gonzalez I (2010) Noise emission evolution on construction sites-measurement for controlling and assessing its impact on the people and on the environment. Build Environ 45:711–717
11. British Standards Institution (2008) BS 5228—Part 1: 2009 code of practice for noise and vibration control on construction and open sites. BSI Group Headquarters, The United Kingdom
12. Idris N (2012) Automated prediction of noise from construction site using stochastic approach. Master thesis, Universiti Teknologi Malaysia
13. Jahya Z (2014) Automated construction noise prediction by considering the variability of noise sources and outdoor sound propagation. Master thesis, Universiti Teknologi Malaysia
14. Haron Z (2007) Probability approach in environmental acoustics. Ph.D. thesis, University of Liverpool

Flame Retardancy of Polymeric Building Material with Recycled Expanded Polystyrene Filler

Syed Anas Bin Syed Mustafa, Rahmah Mohamed, Wan Bahira and K. Rasidan

Abstract Flame retardancy is an important characteristic in the flame and fire safety field to both prevent and limit the effects of material ignition and flame consumption. The preparation and characterization of Unsaturated Polyester Resin (UPR) filled recycled Expanded Polystyrene (EPS) composite systems were systematically investigated. Additives such as organic nanocrystal (ONC) and silica aerogel powder were added to the composite for imparting suitable characteristics to the composite. The result obtained from comparing the current study against its predecessor reveals Liquid Natural Rubber (LNR) is significantly more flammable than Styrene Butadiene Rubber (SBR) by more than doubled but aerogel utilization can negate this effect. It can be concluded that the use of additive could affect flame retardancy and thermal properties. Further studies of the material can be done in determining and/or confirming the actual mechanisms which ONC and aerogel produces this effect on flammability.

Keywords EPS · UPR · Recycled composite · Flame retardancy · Thermal characteristics

S.A.B. Syed Mustafa (✉) · W. Bahira · K. Rasidan
Faculty of Applied Science (FSG), Universiti Teknologi MARA (UiTM), 40450 Shah Alam, Selangor, Malaysia
e-mail: ssyed_anas@hotmail.com

R. Mohamed
GREEN Polymer Research Group, Faculty of Applied Science (FSG), Universiti Teknologi MARA (UiTM), 40450 Shah Alam, Selangor, Malaysia
e-mail: greenkayangan@gmail.com

M.A. Hashim (ed.), *ICGSCE 2014*, DOI 10.1007/978-981-287-505-1_47

Introduction

Unsaturated Polyester (UPE) matrix composites have been used for many years in a broad technology fields such as naval construction, offshore applications, water-lines, and building construction. UPE is an economical thermo-set material that is widely used due to its excellent process ability, good cross-linking tendency as well as mechanical properties when cured. Expanded Polystyrene (EPS) sheet has been used as core material for sandwich core composite door shutter to replace wooden door shutters in building [1]. Studies had been done on use of EPS incorporated into Unsaturated Polyester Resin (UPR) with relevant diluents. Toughening of the unsaturated polyester resin is very important to increase impact performance especially for building structures. [2].

Flammability could be defined as the capability of a material sustaining a flame when it is ignited. Flammability poses an ever-present concern in fire safety and prevention; where reducing the flammability, or increasing the flame retardancy, of materials are a priority to help prevent fires by hardening them against ignition, and reduces a flames spread once it occurs.

The preparation and characterization of UPR filled recycled EPS composite systems were systematically investigated. Additives such as ONC and silica aerogel powder were added to the UPR-EPS blend for imparting suitable characteristics to the composite. The effect of differing weight percentage of additives on the thermal properties and flammability of the material were determined. The fabricated composites undergoes linear flame propagation test to determine its level of flammability.

The use of rubber as a toughening agent with UPE had been investigated [2, 3] and use of aerogel in paint for increasing thermal insulation and flammability had also been performed [4–6]. Integration of nanocrystal and aerogel to UPE/EPS blend was found to lower the thermal conductivity of the material, which is a characteristic required for insulating panels. From our previous paper [7], we found that the thermal conductivity, k value had reduced by more than half upon adding the additives as prescribed.

This paper emphasises on comparing the flammability of various UPR/EPS composite systems impregnated with some selected additives by evaluating the changes in flame propagation speed as effect of the additives utilized.

Experimental

Materials and Method, The selected thermo-set matrix material used for fabricating these composite systems is the Reversol P9780 UPR resin, consisting of vinyl ester oligomers having density of 1.12 g/cc, viscosity of 450–600 cps with 41–44 % styrene content. Methyl Ethyl Ketone Peroxide (MEKP) and Cobalt solution used as initiator and promoter each, were supplied together with the UPE resin by Revertex Sdn Bhd. The EPS or Styrofoam filler was obtained from waste material. These ingredients are similar to materials used by Rashidan et al. [1] and Wan Bahira [8].

This paper was produced by applying an analytical quantitative methodology where the sample produced are subjected repeated testing on its flammability via ASTM D635-10 method, and measurement of its thermo-physical properties via a KD-2 Pro thermal probe and a digital densitometer.

Composite Fabrication and Testing, Recycled EPS, fixed at 10 % parts by weight (%wt) of UPR resin, was blended utilizing a high speed agitation mixer until complete dissolution was attained. This is to enable the gaseous and solid contaminates present are easily eliminated from the mixture via gravity settling for 24 h.

Additives were added before the samples were prepared utilizing hand lay-up processes onto an aluminium mould. The additives such as ONC content was added at specific quantities as a percentage of blend weight. The process was followed by adding 3 %wt of MEKP initiator and suitable amount of cobalt as accelerant to cure the EPS/UPE composite. Gelation time was set to be about 5 min. Once fully gelled, complete curing was done under compression for 1 h at 70 °C. These steps are also similar to the process used by Rashidan et al. [1] and Wan Bahira [8].

The produced sample were tested by using ASTM D635-10 method, where samples were lit while suspended in a horizontal position to which flame characteristic and propagation speed were noted. The samples were also subjected to thermo-physical measurements via a KD-2 Pro thermal probe and density determination via a digital densitometer.

Result and Discussion

The additives utilized across the samples are elastomers (SBR or LNR), antioxidant (AO), flame retardant (FR), silica aerogel (Ag) and organic nanocrystal (ONC). Table 1 shows the time required for burning the samples at the fixed distance, indicated as elapsed time (s), and the rate of combustion is shown by the linear burning formulation, V(mm/min),

$$\text{Linear Burning,} \quad V = 60\,L/t$$

where

L the burn length, in millimetres between reference marks (100 − 25 = 75 mm)
t the time(s) (elapsed time from the 25 to 100 mm reference mark)

Both Fig. 1 and Table 1 shows a few key trends; an integration of elastomer into the system promotes flame propagation, a composite utilizing SBR is shown to be

Table 1 Comparison of combustion elapsed time in different additive systems

Additive in UPE/EPS composite (per 100 % weight mixed UPE/EPS)	Designation	Elapsed time, t (s)	Linear burning, V (mm/min)
20 % SBR − 0.5 % AO	S1	85	52.94
20 %SBR − 0.5 % AO − 0.5 % FR	S2	105	42.86
4 % SBR − 0.5 % AO	SA1	511	8.81
4 % SBR − 0.5 % AO − 2.5 % FR	SA2	814	5.53
4 % SBR − 0.5 % AO − 5 % FR	SA3	1433	3.14
4 % LNR − 0.5 % AO − 5 % FR	L1	542	8.30
4 % LNR − 0.5 % AO − 5 % FR − 0.2 % (ONC)	L2	617	7.29
4 % LNR − 0.5 % AO − 5 % FR − 0.2 % Ag	L3	1196	3.76
4 % LNR − 0.5 % AO − 5 % FR − 0.2 % ONC − 0.2 % Ag	L4	698	6.45

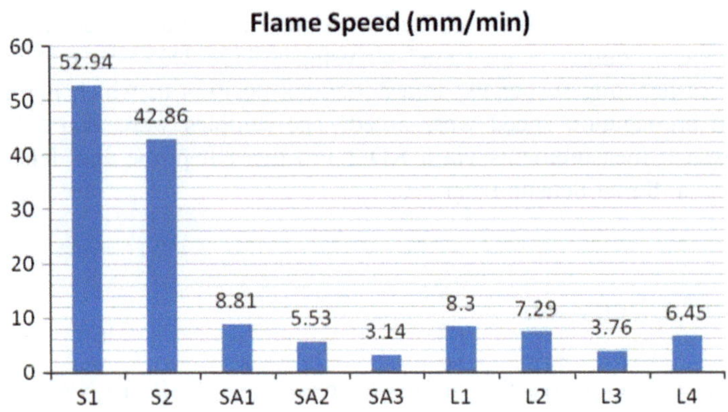

Fig. 1 Flame Propagation Speed for all Systems

more flame retardant compared to the group utilizing LNR, and the use of aerogel and ONC additives have mixed effect on flame propagation speed.

By going along the set definition of flammability, faster flame propagation speeds correlates to better flammability or lower flame retardancy of a material. Hence, a 75 % reduction of SBR content had reduced the flame speed by 80 % as shown by comparing sample S1 and SA1 of similar content. This could also be inferred that excessive amount of elastomer might be a preferred fuel for the flame as shown by how slightly the flame retardant tempers the flame propagation speed at higher elastomer content; at 20 wt% SBR, 5 wt% FR only reduces the flame speed by 19 %, whereas at 4 wt% SBR, similar additions reduces the flame speed by 64 %.

On another note, comparison between sample SA3 and L1 which differs only on the type of elastomer used, it is inferred that SBR is the more flame-retardant of the two. This is shown by monumental increase in flame speed reported from the L1 sample despite using similar flame retardant amount. This phenomena could be attributed to the different chemical composition between the two elastomer as resulted from the manufacturing processes.

The addition of aerogel and ONC into the composite systems had some effects on the flammability of the materials. The aerogel integrated L3 sample shows significant reduction in flame speed (−55 %), while the ONC integrated L2 sample report slight reduction (−12 %). The L4 sample containing both shows value slightly above the expected average of the L2 and L3 (actual speed of 6.45 mm/min against the expected 5.53 mm/min).

The significant reduction imparted by aerogel might be inferred to be attributed to its inherent Silicon, Si based mineral which also serves as thermal insulating material. Silica as a mineral is fairly resistant to ignition and burning which might inhibit flame propagation. The excellent thermal insulating property of aerogel was found to inhibit flame propagation by shielding the more flammable contents from the high temperature and heat necessary for ignition.

In order to further determine the suitability of utilizing these additives systems, these additive systems should be investigated based on another avenue, namely their effects upon the thermo-physical properties of the resultant material. The chosen measured thermo-physical properties are their thermal conductivity, thermal diffusivity, volumetric specific heat, and overall density.

As Table 2 suggested, adding only the flame retardant as in L1 reduces the thermal conductivity and specific heat, thus maintaining the diffusivity which is the ratio of conductivity to specific heat. Diffusivity is an important ratio in due part to a higher value denotes a material which transfer heat better than retaining it, and maybe suitable for certain applications.

Introducing ONC into the system as in L2 reduces all three thermal properties. Adding aerogel instead such as L3 increases them instead. Finally, adding both additives into L4 balances the two systems preceding it.

The results shows that aerogel enhances specific heat capacity as is expected of a material commonly used as a heat insulator, despite going against the same notion by also increasing its thermal conductivity as well. It could be theorized that this is

Table 2 Comparison of thermal properties of different additive systems

Samples	Conductivity (W/m K)	Diffusivity volumetric (mm^2/s)	Specific heat (MJ/m^3 K)	Density (g/cm^3)
UPR/EPS (L0)	0.165	0.093	1.764	1.221
L1	0.099	0.092	1.112	1.230
L2	0.090	0.089	0.983	1.217
L3	0.254	0.122	1.810	1.205
L4	0.227	0.102	1.657	1.240

caused by the silica within the cured material, which as a material, has a fairly good thermal retention. With its minute particle size and proper dispersion across the material, this in turn helps enhance heat transfer through the prompt heat adsorption of adjacent aerogel particles across the material.

The slight thermal properties reduction effect of ONC might be explained by its proposed use as an Infra-Red (IR) radiation reflector. IR radiation generated by a heating source were reflected from the sample surface and thus not absorbed by the material, which in turn reduces the amount of heat that could be measured. This reduction was then manifested as a reduction in the specific heat and thermal conductivity.

Density was also affected by the additives system as shown above. Inclusion of FR which was used in liquid form might had increased the density by due of the denser liquid integrated. Both ONC and aerogel however might had reduced the density by having some of material volume displaced by the suspension solvent used to integrate either additive into the material.

Aerogel is also quite significantly the least dense additive used, hence explaining further density reduction upon its integration. However, the integration of both additives had slight increased the density instead, which may be caused by an undetermined mechanism or reaction to enable the integration, thus requiring further study.

Despite all explanations given, these descriptions were merely deductions that are speculated and interpolated from the inherent properties of the material and additives used. In short, further studies are required to fully determine whether the proposed explanations given are true or another different mechanism are at work.

Conclusion

It could be concluded from the results that incorporation of SBR instead of NR reduced the flammability of the composite; significant reduction in flammability could also be brought by reduced amount of SBR. The increase in FR used proved to inhibit flammability while the use of additives such as ONC and Aerogel have some mixed effects on flame retardancy; the incorporation of aerogel reduces flame retardancy significantly compared to ONC. The integration of aerogel also improves volumetric specific heat while reducing overall density. ONC integrations have slight effect on overall thermal properties. Combining both additives however produces slightly optimized properties. Further research is required to confirm the proposed explanation of the result.

Acknowledgment I would like to acknowledge Universiti Teknologi MARA (UiTM) for their research funding under the Research Intensive Faculty (RIF) fund [File:600-RMI/DANA 5/3/RIF (64/2012)]

References

1. Rashidan K, Rahmah M, Nor ZIZ, Mohd HM (2009). Effect of antioxidant and flame retardant additives on thermal degradation and flammability of SBR filled EPS/UPR composite. In: Proceedings of Malaysia polymer international conference (MPIC 2009)
2. Bonnia NN, Surip SN, Ratim S, Mahat MM (2012) Mechanical performance of hybrid polyester composites reinforced Cloisite 30B and kenaf fibre. In: Proceedings of 2nd ASEAN APCTP workshop on advanced materials science and nanotechnology
3. Redzuan AA, Bonnia NN, Siti Shakirah S, Mohamed MRH, Surip SN (2013) Mechanical properties of rubber toughened polyester filled carbon black. Adv Mater Res 701:37–418
4. Cuce E, Cuce PM, Wood CJ, Riffat SB (2014) Optimizing insulation thickness and analysing environmental impacts of aerogel-based thermal superinsulation in buildings. Energy Build 77:28–39
5. Papadopoulus AM (2005) State of the art in thermal insulation materials and aims for future developments. Energy Build 37(1):77–86
6. Abdou OA, Murali K, Morsi A (1996) Thermal performance evaluation of a prefabricated fiber-reinforced plastic building envelope system. Energy Build 24(1):77–83
7. Mustafa SABS, Mohamed R, Zakaria NZI, Rustam HB (2013) Thermal characterisation of thermoset polyester resin filled recycle expanded polystyrene composite with aerogel and alumina additives. Adv Mater Res 664:600–604
8. Bahirah WN (2011) Effect of curing and flame retardant system on properties of polyester-filled EPS composite (Master's thesis). March

Grey Model for Accident Prediction in Data-Scarce Environment

Ali Al-shanini, Arshad Ahmad and Faisal Khan

Abstract Effective accident prediction is needed in the chemical process industries to facilitate risk management during plant operations. The unavailability of data impeded the process for accident modelling. This article discusses the use of grey model in predicting accidents in data-scarce environment. The model used is a first order single variable, i.e., G(1, 1) and is applied to a vessel and a CSTR case studies. The results obtained are compared to that of Poisson–Gamma model. Results show that G(1, 1) is more accurate compared to the Poisson model that exhibited underestimation in prediction. It is concluded that the use of G(1, 1) can represent the CPI deviations with better accuracy, and it can replace the use of Poisson model especially when some of its weaknesses are overcome.

Keywords CPI accident prediction · Poisson–Gamma model · First order grey model · Precursor data

---

A. Al-shanini · A. Ahmad (✉)
Institute of Hydrogen Economy, Universiti Teknologi Malaysia,
81310 Johor Bahru, Malaysia
e-mail: arshad@cheme.utm.my

A. Al-shanini
e-mail: ali_alshanini@yahoo.com.my

A. Al-shanini · A. Ahmad
Faculty of Chemical Engineering, Universiti Teknologi Malaysia,
81310 Johor Bahru, Malaysia

F. Khan
Faculty of Engineering and Applied Science, Memorial University of Newfoundland,
St. John's, NL A1B 3X5, Canada
e-mail: fkhan@eng.mun.ca

M.A. Hashim (ed.), *ICGSCE 2014*, DOI 10.1007/978-981-287-505-1_48

Introduction

Efforts towards minimizing risks of accidents in the chemical process industry (CPI) can be effectively facilitated by the ability to predict the expected future incidents. These predictions provide valuable information to support risk-based decision to prioritize plant maintenance and management of change needed during operations. A common approach is to use probabilistic methodology by mapping data points onto a distribution function that is assumed to represent the process behavior to generate predictions. These stochastic models e.g. Poisson model is widely applied to predict accidents in the CPI [1–4] since only one observed data is needed to predict the future. However, it has low sensitivity to the observed data which leads to poor prediction with highly significant error when observed data have increasing or decreasing trends [3, 5, 6].

Alternatively, it is also common to classify process behavior according to the order of the process. In this case, the CPI is often represented by a first order, second order or higher order, the latter typically a combination of first order processes. Mathematically, a first order process is modeled by a first order differential equation, and its salient features is embedded in the model parameters. In this case, the challenge is to obtain accurate model parameters with limited amount of data.

This article discusses accident prediction issues in data-scarce environment using grey model. The model is based on a first order differential equation with specific parameter estimation method using only four data points. The scheme is implemented on two case studies and the results obtained are compared to that of Poisson-Gamma model.

Prediction Models

Poisson–Gamma Model

Poisson–Gamma model is a class of stochastic forecasting models that is widely used to predict the number of unwanted consequences in the CPI. The model predicts well when the mean and the variance of the data are in close proximity, whereas in cases where variance exceeds the mean, negative binomial distribution models can provide better prediction [7]. Details about Poisson model can be obtained in the literature e.g., Refs. [3–5, 8].

Time Series Grey Forecasting Model

First introduced by Deng [9], grey model is an M-order partial differential equation of N variables that belongs to time series forecasting family, denoted by G(M, N).

The name of "Grey" represents that the model consists of both components that are founded on fundamental of science and those that have no theoretically explanation, often fitted using experimental data. The model has been applied in various applications including agriculture, environmental, and electronic industry due to its capability to reasonably forecast the unknown events using as few as four data points [10–14].

Grey model has the ability to enlarge few data points into a monotonically increasing data series using an accumulated generating operation (AGO), which approximately would exhibit the exponential behavior. This leads to reduce the noise from the original data series [15]. The solution of the differential equation of Grey model has an exponential form similar to the AGO, and it is used to model the data series from AGO and then to forecast the future events.

First-Order Grey Model G (1, 1)

Although various forms of models are possible, G(1, 1) is the most commonly adopted [10]. It is a single variable first-order grey model is used to predict future events for non-negative data points series $X^{(0)} = \{x^{(0)}(1), x^{(0)}(2), \ldots x^{(0)}(n)|n \geq 4\}$ [16] that is taken in consecutive order and at equal time interval, where $X^{(0)}$ is the time series data at time *i*, and *n* is the number of observation that must be equal or greater than 4 observations. Applying G(1, 1) to predict the future value $x^{(0)}(n + k)$ for $k \geq 1$ is summarized as in the following steps [17–19].

Step 1: the new sequence of data $X^{(1)}$ is generated by AGO and is represented as:

$$X^{(1)} = \left\{x^{(1)}(1), x^{(1)}(2), \ldots, x^{(1)}(n)\right\} \quad (1)$$

Here, $X^{(1)}(1) = X^{(0)}(1)$, and $X^{(1)}(k) = \sum_{i=1}^{k} x^{(0)}(i)$, $k = 2, 3, \ldots, n$. There is a constraint associated with the original historical data that the ratio test $\sigma = [\sigma(2), \sigma(3), \ldots, \sigma(n)]$, hence $\sigma(k) = x^{(0)}(k-1)/x^{(0)}(k), for\, k = 2, 3, \ldots, n$ has to fall within 0.1345–7.389, and this indicates that $X^{(0)}$ satisfy the grey model [20].

Step 2: establish the first-order grey differential equation as follows:

$$x^{(0)}(k) = -aZ^{(1)}(k) + u, \quad k = 2, 3, \ldots, n \quad (2)$$

where, *a* is the developing coefficient, *u* is the grey input, and $z^{(1)}(k) = 0.5x^{(1)}(k) + 0.5x^{(1)}(k-1)$, for k= 2, 3, …, *n*. Equation 2 can be rewritten to a matrix form as follows:

$$\mathrm{X_n} = \mathrm{B}\,\hat{\mathrm{a}}, \quad \text{where } B = \begin{bmatrix} -Z^{(1)}(2) & 1 \\ -Z^{(1)}(3) & 1 \\ \vdots & \vdots \\ -Z^{(1)}(n) & 1 \end{bmatrix}, \tag{3}$$

$$X_n = \left[x^{(0)}(2), x^{(0)}(3), \ldots, x^{(0)}(n)\right]^T, \quad \text{and} \quad \hat{\mathrm{a}} = [\mathrm{a}, \mathrm{u}]^{\mathrm{T}}$$

Step 3: estimating a and u coefficients in Eq. 3 using the ordinary least-squares (OLS) method which is represented as:

$$[\hat{a}, \hat{u}]^T = \left(B^T B\right)^{-1} B^T X_n \tag{4}$$

Step 4: form the first-order grey reflection differential equation as follows:

$$\frac{dx^{(1)}(k)}{dt} = -ax^{(1)}(k) + u \tag{5}$$

Equation 5 is called the shadow equation of the source model $x^{(0)}(k) = -aZ^{(1)}(k) + u$. By solving this equation, with initial condition of $X^{(1)}(1) = X^{(0)}(1)$, the following is obtained:

$$\overset{\Lambda(1)}{x}(k) = \left[x^{(0)}(1) - \frac{u}{a}\right] e^{-a(k-1)} + \frac{u}{a}, \quad k = 2, 3, \ldots, n \tag{6}$$

Here, the ^ donates the grey prediction value.

Step 5: restore the demand prediction value by inverse AGO which is defined as:

$$x^{(0)}(k+1) = x^{(1)}(k+1) - x^{(1)}(k) \tag{7}$$

Substituting Eq. 6 into Eq. 7, the grey prediction of future events will be obtained as follows:

$$\overset{\Lambda(0)}{x}(k) = \left[x^{(0)}(1) - \frac{u}{a}\right] \left(1 - e^{a}\right) e^{-a(k-1)}, \quad k = 2, 3, \ldots, n \tag{8}$$

Grey Model Configurations

The models are applied to two case studies, i.e., a CSTR and a Process Vessel [1, 21]. The data sets employed are shown in Table 1. Two different configurations of Grey prediction models applied with these data sets which are as follows:

Table 1 Historical real time data of the case studies

Time interval	System 1	System 2
	CSTR precursor data	Vessel precursor data
1	6	7
2	9	7
3	12	8
4	13	8
5	15	5
6	11	12
7	16	14
8	10	16

Data of this table has been taken from [1] and [21] for CSTR and vessel processing respectively

- One step prediction using a moving window technique that considers only the last four observed data.
- One step prediction using observed data population increasing technique in which after the predicted event becomes known, it is added to the data points to predict the next event.

However, the second configuration is applied in this article due to space limitation.

Result and Discussion

According to prediction criteria introduced by Lewis [22], typically a prediction is considered excellent if (MAPE < 10), good if (10 < MAPE < 20), reasonable if (20 < MAPE < 50), and incorrect if (MAPE > 50). The results obtained in revealed that both models are considered reasonable with MAPE of 20–30 %. These are shown in Tables 2 and 3. The results also illustrated the better prediction provided by the G(1, 1) model for both the CSTR and the process vessel data.

Table 2 Predictive models' outputs and errors CSTR process with observed data population configuration

Time interval	Actual no. of consequences	Poisson model		G(1, 1)	
		Model value	Absolute error (%)	Model value	Absolute error (%)
1	6	–	–	–	–
2	9	6	33.33	9	0
3	12	7	41.67	11	8.33
4	13	9	30.77	13	0
5	15	10	33.33	16	6.67
6	11	11	0	18	63.64
7	16	11	31.25	14	12.5
8	10	12	20	16	60
MAPE (%)		–	27.2	–	21.6

Table 3 Predictive models' outputs and errors of tank process with observed data population configuration

Time interval	Actual no. of consequences	Poisson model		G(1, 1)	
		Model value	Absolute error (%)	Model value	Absolute error (%)
1	7	–	–	–	–
2	7	7	0	7	0
3	8	7	12.5	8	0
4	8	7	12.5	8	0
5	5	7	40	9	80
6	12	6	50	6	50
7	14	8	42.86	11	21.43
8	16	9	43.75	15	6.25
MAPE (%)		–	28.8	–	22.5

Figures 1 and 2 show the validation of the prediction models to the actual observed data. The lack of sensitivity of the Poisson model to the observed data has led to underestimation especially when the observed data are more excited. The Grey model has some difficulties in dealing with fast changes. This has caused large prediction error on 6th and 8th intervals for the G(1, 1). The 6th interval prediction output has extended the data trend from the 1st to the 5th interval. Likewise, the 8th interval prediction also followed the decreasing trend, which indicate consistent

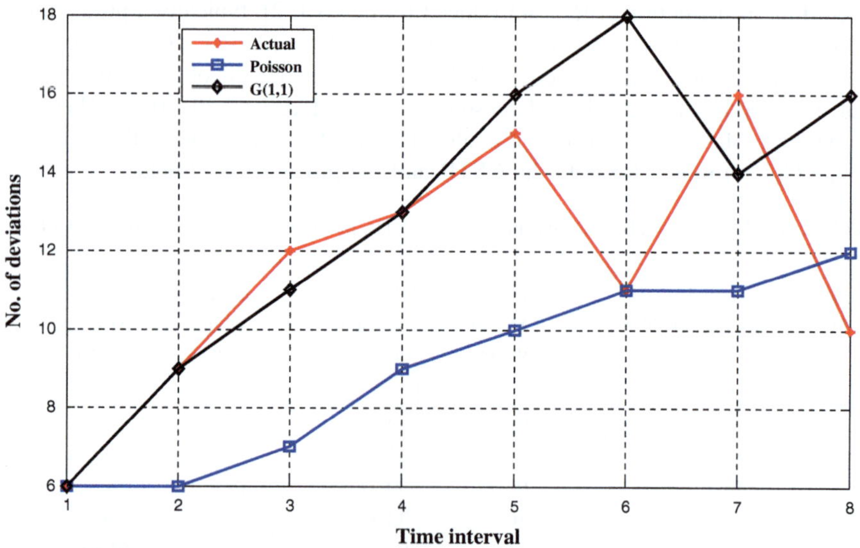

Fig. 1 Prediction models validation of CSTR case study

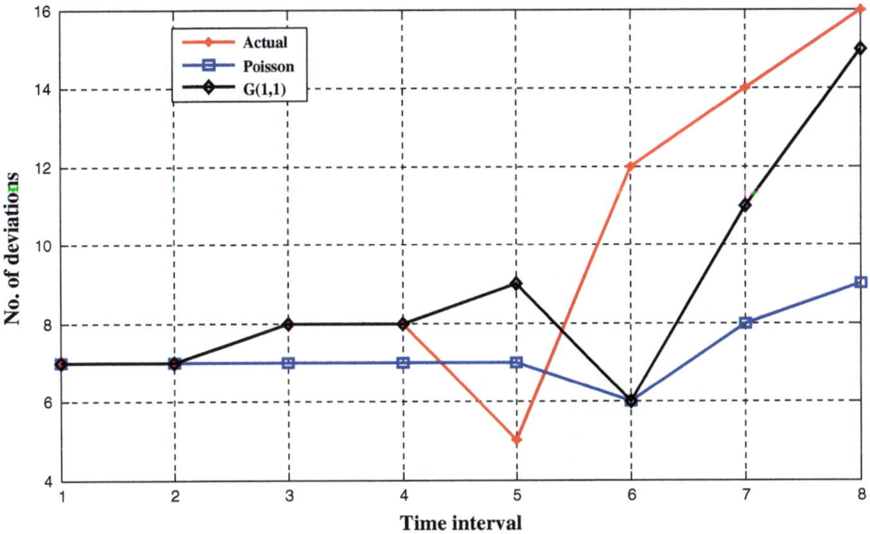

Fig. 2 Prediction models validation of vessel tank case study

weakness on the G(1, 1) model. In the case of the Poisson model, in all cases it is underestimating the prediction. It consistently showed weak response in the manner of excessive filtering of all observed data.

Conclusion

Grey model has been studied to predict CPI consequences in scarce data environment. In both case studies considered, the Grey model based on G(1, 1) configuration produced higher prediction accuracy than the commonly used Poisson model. However, the implementation of Grey model requires a minimum of four data points. The capability of the original G(1, 1) model in providing reasonably good prediction has inspired further extension by using hybrid approach using Bayesian network, which is currently on-going. It is believed that by producing better predictions, effective risk based decision supports methodology can be established to facilitate the maintenance planning and management of change.

References

1. Meel A, Seider WD (2006). Plant-Specific Dynamic Failure Assessment Using Bayesian Theory. Chem Eng Sci 61:7036–7056
2. Rathnayaka S, Khan F, Amyotte P (2011). SHIPP methodology: Predictive accident modeling approach. Part I: Methodology and model description. Process Saf Environ Prot 89:151–164

3. Rathnayaka S, Khan F, Amyotte P (2012). Accident modeling approach for safety assessment in an LNG processing facility. J Loss Prev Process Ind 25:414–423
4. Meel A, O'Neill L, Levin J, Seider WD, Oktem U, Keren N (2007). Operational risk assessment of chemical industries by exploiting accident databases. J Loss Prev Process Ind 20:113–127
5. Rathnayaka S, Khan F, Amyotte P (2011). SHIPP methodology: Predictive accident modeling approach. Part II. Validation with case study. Process Saf Environ Prot 89:75–88
6. Al-shanini A, Ahmad A, Khan F (2014). Accident Modelling and Analysis in Process Industries. J Loss Prev Process Ind 32:319–334. doi:10.1016/j.jlp.2014.09.016
7. Bradlow ET, Hardie BG, Fader PS (2002). Bayesian inference for the negative binomial distribution via polynomial expansions. J Comput Graph Stat 11:1
8. Al-shanini A, Ahmad A, Khan F (2014). Accident Modelling and Safety Measure Design of a Hydrogen Station. Int J Hydrogen Energy 39(35):20362–20370. doi:10.1016/j.ijhydene.2014.05.044
9. Ju-Long D (1982). Control problems of grey systems. Syst Control Lett 1:288–294
10. Deng J-L (1989). Introduction to grey system theory. J Grey Syst 1:1–24
11. Tseng F-M, Yu H-C, Tzeng G-H (2001). Applied hybrid grey model to forecast seasonal time series. Technol Forecast Soc Chang 67:291–302
12. Hsu L-C (2003). Applying the grey prediction model to the global integrated circuit industry. Technol Forecast Soc Chang 70:563–574
13. Lin C-T, Yang S-Y (2003). Forecast of the output value of Taiwan's opto-electronics industry using the Grey forecasting model. Technol Forecast Soc Chang 70:177–186
14. Chang N-B, Tseng C (1999). Optimal evaluation of expansion alternatives for existing air quality monitoring network by grey compromise programing. J Environ Manage 56:61–77
15. Huang Y-P, Wang S-F (1997). The identification of fuzzy grey prediction system by genetic algorithms. Int J Syst Sci 28:15–24
16. Zhou P, Ang B, Poh K (2006). A trigonometric grey prediction approach to forecasting electricity demand. Energy 31:2839–2847
17. Hsu C-C, Chen C-Y (2003). Applications of improved grey prediction model for power demand forecasting. Energy Convers Manag 44:2241–2249
18. Yao AW, Chi S, Chen J (2003). An improved grey-based approach for electricity demand forecasting. Electr Power Syst Res 67:217–224
19. Hsu L-C, Wang C-H (2007). Forecasting the output of integrated circuit industry using a grey model improved by the Bayesian analysis. Technol Forecast Soc Chang 74:843–853
20. Wang Y-F (2002). Predicting stock price using fuzzy grey prediction system. Expert Syst Appl 22:33–38
21. Kalantarnia M, Khan F, Hawboldt K (2009). Dynamic risk assessment using failure assessment and Bayesian theory. J Loss Prev Process Ind 22:600–606
22. Lewis CD (1982) Industrial and business forecasting methods: a practical guide to exponential smoothing and curve fitting. Butterworth Scientific, London

Index

M.A. Hashim (ed.), *ICGSCE 2014*, DOI 10.1007/978-981-287-505-1

MIX
Papier aus verantwortungsvollen Quellen
Paper from responsible sources
FSC® C105338
FSC
www.fsc.org

If you have any concerns about our products,
you can contact us on
ProductSafety@springernature.com

In case Publisher is established outside the EU,
the EU authorized representative is:
Springer Nature Customer Service Center GmbH
Europaplatz 3, 69115 Heidelberg, Germany

Printed by Libri Plureos GmbH
in Hamburg, Germany